MEDICAL NUTRITION THERAPY

A CASE STUDY APPROACH

MEDICAL NUTRITION THERAPY

A CASE STUDY APPROACH

FOURTH EDITION

MARCIA NAHIKIAN NELMS, PhD, RD, LD, CNSC
OHIO STATE UNIVERSITY

SARA LONG ROTH, PhD, RD, LD
SOUTHERN ILLINOIS UNIVERSITY

CENGAGE
Learning·

Australia • Brazil • Japan • Korea • Mexico • Singapore • Spain • United Kingdom • United States

CENGAGE
Learning·

Medical Nutrition Therapy: A Case Study Approach, Fourth Edition
Marcia Nahikian Nelms and Sara Long Roth

Publisher: Yolanda Cossio

Acquisitions Editor: Peggy Williams

Developmental Editor: Elesha Feldman

Editorial Assistant: Kellie Petruzzelli

Media Editor: Miriam Myers

Brand Manager: Elisabeth Rhoden

Market Development Manager: Kara Kindstron

Manufacturing Planner: Karen Hunt

Rights Acquisitions Specialist: Roberta Broyer

Text Researcher: Sarah D'Stair

Production Management and Composition: Teresa Christie, MPS Limited

Copy Editor: William Clark

Art and Design Direction: Carolyn Deacy, MPS Limited

Cover Designer: Kathleen Cunningham

Cover Image: © Christoph Rosenberger/ GettyImages

For product information and technology assistance, contact us at **Cengage Learning Customer & Sales Support, 1-800-354-9706.**

For permission to use material from this text or product, submit all requests online at **www.cengage.com/permissions** Further permissions questions can be e-mailed to **permissionrequest@cengage.com**

Library of Congress Control Number: 2013931079
ISBN-13: 978-1-133-59315-7
ISBN-10: 1-133-59315-1

Cengage Learning
200 First Stamford Place, 4th Floor
Stamford, CT 06902
USA

Cengage Learning is a leading provider of customized learning solutions with office locations around the globe, including Singapore, the United Kingdom, Australia, Mexico, Brazil, and Japan. Locate your local office at **www.cengage.com/global**

Cengage Learning products are represented in Canada by Nelson Education, Ltd.

To learn more about Wadsworth, visit **www.cengage.com/wadsworth**

Purchase any of our products at your local college store or at our preferred online store **www.CengageBrain.com**

Printed in the United States of America
2 3 4 5 6 7 17 16 15 14 13

DEDICATION

Dedicated to our students—past and present—who continue to challenge us, teach us, and guide us as we strive to enhance clinical education.

CONTENTS

Preface xi

Acknowledgments xv

About the Authors xvii

Introducing Case Studies, or Finding Your Way Through a Case Study xix

Unit One

ENERGY BALANCE AND BODY WEIGHT 1

1 Pediatric Weight Management 3
2 Bariatric Surgery for Morbid Obesity 11
3 Malnutrition Associated with Chronic Disease 23

Unit Two

NUTRITION THERAPY FOR CARDIOVASCULAR DISORDERS 33

4 Hypertension and Cardiovascular Disease 35
5 Myocardial Infarction 47
6 Heart Failure with Resulting Cardiac Cachexia 59

Unit Three

NUTRITION THERAPY FOR UPPER GASTROINTESTINAL DISORDERS 69

7 Gastroesophageal Reflux Disease 71
8 Ulcer Disease: Medical and Surgical Treatment 81

Unit Four

NUTRITION THERAPY FOR LOWER GASTROINTESTINAL DISORDERS 91

9 Celiac Disease 93
10 Irritable Bowel Syndrome (IBS) 103
11 Inflammatory Bowel Disease: Crohn's Disease 115

Unit Five

NUTRITION THERAPY FOR HEPATOBILIARY AND PANCREATIC DISORDERS 127

12 Cirrhosis of the Liver 129
13 Acute Pancreatitis 141

Unit Six

NUTRITION THERAPY FOR ENDOCRINE DISORDERS 153

14 Pediatric Type 1 Diabetes Mellitus 155
15 Type 1 Diabetes Mellitus in the Adult 169
16 Type 2 Diabetes Mellitus—Pediatric Obesity 181
17 Adult Type 2 Diabetes Mellitus: Transition to Insulin 193

Unit Seven

NUTRITION THERAPY FOR RENAL DISORDERS 205

18 Chronic Kidney Disease (CKD) Treated with Dialysis 207
19 Chronic Kidney Disease: Peritoneal Dialysis 219
20 Acute Kidney Injury (AKI) 229

Unit Eight

NUTRITION THERAPY FOR HEMATOLOGICAL DISORDERS 239

21 Anemia in Pregnancy 241
22 Folate and Vitamin B_{12} Deficiencies 253

Unit Nine

NUTRITION THERAPY FOR NEUROLOGICAL DISORDERS 265

23 Ischemic Stroke 267
24 Progressive Neurological Disease: Parkinson's Disease 279
25 Alzheimer's Disease 289

Unit Ten

NUTRITION THERAPY FOR PULMONARY DISORDERS 299

26 Chronic Obstructive Pulmonary Disease 301
27 COPD with Respiratory Failure 313

Unit Eleven

NUTRITION THERAPY FOR METABOLIC STRESS AND CRITICAL ILLNESS 325

28 Pediatric Traumatic Brain Injury: Metabolic Stress with Nutrition Support 327
29 Metabolic Stress and Trauma: Open Abdomen 339
30 Nutrition Support for Burn Injury 351
31 Nutrition Support in Sepsis and Morbid Obesity 365

Unit Twelve

NUTRITION THERAPY FOR NEOPLASTIC DISEASE 377

32 Acute Lymphoblastic Leukemia Treated with Hematopoietic Cell Transplantation 379
33 Esophageal Cancer Treated with Surgery and Radiation 391

Unit Thirteen

NUTRITION THERAPY FOR HIV/AIDS 401

34 AIDS 403

Appendices

A Common Medical Abbreviations 413
B Normal Values for Physical Examination 417
C Routine Laboratory Tests with Nutritional Implications 419
D Exchange Lists for Diabetes Meal Plan Form and Food Lists Nutrient Information 421

Index 423

PREFACE

In teaching, we seek to promote the fundamental values of humanism, democracy, and the sciences—that is, a curiosity about new ideas and enthusiasm for learning, a tolerance for the unfamiliar, and the ability to critically evaluate new ideas.

We wish to provide the environment that will support students in their quest for integration of knowledge and support the development of critical thinking skills. Thus, we strive to develop these "laboratories" and "real-world" situations that mimic the professional community to build that bridge to clinical practice.

The idea for this book actually began more than fifteen years ago as we began teaching medical nutrition therapy for dietetic students, and now as this fourth edition publishes we hope that these cases reflect the most recent changes in nutrition therapy practice. Entering the classroom after being clinicians for many years, we knew we wanted our students to experience nutritional care as realistically as possible. We wanted the classroom to actually be the bridge between the textbook and the clinical setting. In fashioning one of the tools used to build that bridge, we relied heavily on our clinical experience to develop what we hoped would be realistic clinical applications. Use of a clinical application or case study is not a new concept; the use of case studies in nutrition, medicine, nursing, and many other allied health fields is commonplace. The case study places the student in a situation that forces integration of knowledge from many sources; supports use of previously learned information; puts the student in a decision-making role; and nurtures critical thinking.

What makes this text different, then, from a simple collection of case studies? The pedagogy we have developed with each case takes the student one step closer as he or she moves from the classroom to the real world. The cases represent the most common diagnoses that rely on nutrition therapy as an essential component of the medical care. Therefore, we believe these cases represent the type of patient with which the student will most likely be involved. The concepts presented in these cases can apply to many other medical conditions that may not be presented here. Furthermore, the instructor can choose a variety of questions from each case even if he or she chooses not to have the student complete the entire case. The cases represent both introductory and advanced-level practice and, therefore, use of this text allows faculty to choose among many cases and questions that fit the students' level of expertise.

The cases cross the life span, allowing the student to see the practice of nutrition therapy during pregnancy, childhood, adolescence, and adulthood through the elder years. We have tried to represent the diversity of individual patients we encounter today. Placing nutrition therapy and nutrition education within the appropriate cultural context is crucial.

The medical record provides the structure for each case. With the fourth edition, our format has changed to reflect the components of the electronic medical record. The student will seek information to solve the case by using the exact tools he or she will need to use in the clinical setting. As the student moves from the admission or outpatient visit record to the physician's history and physical, to laboratory data, and to documentation of daily care, the student will need to discern the relevant information from the medical record.

Questions for each case are organized using the nutrition care process, beginning with items introducing the pathophysiology and principles of nutrition therapy for the case and then proceeding through each component of the process. Questions prompt the student to identify nutrition problems and then synthesize a PES statement. It will be helpful to begin by orienting the student to the components of a case. We have provided an outline of this introduction below (see "Introducing Case Studies"). Teaching needs to be purposeful. If the instructor takes the responsibility of teaching students how to use this book seriously, it is much more likely that student autonomy will be the end result.

To be consistent with the philosophy of the text, each case requires that the student seek information from multiple resources to complete the case. Many of the articles and online sites provide essential data regarding diagnosis and treatment within that case. We have found that when students learn how to research the case, their expertise grows exponentially.

The cases lend themselves to be used in several different teaching situations. They fit easily into a problem-based learning curriculum, and also can be used as a summary for classroom teaching of the pathophysiology and nutrition therapy for each diagnosis. The cases can be integrated into the appropriate rotation for a dietetic internship, medical school, or nursing school. Furthermore, these cases can be successfully used to develop standardized patient and simulation experiences.

Objectives for student learning within each case are built around the nutrition care process and competencies for dietetic education. This allows an additional path for nutrition and dietetic faculty to document student performance as part of program assessment.

New to the Fourth Edition

Several important factors have prompted the changes to this fourth edition. The first is the transition to the electronic medical record (EMR). For this edition, the case components are formatted to mimic the "screens" one might find in the electronic medical record (EMR). Though the EMRs used in clinics, physician's offices, and hospitals vary, these cases capture the primary sources of information that the clinician will access to provide a thorough nutrition assessment for her or his patient. The setting for some of the cases has also been changed to reflect outpatient care within the patient-centered medical home.

Secondly, our reviewers requested that the cases be shortened in length. We have streamlined all of the cases so that questions are more precise. Finally, even within a two- to three-year period, medical and nutritional care can change dramatically. These cases reflect the most recent research and evidenced-based literature so that the student moves toward higher levels of practice.

The fourth edition introduces sixteen totally new cases:

Case 1 Pediatric Weight Management
Case 2 Bariatric Surgery for Morbid Obesity
Case 3 Malnutrition Associated with Chronic Disease
Case 10 Irritable Bowel Syndrome (IBS)
Case 13 Acute Pancreatitis
Case 14 Pediatric Type 1 Diabetes Mellitus
Case 15 Type 1 Diabetes Mellitus in the Adult
Case 16 Type 2 Diabetes Mellitus—Pediatric Obesity
Case 17 Adult Type 2 Diabetes Mellitus: Transition to Insulin
Case 19 Chronic Kidney Disease: Peritoneal Dialysis
Case 20 Acute Kidney Injury (AKI)
Case 21 Anemia in Pregnancy
Case 22 Folate and Vitamin B_{12} Deficiencies
Case 24 Progressive Neurological Disease: Parkinson's Disease
Case 30 Nutrition Support for Burn Injury
Case 31 Nutrition Support in Sepsis and Morbid Obesity
Case 32 Acute Lymphoblastic Leukemia Treated with Hematopoietic Cell Transplantation

The additional cases you will find in this edition, though they have been included in previous editions, have also been changed to reflect current medical care with appropriate changes in drugs, procedures, and nutrition interventions. For example, the presenting signs and symptoms in the celiac disease case have been changed so they are not the classic gastrointestinal complaints traditionally associated with this disorder. The heart failure case addresses the risk of micronutrient deficiency that is often seen in these patients. Within the open abdomen surgical case, the most recent literature about assessment of these critically ill patients has been incorporated, and the use of nutrition support has been altered to reflect current practice. Incorporation of evidence-based guidelines is encouraged throughout each of the cases and the questions are designed to not only follow the nutrition care process but also require the student to evaluate the most current literature.

TEACHING STRATEGIES

You can find cases to emphasize specific topics that are part of the curriculum for pathophysiology and medical nutrition therapy (a list of cases by topic is provided below). We have found that when specific questions are selected for each case, they can be modified to assist in the pedagogy for other classes as well.

Nutrition Assessment: Case 1 Pediatric Weight Management; Case 4 Hypertension and Cardiovascular Disease

Fluid Balance/Acid-Base Balance: Case 27 COPD with Respiratory Failure; Case 29 Metabolic Stress and Trauma: Open Abdomen

Genetics/Immunology/Infectious Process: Case 9 Celiac Disease; Case 12 Cirrhosis of the Liver; Case 14 Pediatric Type 1 Diabetes Mellitus; Case 15 Type 1 Diabetes Mellitus in the Adult

Hypermetabolism/Metabolic Stress: Case 13 Acute Pancreatitis; Case 20 Acute Kidney Injury (AKI); Case 28 Pediatric Traumatic Brain Injury: Metabolic Stress with Nutrition Support; Case 29 Metabolic Stress and Trauma: Open Abdomen; Case 30 Nutrition Support for Burn Injury; Case 31 Nutrition Support in Sepsis and Morbid Obesity

Dysphagia: Case 23 Ischemic Stroke; Case 24 Progressive Neurological Disease: Parkinson's Disease; Case 25 Alzheimer's Disease; Case 33 Esophageal Cancer Treated with Surgery and Radiation

Nutritional Needs of the Elderly: Case 3 Malnutrition Associated with Chronic Disease; Case 25 Alzheimer's Disease; Case 27 COPD with Respiratory Failure

Malnutrition: Case 3 Malnutrition Associated with Chronic Disease; Case 6 Heart Failure with Resulting Cardiac Cachexia; Case 29 Metabolic Stress and Trauma: Open Abdomen; Case 30 Nutrition Support for Burn Injury; Case 31 Nutrition Support in Sepsis and Morbid Obesity; Case 33 Esophageal Cancer Treated with Surgery and Radiation

Pediatrics: Case 1 Pediatric Weight Management; Case 14 Pediatric Type 1 Diabetes Mellitus; Case 16 Type 2 Diabetes Mellitus—Pediatric Obesity; Case 28 Pediatric Traumatic Brain Injury: Metabolic Stress with Nutrition Support

Nutrition Support: Case 11 Inflammatory Bowel Disease: Crohn's Disease; Case 13 Acute Pancreatitis; Case 24 Progressive Neurological Disease: Parkinson's Disease; Case 28 Pediatric Traumatic Brain Injury: Metabolic Stress with Nutrition Support; Case 29 Metabolic Stress and Trauma: Open Abdomen; Case 30 Nutrition Support for Burn Injury; Case 31 Nutrition Support in Sepsis and Morbid Obesity; Case 32 Acute Lymphoblastic Leukemia Treated with Hematopoietic Cell Transplantation; Case 33 Esophageal Cancer Treated with Surgery and Radiation

ACKNOWLEDGMENTS

We first need to thank our developmental editor—Elesha Feldman—who provided expert guidance in all steps of this revision. We would like to thank the following Ohio State University medical dietetic students who provided input to the cases and the answer guide: Elizabeth Bastian, Ann Barrett, Chelsea Britton, and Jennifer Galvin. We also have three contributors to new cases and we are fortunate to benefit from the expertise of these outstanding clinicians: Dawn Scheiderer, RD, LD; Sheela Thomas, MS, RD, CNSC; and Kimberlee Orben, MS, RD, CSO. We would also like to thank the following reviewers who provided invaluable suggestions for updating the existing cases and refining the new ones:

Susan N. Hawk, PhD, RD, Central Washington University

Magaly Hernandez, MPH, RD, Andrews University

Lisa Herzig, PhD, RD, CDE, California State University, Fresno

Norman G. Hord, PhD, MPH, RD, Michigan State University

Maria T. Spicer, Florida State University

Jillian Trabulsi, University of Delaware

ABOUT THE AUTHORS

Marcia Nahikian-Nelms, PhD, RD, LD, CNSC

Dr. Nahikian-Nelms is currently a professor of clinical health and rehabilitation sciences and director of the dietetic internship in the Division of Medical Dietetics – College of Medicine at The Ohio State University. She has practiced as a dietitian and public health nutritionist for over 25 years. She is the lead author for the textbooks *Nutrition Therapy and Pathophysiology; Medical Nutrition Therapy: A Case Study Approach*; and *Food and Culture*. Additionally, she has contributed to the Academy of Nutrition and Dietetics *Nutrition Care Manual* sections on gastrointestinal disorders and is the author of numerous peer-reviewed journal articles and chapters for other texts. The focus of her clinical expertise is the development and practice of evidence-based nutrition therapy for a variety of conditions including diabetes, gastrointestinal disease, and hematology/oncology for both pediatric and adult populations, as well as the development of alternative teaching environments for students receiving their clinical training. Dr. Nahikian-Nelms has received the Governor's Award for Outstanding Teaching for the State of Missouri, Outstanding Dietetic Educator in Missouri and Ohio, and the PRIDE award from Southeast Missouri State University in recognition of her teaching.

Sara Long Roth, PhD, RD

Dr. Long is a professor and director of the Didactic Program in Dietetics in the Department of Animal Science, Food and Nutrition at Southern Illinois University Carbondale. Prior to obtaining her PhD in health education, she practiced as a clinical dietitian for 11 years. Dr. Long also served as the nutrition education/counseling consultant for Carbondale Family Medicine for 18 years. She has been an active leader in national, state, and district dietetic associations, where she has served in numerous elected and appointed positions including President of the Illinois Dietetic Association, Council on Professional Issues Delegate (Education) in the Academy of Nutrition and Dietetics House of Delegates, member of the Commission on Accreditation for Dietetics Education, member of the Commission on Dietetic Registration, and member of the Illinois Dietetic and Nutrition Services Practice Board.

Dr. Long is co-author of *Nutrition Therapy and Pathophysiology* and three other nutrition texts. She has received various awards and honors for teaching, including Outstanding Dietetic Education (Academy of Nutrition and Dietetics), SIUC Undergraduate Student Government Outstanding Educator for the College of Agriculture, Outstanding Educator for the College of Agricultural Sciences, and Inter-Greek Council of SIUC Professor of the Year.

INTRODUCING CASE STUDIES, OR FINDING YOUR WAY THROUGH A CASE STUDY

Have you ever put together a jigsaw puzzle or taught a young child how to complete a puzzle?

Almost everyone has at one time or another. Recall the steps that are necessary to build a puzzle. You gather together the straight edges, identify the corner pieces, and match the like colors. There is a method and a procedure to follow that, when used persistently, leads to the completion of the puzzle.

Finding your way through a case study is much like assembling a jigsaw puzzle. Each piece of the case study tells a portion of the story. As a student, your job is to put together the pieces of the puzzle to learn about a particular diagnosis, its pathophysiology, and the subsequent medical and nutritional treatment. Although each case in the text is different, the approach to working with the cases remains the same, and with practice, each case study and each medical record becomes easier to manage. The following steps provide guidance for working with each case study.

1. Identify the major parts of the case study.
   - Admitting history and physical
   - Documentation of MD orders, nursing assessment, and results from other care providers
   - Laboratory data
   - Bibliography

2. Read the case carefully.
   - Get a general sense of why the person has been admitted to the hospital.
   - Use a medical dictionary to become acquainted with unfamiliar terms.
   - Use the list of medical abbreviations provided in Appendix A to define any that are unfamiliar to you.

3. Examine the admitting history and physical for clues.
   - Height, weight
   - Vital signs (compare to normal values for physical examination in Appendix B)
   - Chief complaint
   - Patient and family history
   - Lifestyle risk factors

4. Review the medical record.
   - Examine the patient's vital statistics and demographic information (e.g., age, education, marital status, religion, ethnicity).
   - Read the patient history (remember, this is the patient's subjective information).

5. Use the information provided in the physical examination.
   - Familiarize yourself with the normal values found in Appendix B.
   - Make a list of those things that are abnormal.
   - Now compare abnormal values to the pathophysiology of the admitting diagnosis. Which are consistent? Which are inconsistent?

6. Evaluate the nutrition history.
   - Note appetite and general descriptions.
   - Evaluate the patient's dietary history: calculate average kcal and protein intakes and compare to population standards and recommendations such as the USDA Food Patterns.
   - Is there any information regarding physical activity?
   - Find anthropometric information.
   - Is the patient responsible for food preparation?
   - Is the patient taking a vitamin or mineral supplement?

7. Review the laboratory values.
   - Hematology
   - Chemistry
   - What other reports are present?
   - Compare the values to the normal values listed. Which are abnormal?

Highlight those and then compare to the pathophysiology. Are they consistent with the diagnosis? Do they support the diagnosis? Why?

8. Use your resources.
   - Use the bibliography provided for each case.
   - Review your nutrition textbooks.
   - Use any books on reserve.
   - Access information on the Internet but choose your sources wisely: stick to

government, not-for-profit organizations, and other legitimate sites. A list of reliable Internet resources is provided for each case.

Mindmap

A mindmap is a graphic representation of the elements of the case study and the steps in its analysis. This organization can assist in connecting bodies of information and allow for further development of critical thinking skills.

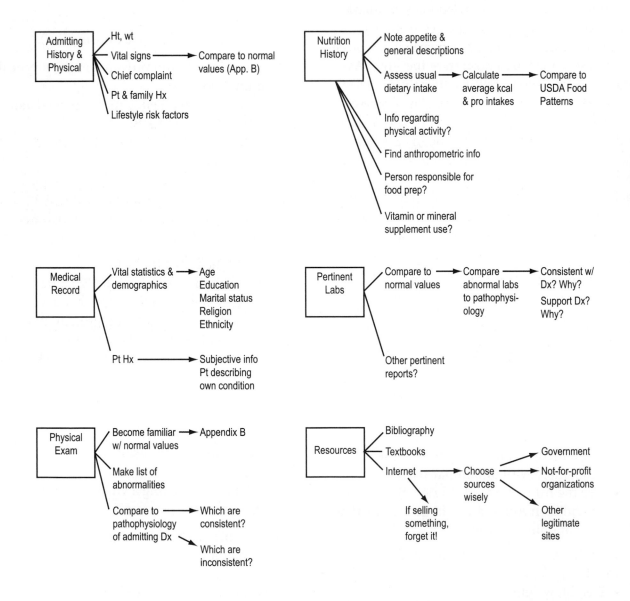

Unit One

ENERGY BALANCE AND BODY WEIGHT

Unit One introduces nutrition therapy for treatment of disorders of weight balance and draws our attention to these major public health concerns in the United States. The first case uses pediatric obesity as a springboard for a discussion of the implications of the rapidly rising rate of childhood obesity. The incidence of childhood obesity has more than tripled over the past three decades, and the prevalence for children ages 6–11 in the U.S. has risen to 20%. The child featured in Case 1 is representative of children ages 6–11. Pediatric obesity treatment requires complex interventions to address family, environmental, and economic concerns. This case allows the student to explore the current research and the use of evidence-based guidelines to determine appropriate nutrition therapy.

Case 2 uses the record of a bariatric surgery patient as an opportunity to learn about morbid obesity. More than 3 million individuals in the U.S. are considered to be morbidly obese—this is also referred to as Class III obesity or a body mass index (BMI) >40.0. Health consequences of untreated morbid obesity include type 2 diabetes mellitus, coronary heart disease and hypertension, cancer, sleep apnea, and even premature death. Individuals who have failed to lose weight by less invasive means, and who meet the medical criteria, may consider bariatric surgery as a treatment method for weight control. This case allows the student to research the surgical options used for bariatric surgery and to begin to understand the progression of nutrition therapy used postoperatively.

Case 3 explores the diagnosis of malnutrition. As early as 1979, Charles Butterworth attempted to raise awareness of the increasing incidence of malnutrition in the U.S. health care system with his classic article, "The Skeleton in the Hospital Closet." Unfortunately, the rate of malnutrition is still considered to be significant today—and is associated with increased hospital costs, increased morbidity and mortality, and decreased quality of life for these individuals. Recently, new definitions of malnutrition have been proposed by the Academy of Nutrition and Dietetics (AND) and the Association for Parenteral and Enteral Nutrition (ASPEN) in an effort to more consistently identify those individuals who are at risk for malnutrition and who are malnourished, so that expedient interventions may occur. This case uses the most recent literature to provide the opportunity to recognize and apply the newly proposed diagnostic criteria for malnutrition.

Case 1

Pediatric Weight Management

Objectives

After completing this case, the student will be able to:

1. Describe the physiological effects of overweight/obesity in the pediatric population.
2. Interpret laboratory parameters for nutritional implications and significance.
3. Analyze nutrition assessment data to evaluate nutritional status and identify specific nutrition problems.
4. Determine nutrition diagnoses and write appropriate PES statements.
5. Prescribe appropriate nutrition therapy.
6. Develop a nutrition care plan with appropriate measurable goals, interventions, and strategies for monitoring and evaluation consistent with the nutrition diagnoses of this case.

Jamey Whitmer is taken to see her pediatrician by her parents, who have noticed she appears to stop breathing while sleeping. She is diagnosed with sleep apnea related to her weight and referred to the registered dietitian for nutrition counseling.

Whitmer, Jamey, Female, 10 y.o.
Allergies: No known allergies   **Code:** FULL   **Isolation:** None
Pt. Location: University Clinic   **Physician:** Lambert, S. David   **Appointment Date:** 9/22

Patient Summary: 10-year-old female is here with parents who describe concerns that their daughter appears to stop breathing while she is sleeping.

History:
Onset of disease: Parents describe sleep disturbance in their daughter for the past several years, including: sleeping with her mouth open, cessation of breathing for at least 10 seconds (per episode), snoring, restlessness during sleep, enuresis, and morning headaches. They also mention that Jamey's teacher reports difficulty concentrating in school and a change in her performance. She is the second child born to these parents—full-term infant with birthweight of 10 lbs 5 oz; 23" length. Actual date of onset unclear, but parents first noticed onset of the above-mentioned symptoms about one year ago.
Medical history: None
Surgical history: None
Family history: What? Possible gestational diabetes; type 2 DM; Who? Mother and grandmother

Demographics:
Years education: Third grade
Language: English only
Occupation: Student
Household members: Father age 36, mother age 35, sister age 5
Ethnicity: Caucasian
Religious affiliation: Presbyterian

MD Progress Note:
Review of Systems
Constitutional: Negative
Skin: Negative
Cardiovascular: Negative
Respiratory: Negative
Gastrointestinal: Negative
Neurological: Negative
Psychiatric: Negative

Physical Exam
Constitutional: Somewhat tired and irritable 10-year-old female
Cardiovascular: Regular rate and rhythm, heart sounds normal
HEENT: Eyes: Clear
           Ears: Clear
           Nose: Normal mucous membranes
           Throat: Dry mucous membranes, no inflammation, tonsillar hypertrophy
Genitalia: SMR (Tanner) pubic hair stage 3, genital stage 3

Whitmer, Jamey, Female, 10 y.o.
Allergies: No known allergies
Pt. Location: University Clinic

Code: FULL
Physician: Lambert, S. David

Isolation: None
Appointment Date: 9/22

Neurologic: Alert, oriented × 3
Extremities: No joint deformity or muscle tenderness, but patient complains of occasional knee pain. No edema, strength 5/5.
Skin: Warm, dry; reduced capillary refill (approximately 2 seconds); slight rash in skin folds
Chest/lungs: Clear
Abdomen: Obese

Vital Signs: Temp: 98.5   Pulse: 85   Resp rate: 27
                   BP: 123/80   Height: 57"   Weight: 115 lbs   BMI: 24.9

Assessment and Plan:

10-year-old female here with parents c/o of breathing difficulty at night. Child has steadily gained weight over previous several years— >10 lbs per year

Dx: R/O Obstructive sleep apnea (OSA) secondary to obesity and physical inactivity

Medical Tx plan: Polysomnography to diagnose OSA, FBG, HbA$_{1C}$, lipid panel (total cholesterol, HDL-C, LDL-C, triglycerides), psychological evaluation, nutrition assessment

.... SD Lambert, MD

Nutrition:

General: Very good appetite with consumption of a wide variety of foods. Jamey's physical activity level is generally low. Her elementary school discontinued physical education, art, and music classes due to budget cuts five years ago. She likes playing video games and reading.

24-hour recall:

AM:	2 breakfast burritos, 8 oz whole milk, 4 oz apple juice, 6 oz coffee with ¼ c cream and 2 tsp sugar
Lunch:	2 bologna and cheese sandwiches with 1 tbsp mayonnaise each, 1-oz pkg Fritos corn chips, 2 Twinkies, 8 oz whole milk
After school snack:	Peanut butter and jelly sandwich (2 slices enriched bread with 2 tbsp crunchy peanut butter and 2 tbsp grape jelly), 12 oz whole milk
Dinner:	Fried chicken (2 legs and 1 thigh), 1 c mashed potatoes (made with whole milk and butter), 1 c fried okra, 20 oz sweet tea
Snack:	3 c microwave popcorn, 12 oz Coca-Cola

Food allergies/intolerances/aversions: NKA
Previous nutrition therapy? No
Food purchase/preparation: Parent(s)
Vitamin intake: Flintstones vitamin daily

Whitmer, Jamey, Female, 10 y.o.
Allergies: No known allergies   **Code:** FULL   **Isolation:** None
Pt. Location: University Clinic   **Physician:** Lambert, S. David   **Appointment Date:** 9/22

Laboratory Results

	Ref. Range	9/22
Chemistry		
Sodium (mEq/L)	136–145	142
Potassium (mEq/L)	3.5–5.5	4.3
Chloride (mEq/L)	95–105	101
Carbon dioxide (CO_2, mEq/L)	23–30	25
BUN (mg/dL)	8–18	8
Creatinine serum (mg/dL)	0.6–1.2	0.6
Glucose (mg/dL)	70–110	112 !↑
Calcium (mg/dL)	9–11	9.2
Bilirubin total (mg/dL)	≤1.5	0.1
Protein, total (g/dL)	6–8	6.2
Albumin (g/dL)	3.5–5	4.8
Prealbumin (mg/dL)	16–35	33
Ammonia (NH_3, μmol/L)	9–33	9
Alkaline phosphatase (U/L)	30–120	99
ALT (U/L)	4–36	5
AST (U/L)	0–35	6
CPK (U/L)	30–135 F 55–170 M	72
Lactate dehydrogenase (U/L)	208–378	220
Cholesterol (mg/dL)	<170	165
HDL-C (mg/dL)	>55 F, >45 M	34 !↓
LDL (mg/dL)	<110	110
LDL/HDL ratio	<3.22 F <3.55 M	3.23 !↑
Triglycerides (mg/dL)	≤150	114
T_4 (μg/dL)	4–12	5
T_3 (μg/dL)	75–98	78
HbA_{1C} (%)	3.9–5.2	4.9
Hematology		
Transferrin (mg/dL)	250–380 F 215–365 M	254

Case Questions

I.   Understanding the Disease and Pathophysiology

1.   Current research indicates that the cause of childhood obesity is multifactorial. Briefly outline the roles of genetics, environment, and nutritional intake in development of obesity in children.

2.   Describe health consequences of overweight and obesity for children.

3.   Jamey has been diagnosed with obstructive sleep apnea. Define *sleep apnea*.

4.   Explain the relationship between sleep apnea and obesity.

II.   Understanding the Nutrition Therapy

5.   What are the goals for weight loss in the pediatric population?

6.   Under what circumstances might weight loss in overweight children not be appropriate?

7.   What would you recommend as the current focus for nutritional treatment of Jamey's obesity?

III.   Nutrition Assessment

8.   Evaluate Jamey's weight using the CDC growth charts provided (p.8): What is Jamey's BMI percentile? How is her weight status classified? Use the growth chart to determine Jamey's optimal weight for height and age.

9.   Identify two methods for determining Jamey's energy requirements other than indirect calorimetry, and then use them to calculate Jamey's energy requirements.

10.   Dietary factors associated with increased risk of overweight are increased dietary fat intake and increased calorie-dense beverages. Identify foods from Jamey's diet recall that fit these criteria.

11.   Calculate the percent of kcal from each macronutrient and the percent of kcal provided by fluids for Jamey's 24-hour recall.

12.   Increased fruit and vegetable intake is associated with decreased risk of overweight. What foods in Jamey's diet fall into these categories?

13.   Use the ChooseMyPlate online tool (available from www.choosemyplate.gov; click on "Daily Food Plans" under "SuperTracker and Other Tools") to generate a customized daily food plan. Using this eating pattern, plan a 1-day menu for Jamey.

Stature-for-Age and Weight-for-Age Percentiles: Girls, 2 to 20 Years

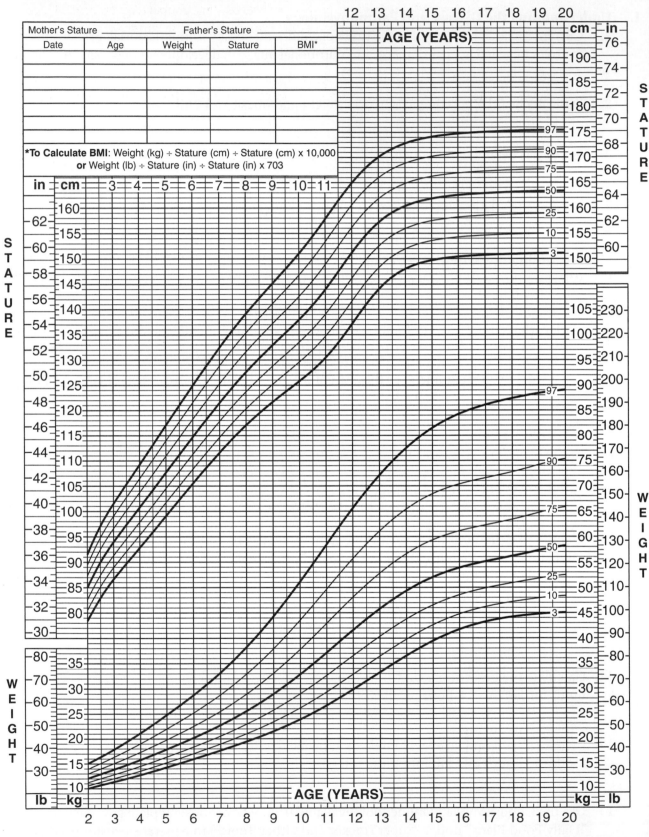

Source: Centers for Disease Control and Prevention. National Center for Health Statistics. 2000 CDC Growth Charts: United States. Available at http://www.cdc.gov/growthcharts. Accessed April 10, 2008.

14. Now enter and assess the 1-day menu you planned for Jamey using the MyPlate SuperTracker online tool (http://www.choosemyplate.gov/supertracker-tools/supertracker.html). Does your menu meet macro- and micronutrient recommendations for Jamey?

15. Why did Dr. Lambert order a lipid profile and blood glucose tests? What lipid and glucose levels are considered altered (i.e., outside of normal limits) for the pediatric population? Evaluate Jamey's lab results.

16. What behaviors associated with increased risk of overweight would you look for when assessing Jamey's and her family's diets? What aspects of Jamey's lifestyle place her at increased risk for overweight?

17. You talk with Jamey and her parents, who are friendly and cooperative. Jamey's mother asks if it would help for them to not let Jamey snack between meals and to reward her with dessert when she exercises. What would you tell them?

18. Identify one specific physical activity recommendation for Jamey.

IV. Nutrition Diagnosis
19. Select two nutrition problems and complete PES statements for each.

V. Nutrition Intervention
20. For each PES statement written, establish an ideal goal (based on signs and symptoms) and an appropriate intervention (based on etiology).

21. Mr. and Mrs. Whitmer ask about using over-the-counter diet aids, specifically Alli (orlistat). What would you tell them?

22. Mr. and Mrs. Whitmer ask about gastric bypass surgery for Jamey. Using the EAL, what are the recommendations regarding gastric bypass surgery for the pediatric population?

VI. Nutrition Monitoring and Evaluation
23. What is the optimal length of weight management therapy for Jamey?

24. Should her parents be included? Why or why not?

25. What would you assess during this follow-up counseling session?

Bibliography

Academy of Nutrition and Dietetics Evidence Based Library. Balanced Macronutrient Diet and Treating Childhood Obesity in Children Ages 6-12. ADA Evidence Analysis Library. http://www.adaevidencelibrary.com/evidence.cfm?format_tables=0&evidence_summary_id=250036. Accessed 01/03/12.

Academy of Nutrition and Dietetics Evidence Based Library. Effectiveness of using balanced macronutrient, reduced calorie (>1200 kcal-DRI per day) dietary in children ages 6-12. ADA Evidence Analysis Library.http://www.adaevidencelibrary.com/evidence.cfm?format_tables=0&evidence_summary_id=250036. Accessed 01/03/12.

Academy of Nutrition and Dietetics Evidence Based Library. Effectiveness of using a program to increase physical activity as a part of an intervention program to treat childhood overweight. ADA Evidence Analysis Library. http://www.adaevidencelibrary.com/conclusion.cfm?format_tables=0&conclusion_statement_id=105. Accessed 01/03/12.

Academy of Nutrition and Dietetics Evidence Based Library. Increasing physical activity as a part of an intervention program to treat childhood obesity. ADA Evidence Analysis Library. http://www.adaevidencelibrary.com/evidence.cfm?evidence_summary_id=55. Accessed 01/03/12.

Academy of Nutrition and Dietetics. Pediatric Nutrition Care Manual. http://peds.nutritioncaremanual.org/welcome.cfm. Accessed 02/01/12.

Academy of Nutrition and Dietetics Evidence Based Library. Pediatric Weight Management (PWM) Comprehensive, Multicomponent Weight Program for Treating Childhood Obesity. ADA Evidence Analysis Library. http://www.adaevidencelibrary.com/template.cfm?format_tables=0&template=guide_summary&key=1284, Accessed 01/03/12.

Academy of Nutrition and Dietetics Evidence Based Library. Pediatric Weight Management Evidence-Based Nutrition Practice Guideline. ADA Evidence Analysis Library. http://www.adaevidencelibrary.com/topic.cfm?format_tables=0&cat=2721. Accessed 12/24/11.

Academy of Nutrition and Dietetics Evidence Based Library. Pediatric Weight Management (PWM) Physical Activity in the Treatment of Childhood and Adolescent Obesity. ADA Evidence Analysis Library. http://www.adaevidencelibrary.com/template.cfm?format_tables=0&template=guide_summary&key=1224&highlight=childhoodobesity&home=1. Accessed 01/03/12.

Academy of Nutrition and Dietetics Evidence Based Library. The Traffic Light Diet and Treating Childhood Obesity. ADA Evidence Analysis Library. http://www.adaevidencelibrary.com/evidence.cfm?format_tables=0&evidence_summary_id=250033. Accessed 01/03/12.

Centers for Disease Control and Prevention, National Center for Health Statistics. CDC growth charts: United States. http://www.cdc.gov/growthcharts/. Accessed 01/03/12.

Lee RD. Energy balance and body weight. In: Nelms M, Sucher K, Lacey K, Long S. *Nutrition Therapy and Pathophysiology*. 2nd ed. Belmont, CA: Wadsworth, Cengage Learning; 2011:323–369.

Let's Move! http://www.letsmove.gov/. Accessed 01/03/12.

Lucas BL, Feucht SA, Grieger LE. *Children with Special Health Care Needs: Nutrition Care Handbook*. Chicago, IL: American Dietetic Association; 2004.

National Academy of Sciences Food and Nutrition Board. *Dietary Reference Intakes for Energy, Carbohydrate, Fiber, Fat, Fatty Acids, Cholesterol, Protein, and Amino Acids*. Washington, DC: The National Academies Press; 2005.

United States Department of Agriculture. Information for Health Care Professionals. MyPlate. http://www.choosemyplate.gov/information-healthcare-professionals.html. Accessed 01/03/12.

Internet Resources

American Academy of Pediatrics: http://www.aap.org/obesity/

American Sleep Apnea Association: http://www.sleepapnea.org/

Baylor College of Medicine Children's Nutrition Research Center: http://www.bcm.edu/cnrc/healthyeatingcalculator/eatingCal.html.

Centers for Disease Control and Prevention. Defining Obesity and Overweight: http://www.cdc.gov/nccdphp/dnpa/bmi/childrens_BMI/about_childrens_BMI.htm

Centers for Disease Control and Prevention. Growth Charts: http://www.cdc.gov/growthcharts/

Centers for Disease Control. BMI for Children and Teens: http://www.cdc.gov/nccdphp/dnpa/bmi/childrens_BMI/about_childrens_BMI.htm

Childhood Sleep Apnea: http://www.stanford.edu/~dement/childapnea.html

eMedicine: http://www.emedicine.com/neuro/topic566.htm

Mayo Clinic: http://www.mayoclinic.com/health/childhood-obesity/DS00698

Medline Plus: http://www.nlm.nih.gov/medlineplus/ency/article/003932.htm

Medline Plus: http://www.nlm.nih.gov/medlineplus/obesityinchildren.html

Bariatric Surgery for Morbid Obesity

After completing this case, the student will be able to:

1. Identify criteria that allow for an individual to qualify as a candidate for bariatric surgery.
2. Research and outline the health risks associated with morbid obesity.
3. Identify the current surgical procedures used for bariatric surgery.
4. Describe the potential physiological changes and nutrition problems that may occur after bariatric surgery.
5. Interpret nutrition assessment data to assist with the design of measurable goals, interventions, and strategies for monitoring and evaluation that address the nutrition diagnoses for the patient.

6. Understand current nutrition therapy guidelines for progression of oral intake after bariatric surgery.

Mr. McKinley is admitted for a Roux-en-Y gastric bypass surgery. He has suffered from type 2 diabetes mellitus, hyperlipidemia, hypertension, and osteoarthritis. Mr. McKinley has weighed over 250 lbs since age 15 with steady weight gain since that time. He has attempted to lose weight numerous times but the most weight he ever lost was 75 lbs, which he regained over a two-year period. He had recently reached his highest weight of 434 lbs, but since beginning the preoperative nutrition education program he has lost 24 lbs.

McKinley, Chris, Male, 37 y.o.
Allergies: NKA   **Code:** FULL   **Isolation:** None
Pt. Location: RM 703   **Physician:** P Walker   **Admit Date:** 2/23

Patient Summary: Patient is a morbidly obese 37-year-old white male who is admitted for Roux-en-Y gastric bypass surgery tomorrow morning. Patient has been obese his entire adult life with highest weight 6 months ago at 434 lbs. He has lost 24 lbs since that time as he has been attending the preoperative nutrition program at our clinic.

History:
Onset of disease: Lifelong obesity
Medical history: Type 2 diabetes mellitus, hypertension, hyperlipidemia, osteoarthritis
Surgical history: R total knee replacement 3 years previous
Medications at home: Metformin 1000 mg/twice daily; 35 u Lantus pm; Lasix 25 mg/day; Lovastatin 60 mg/day
Tobacco use: None
Alcohol use: Socially, 2–3 beers per week
Family history: Father: Type 2 DM, CAD, Htn, COPD; Mother: Type 2 DM, CAD, osteoporosis

Demographics:
Marital status: Single
Number of children: 0
Years education: Associate's degree
Language: English only
Occupation: Office manager for real estate office
Hours of work: 8-5 daily—sometimes on weekend
Household members: Lives with roommate
Ethnicity: Caucasian
Religious affiliation: None stated

Admitting History/Physical:
Chief complaint: "I am here for weight-loss surgery."
General appearance: Obese white male

Vital Signs: Temp: 98.9   Pulse: 85   Resp rate: 23
                    BP: 135/90   Height: 5'10"   Weight: 410 lbs
Heart: Normal rate, regular rhythm, normal heart; diminished distal pulses. Exam reveals no gallop and no friction rub.
HEENT: Head: WNL
           Eyes: PERRLA
           Ears: Clear
           Nose: WNL
           Throat: Moist mucous membranes without exudates or lesions
Genitalia: Normally developed 37-year-old male

McKinley, Chris, Male, 37 y.o.
Allergies: NKA
Pt. Location: RM 703

Code: FULL
Physician: P Walker

Isolation: None
Admit Date: 2/23

Neurologic: Alert and oriented
Extremities: Ecchymosis, abrasions, petechiae on lower extremities, 2+ pitting edema
Skin: Warm, dry to touch
Chest/lungs: Respirations WNL, clear to auscultation and percussion
Peripheral vascular: Diminished pulses bilaterally
Abdomen: Obese, rash present under skinfolds

Nursing Assessment	2/23
Abdominal appearance (concave, flat, rounded, obese, distended)	obese
Palpation of abdomen (soft, rigid, firm, masses, tense)	soft
Bowel function (continent, incontinent, flatulence, no stool)	continent
Bowel sounds (P=present, AB=absent, hypo, hyper)	
RUQ	P
LUQ	P
RLQ	P
LLQ	P
Stool color	lt brown
Stool consistency	formed
Tubes/ostomies	NA
Genitourinary	
Urinary continence	NA
Urine source	NA
Appearance (clear, cloudy, yellow, amber, fluorescent, hematuria, orange, blue, tea)	clear, yellow
Integumentary	
Skin color	pale
Skin temperature (DI=diaphoretic, W=warm, dry, CL=cool, CLM=clammy, CD+=cold, M=moist, H=hot)	W, M
Skin turgor (good, fair, poor, TENT=tenting)	good
Skin condition (intact, EC=ecchymosis, A=abrasions, P=petechiae, R=rash, W=weeping, S=sloughing, D=dryness, EX=excoriated, T=tears, SE=subcutaneous emphysema, B=blisters, V=vesicles, N=necrosis)	EC, A, R
Mucous membranes (intact, EC=ecchymosis, A=abrasions, P=petechiae, R=rash, W=weeping, S=sloughing, D=dryness, EX=excoriated, T=tears, SE=subcutaneous emphysema, B=blisters, V=vesicles, N=necrosis)	intact
Other components of Braden score: special bed, sensory pressure, moisture, activity, friction/shear (>18 = no risk, 15–16 = low risk, 13–14 = moderate risk, ≤12 = high risk)	15

McKinley, Chris, Male, 37 y.o.
Allergies: NKA   **Code:** FULL   **Isolation:** None
Pt. Location: RM 703   **Physician:** P Walker   **Admit Date:** 2/23

Orders:
Vital Signs, Routine, Every 4 hours
CBC with differential, comprehensive metabolic profile; PT/PTT; EKG; Urinalysis
NPO after midnight

Nutrition:
Meal type: NPO
Intake % of meals: NPO
Fluid requirement: 1800–2000 mL

MD Progress Note:
2/24
Subjective:Chris McKinley's previous 24 hours reviewed
Vitals: Temp: 98.9   Pulse: 78   Resp rate: 24   BP: 115/70
Urine output: 2230 mL Point of Care-Glu: 145

Physical Exam:
HEENT: WNL
Neck: WNL
Heart: WNL
Lungs: Clear to auscultation
Abdomen: Obese, soft, some epigastric tenderness +BS × 4

Assessment/Plan:
POD#1 s/p Roux-en-Y gastric surgery—now with positive bowel sounds. Will progress to Stage 1 Bariatric surgery diet. If tolerated, discharge to home after nutrition consult. Schedule for postoperative visit in one week. P. Walker MD

McKinley, Chris, Male, 37 y.o.
Allergies: NKA
Pt. Location: RM 703

Code: FULL
Physician: P Walker

Isolation: None
Admit Date: 2/23

Intake/Output

Date		2/23 0701–2/24 0700			
Time		0701–1500	1501–2300	2301–0700	Daily total
IN	P.O.	0	60	100	160
	I.V.	680	680	680	2040
	(mL/kg/hr)	(0.45)	(0.45)	(0.45)	(0.45)
	I.V. piggyback	0	0	0	0
	TPN	0	0	0	0
	Total intake	680	740	780	2200
	(mL/kg)	(3.6)	(3.9)	(4.2)	(11.8)
OUT	Urine	700	710	820	2230
	(mL/kg/hr)	(0.47)	(0.47)	(0.55)	(0.50)
	Emesis output	0	0	0	0
	Other	0	0	0	0
	Stool	0	0	0	0
	Total output	700	710	820	2230
	(mL/kg)	(3.7)	(3.8)	(4.4)	(12.0)
Net I/O		−20	+30	−40	−30
Net since admission (2/23)		−20	+10	−30	−30

Laboratory Results

	Ref. Range	2/23
Chemistry		
Sodium (mEq/L)	136–145	138
Potassium (mEq/L)	3.5–5.5	5.8 !↑
Chloride (mEq/L)	95–105	99
Carbon dioxide (CO_2, mEq/L)	23–30	27
BUN (mg/dL)	8–18	15
Creatinine serum (mg/dL)	0.6–1.2	0.9
BUN/Crea ratio	10–20	16.7:1
Glucose (mg/dL)	70–110	145 !↑
Phosphate, inorganic (mg/dL)	2.3–4.7	3.9
Magnesium (mg/dL)	1.8–3	2.0
Calcium (mg/dL)	9–11	9.5
Osmolality (mmol/kg/H_2O)	285–295	289

(Continued)

McKinley, Chris, Male, 37 y.o.

Allergies: NKA	**Code:** FULL
Pt. Location: RM 703	**Physician:** P Walker

Isolation: None
Admit Date: 2/23

Laboratory Results *(Continued)*

	Ref. Range	2/23
Bilirubin total (mg/dL)	≤1.5	0.8
Bilirubin, direct (mg/dL)	<0.3	0.07
Protein, total (g/dL)	6–8	6.8
Albumin (g/dL)	3.5–5	4.2
Prealbumin (mg/dL)	16–35	22
Ammonia (NH_3, µmol/L)	9–33	11
Alkaline phosphatase (U/L)	30–120	118
ALT (U/L)	4–36	21
AST (U/L)	0–35	10
CPK (U/L)	30–135 F 55–170 M	220 !↑
Cholesterol (mg/dL)	120–199	320 !↑
HDL-C (mg/dL)	>55 F, >45 M	32 !↓
VLDL (mg/dL)	7–32	45 !↑
LDL (mg/dL)	<130	232 !↑
LDL/HDL ratio	<3.22 F <3.55 M	7.5 !↑
Triglycerides (mg/dL)	35–135 F 40–160 M	245 !↑
T_4 (µg/dL)	4–12	6.1
T_3 (µg/dL)	75–98	82
Amylase (u/L)	25–125	26
Lipase (u/L)	10–140	11
HbA_{1C} (%)	3.9–5.2	7.2 !↑
Hematology		
WBC ($\times 10^3$/mm^3)	4.8–11.8	10.2
RBC ($\times 10^6$/mm^3)	4.2–5.4 F 4.5–6.2 M	5.5
Hemoglobin (Hgb, g/dL)	12–15 F 14–17 M	14.5
Hematocrit (Hct, %)	37–47 F 40–54 M	42
Mean cell volume (µm^3)	80–96	82
Mean cell Hgb (pg)	26–32	27

McKinley, Chris, Male, 37 y.o.
Allergies: NKA
Pt. Location: RM 703

Code: FULL
Physician: P Walker

Isolation: None
Admit Date: 2/23

Laboratory Results *(Continued)*

	Ref. Range	2/23
Mean cell Hgb content (g/dL)	31.5–36	33
RBC distribution (%)	11.6–16.5	12.3
Platelet count ($\times 10^3$/mm^3)	140–440	261
Transferrin (mg/dL)	250–380 F 215–365 M	279
Ferritin (mg/mL)	20–120 F 20–300 M	210
Vitamin B$_{12}$ (ng/dL)	24.4–100	72
Folate (ng/dL)	5–25	15
Urinalysis		
Collection method	—	clean catch
Color	—	yellow
Appearance	—	clear
Specific gravity	1.003–1.030	1.004
pH	5–7	6.1
Protein (mg/dL)	Neg	Neg
Glucose (mg/dL)	Neg	Neg
Ketones	Neg	Neg
Blood	Neg	Neg
Bilirubin	Neg	Neg
Nitrites	Neg	Neg
Urobilinogen (EU/dL)	<1.1	Neg
Leukocyte esterase	Neg	Neg
Prot chk	Neg	Neg
WBCs (/HPF)	0–5	0
RBCs (/HPF)	0–5	0
Bact	0	0
Mucus	0	0
Crys	0	0
Casts (/LPF)	0	0
Yeast	0	0

Case Questions

I.   Understanding the Diagnosis and Pathophysiology

1. Discuss the classification of morbid obesity.

2. Describe the primary health risks involved with untreated morbid obesity. What health risks does Mr. McKinley present with?

3. What are the standard adult criteria for consideration as a candidate for bariatric surgery? After reading Mr. McKinley's medical record, determine the criteria that allow him to qualify for surgery.

4. By performing an Internet search or literature review, find one example of a bariatric surgery program. Describe the information that is provided for the patient regarding qualification for surgery. Outline the personnel involved in the evaluation and care of the patient in this particular program.

5. Describe the following surgical procedures used for bariatric surgery, including advantages, disadvantages, and potential complications.

    a. Roux-en-Y gastric bypass

    b. Vertical sleeve gastrectomy

    c. Adjustable gastric banding (Lap-Band®)

    d. Vertical banded gastroplasty

    e. Duodenal switch

    f. Biliopancreatic diversion

6. Mr. McKinley has had type 2 diabetes for several years. His physician shared with him that after surgery he will not be on any medications for his diabetes and that he may be able to stop his medications for diabetes altogether. Describe the proposed effect of bariatric surgery on the pathophysiology of type 2 diabetes. What, if any, other medical conditions might be affected by weight loss?

II.   Understanding the Nutrition Therapy

7. On post-op day one, Mr. McKinley was advanced to the Stage 1 Bariatric Surgery Diet. This consists of sugar-free clear liquids, broth, and sugar-free Jell-O. Why are sugar-free foods used?

8. Over the next two months, Mr. McKinley will be progressed to a pureed-consistency diet with 6–8 small meals. Describe the major goals of this diet for the Roux-en-Y patient. How might the nutrition guidelines differ if Mr. McKinley had undergone a Lap-Band procedure?

9. Mr. McKinley's RD has discussed the importance of hydration, protein intake, and intakes of vitamins and minerals, especially calcium, iron, and B_{12}. For each of these nutrients, describe why intake may be inadequate and explain the potential complications that could result from deficiency.

III.   Nutrition Assessment

10. Assess Mr. McKinley's height and weight. Calculate his BMI and % usual body weight. What would be a reasonable weight goal for Mr. McKinley? Give your rationale for the method you used to determine this.

11. After reading the physician's history and physical, identify any signs or symptoms that are most likely a consequence of Mr. McKinley's morbid obesity.

12. Identify any abnormal biochemical indices and discuss the probable underlying etiology. How might they change after weight loss?

13. Determine Mr. McKinley's energy and protein requirements to promote weight loss. Explain the rationale for the method you used to calculate these requirements.

IV.   Nutrition Diagnosis

14. Identify at least two pertinent nutrition problems and the corresponding nutrition diagnoses.

V.   Nutrition Intervention

15. Determine the appropriate progression of Mr. McKinley's post-bariatric-surgery diet. Include recommendations for any supplementation that you would advise.

16. Describe any pertinent lifestyle changes that you would view as a priority for Mr. McKinley.

17. How would you assess Mr. McKinley's readiness for a physical activity plan? How does exercise assist in weight loss after bariatric surgery?

VI.   Nutrition Monitoring and Evaluation

18. Identify the steps you would take to monitor Mr. McKinley's nutritional status postoperatively.

19. From the literature, what is the success rate of bariatric surgery? What patient characteristics may increase the likelihood for success?

20. Mr. McKinley asks you about the possibility of bariatric surgery for a young cousin who is 10 years old. What are the criteria for bariatric surgery in children and adolescents?

21. Write an ADIME note for your inpatient nutrition assessment with initial education for the Stage 1 (liquid) diet for Mr. McKinley.

Bibliography

Academy of Nutrition and Dietetics. Nutrition Care Manual. Bariatric Surgery. http://nutritioncaremanual.org/content.cfm?ncm_content_id=79396. Accessed March 20, 2012.

Charney P, Malone AM. *ADA Pocket Guide to Pediatric Assessment*, 2nd ed. Chicago, IL; The Academy of Nutrition and Dietetics; 2009.

Egberts K, Brown WA, Brennan L, O'Brien PE. Does exercise improve weight loss after bariatric surgery? A Systematic Review. *Obes Surg.* 2012;22:335–341.

Junior WS, Lopes do Amaral J, Nonino-Borges CB. Factors related to weight loss up to 4 years after bariatric surgery. *Obes Surg.* 2011;21:1724–1730.

Laferrère B. Diabetes remission after bariatric surgery: Is it just the incretins? *International Journal of Obesity.* 2011;35:S22–S25.

Lee R. Disorders of weight balance. In: Nelms M, Sucher K, Lacey K, Long S. *Nutrition Therapy and Pathophysiology.* 2nd ed. Belmont, CA: Wadsworth, Cengage Learning; 2011:238–282.

Livhits M, Mercado C, Yermilov I, et al. Preoperative predictors of weight loss following bariatric surgery: Systematic review. *Obes Surg.* 2012;22:70–89.

Nelms M, Sucher KP, Lacey K, Long SR. *Nutrition Therapy and Pathophysiology.* 2nd ed. Wadsworth, Cengage Learning; 2011:364–366.

Padwal R, Klarenbach S, Wiebe N, et al. Bariatric surgery: A systematic review of the clinical and economic evidence. *J Gen Intern Med.* 2011;26:1183–94.

Park C, Torquati A. Physiology of weight loss surgery. *Surg Clin N Am.* 2001;91:1149–1161.

Raftopoulos, I Bernstein, O'Hara K, Ruby JA, Chhatrala R, Carty,J. Protein intake compliance of morbidly obese patients undergoing bariatric surgery and its effect on weight loss and biochemical parameters. *Surgery for Obesity and Related Diseases.* 2011;7:733–742

Internet Resources

American Society for Metabolic and Bariatric Surgery: http://ASMBS.org

Bariatric Eating: www.bariatriceating.com

LapBand: www.lapband.com/en/home

Nutrition Care Manual: http://www.nutritioncaremanual.org

Obesity Help: www.obesityhelp.com

Realize www.realize.com

VerticalSleeveTalk:www.verticalsleevetalk.com

Malnutrition Associated with Chronic Disease

Objectives

After completing this case, the student will be able to:

1. Identify the signs and symptoms associated with malnutrition.
2. Discern the physiological differences among starvation, chronic disease-related malnutrition, and malnutrition associated with acute disease.
3. Develop a nutrition care plan—with appropriate measurable goals, interventions, and strategies for monitoring and evaluation—that addresses the nutrition diagnoses for this case.

Harry Campbell is a 68-year-old male admitted to acute care for possible dehydration, weight loss, generalized weakness, and malnutrition.

Campbell, Harry, Male, 68 y.o.
Allergies: NKA
Pt. Location: RM 1119

Code: FULL
Physician: F. Connors

Isolation: None
Admit Date: 9/22

Patient Summary: Harry Campbell is a 68-year-old male admitted to acute care for possible dehydration, weight loss, generalized weakness, and malnutrition.

History:

Onset of disease: Patient diagnosed with squamous cell carcinoma of tongue five years ago. Patient previously treated with radiation therapy—no treatment $\times$ 3 years
Medical history: Essential hypertension; hyperlipidemia; weight loss; primary tongue squamous cell carcinoma five years previous; peripheral vascular disease
Surgical history: s/p partial glossectomy five years ago
Medications at home: Lipitor 80 mg daily; Capoten 25 mg twice daily
Tobacco use: 1 ppd for 60 plus years
Alcohol use: 1–3 cans of beer per day
Family history: Mother died of pneumonia; father died of lung cancer

Demographics:

Marital status: Married—lives with wife; *Spouse name:* Carol
Number of children: 2—alive, ages 42, 45
Years education: 9 years
Language: English only
Occupation: Meat cutter for 26 years; retired
Hours of work: N/A
Household members: Wife and patient
Ethnicity: Caucasian
Religious affiliation: Baptist

Admitting History/Physical:

Chief complaint: "I just feel weak all over and don't have the energy to do anything."
General appearance: Cachetic, appears older than years

Vital Signs: Temp: 96.6   Pulse: 101   Resp rate: 20
                     BP: 122/77   Height: 6'3"   Weight: 156 lbs
Heart: Regular rate and rhythm
HEENT: Head: Noted temporal wasting
           Eyes: PERRLA
           Ears: Clear
           Nose: Dry mucous membranes with petechiae
           Throat: Dry mucous membranes without exudates or lesions
Genitalia: Deferred
Neurologic: Alert and oriented; strength reduced
Extremities: Decreased muscle tone with normal ROM; loss of lean mass noted quadriceps and gastrocnemius; 1+ pedal edema
Skin: Warm and dry with ecchymoses

Campbell, Harry, Male, 68 y.o.
Allergies: NKA   **Code:** FULL   **Isolation:** None
Pt. Location: RM 1119   **Physician:** F. Connors   **Admit Date:** 9/22

Chest/lungs: Respirations are shallow—clear to auscultation and percussion
Peripheral vascular: Diminished pulses bilaterally
Abdomen: Hypoactive bowel sounds × 4; nontender, nondistended

Nursing Assessment	9/22
Abdominal appearance (concave, flat, rounded, obese, distended)	flat
Palpation of abdomen (soft, rigid, firm, masses, tense)	soft
Bowel function (continent, incontinent, flatulence, no stool)	continent
Bowel sounds (P=present, AB=absent, hypo, hyper)	
RUQ	P, hypo
LUQ	P, hypo
RLQ	P, hypo
LLQ	P, hypo
Stool color	light brown
Stool consistency	soft
Tubes/ostomies	NA
Genitourinary	
Urinary continence	catheter
Urine source	catheter
Appearance (clear, cloudy, yellow, amber, fluorescent, hematuria, orange, blue, tea)	clear, yellow
Integumentary	
Skin color	pale
Skin temperature (DI=diaphoretic, W=warm, dry, CL=cool, CLM=clammy, CD+=cold, M=moist, H=hot)	W, dry
Skin turgor (good, fair, poor, TENT=tenting)	TENT
Skin condition (intact, EC=ecchymosis, A=abrasions, P=petechiae, R=rash, W=weeping, S=sloughing, D=dryness, EX=excoriated, T=tears, SE=subcutaneous emphysema, B=blisters, V=vesicles, N=necrosis)	EC, D, T
Mucous membranes (intact, EC=ecchymosis, A=abrasions, P=petechiae, R=rash, W=weeping, S=sloughing, D=dryness, EX=excoriated, T=tears, SE=subcutaneous emphysema, B=blisters, V=vesicles, N=necrosis)	intact, D, P
Other components of Braden score: special bed, sensory pressure, moisture, activity, friction/shear (>18 = no risk, 15–16 = low risk, 13–14 = moderate risk, ≤12 = high risk)	friction/shear; 17

Orders:

0.9% sodium chloride with potassium chloride 20 mEq 125 mL/hr
Vancomycin 1 g in dextrose 200 mL IVPB

Campbell, Harry, Male, 68 y.o.
Allergies: NKA
Pt. Location: RM 1119

Code: FULL
Physician: F. Connors

Isolation: None
Admit Date: 9/22

Thiamin injection 100 mg daily
Multivitamin capsule 1 Cap daily
Metronidazole 500 mg in NaCl premix IVPB
Docusate capsule 100 mg twice daily
Lipitor 80 mg daily
Lopressor 5 mg every 6 hours

Nutrition:

Meal type: Mechanical soft diet
Intake % of meals: <5%; sips of liquids
Fluid requirement: 2000–2500 mL
History: Patient states that he has lost over 60 lbs in past 1–2 years. He lost some weight when diagnosed with cancer 5 years ago but held steady at approximately 220 lbs even after completing radiation therapy. He states that he gets full really easily and never feels hungry.
Usual intake (for past several months): AM—egg, coffee, few bites of toast; 10 am—½ can Ensure Plus; lunch—soup or ½ sandwich, milk; dinner – few bites of soft meat, potatoes or rice. Tries to drink the other ½ can of Ensure Plus.

Intake/Output

Date		9/22 0701–9/23 0700			
Time		0701–1500	1501–2300	2301–0700	Daily total
IN	P.O.	**sips**	**120**	**240**	**360**
	I.V.	**720**	**720**	**720**	**2,160**
	(mL/kg/hr)	(1.3)	(1.3)	(1.3)	(1.3)
	I.V. piggyback				
	TPN				
	Total intake	**720**	**840**	**960**	**2,520**
	(mL/kg)	(10.2)	(11.8)	(13.5)	(35.5)
OUT	Urine	**480**	**320**	**643**	**1,443**
	(mL/kg/hr)	(0.8)	(0.6)	(1.1)	(0.8)
	Emesis output				
	Other				
	Stool	**1**			
	Total output	**481**	**320**	**643**	**1,444**
	(mL/kg)	(6.8)	(4.5)	(9.1)	(20.4)
Net I/O		**+239**	**+520**	**+317**	**+1,076**
Net since admission (9/22)		**+239**	**+759**	**+1,076**	**+1,076**

Campbell, Harry, Male, 68 y.o.
Allergies: NKA
Pt. Location: RM 1119

Code: FULL
Physician: F. Connors

Isolation: None
Admit Date: 9/22

Laboratory Results

	Ref. Range	9/22 1522
Chemistry		
Sodium (mEq/L)	136–145	150 !↑
Potassium (mEq/L)	3.5–5.5	3.4 !↓
Chloride (mEq/L)	95–105	118 !↑
Carbon dioxide (CO_2, mEq/L)	23–30	29
BUN (mg/dL)	8–18	36 !↑
Creatinine serum (mg/dL)	0.6–1.2	1.27 !↑
BUN/Crea ratio		28
Est GFR, non-Afr-Amer	—	55
Est GFR, Afr-Amer	—	>60
Glucose (mg/dL)	70–110	71
Phosphate, inorganic (mg/dL)	2.3–4.7	2.8
Magnesium (mg/dL)	1.8–3	2.0
Calcium (mg/dL)	9–11	8.4 !↓
Anion gap	—	7
Bilirubin total (mg/dL)	≤1.5	0.6
Bilirubin, direct (mg/dL)	<0.3	0.1
Protein, total (g/dL)	6–8	5.8 !↓
Albumin (g/dL)	3.5–5	1.8 !↓
Prealbumin (mg/dL)	16–35	9 !↓
Ammonia (NH_3, μmol/L)	9–33	11
Alkaline phosphatase (U/L)	30–120	75
ALT (U/L)	4–36	31
AST (U/L)	0–35	24
C-reactive protein (mg/dL)	<1.0	2.4 !↑
Cholesterol (mg/dL)	120–199	92 !↓
Triglycerides (mg/dL)	35–135 F 40–160 M	72
Coagulation (Coag)		
PT (sec)	12.4–14.4	15.1 !↑
INR	0.9–1.1	1.0
PTT (sec)	24–34	25
Hematology		
WBC (× 10^3/mm^3)	4.8–11.8	10.6

(Continued)

Campbell, Harry, Male, 68 y.o.
Allergies: NKA   **Code:** FULL   **Isolation:** None
Pt. Location: RM 1119   **Physician:** F. Connors   **Admit Date:** 9/22

Laboratory Results *(Continued)*

	Ref. Range	9/22 1522
RBC ($\times 10^6$/mm^3)	4.2–5.4 F 4.5–6.2 M	2.4 !↓
Hemoglobin (Hgb, g/dL)	12–15 F 14–17 M	8.1 !↓
Hematocrit (Hct, %)	37–47 F 40–54 M	24.1 !↓
Mean cell volume (µm^3)	80–96	100.6 !↑
Mean cell Hgb (pg)	26–32	33.6 !↑
Mean cell Hgb content (g/dL)	31.5–36	33
RBC distribution (%)	11.6–16.5	18 !↑
Platelet count ($\times 10^3$/mm^3)	140–440	240
Hematology, Manual Diff		
Neutrophil (%)	50–70	55
Lymphocyte (%)	15–45	11 !↓
Monocyte (%)	3–10	4
Eosinophil (%)	0–6	0
Basophil (%)	0–2	0
Blasts (%)	3–10	4
Segs (%)	0–60	45
Bands (%)	0–10	10
Urinalysis		
Color	—	yellow
Appearance	—	clear
Specific gravity	1.003–1.030	1.033
pH	5–7	7.0
Protein (mg/dL)	Neg	100
Glucose (mg/dL)	Neg	Neg
Ketones	Neg	+
Blood	Neg	small
Urobilinogen (EU/dL)	<1.1	0.2
Leukocyte esterase	Neg	small
Prot chk	Neg	+
WBCs (/HPF)	0–5	3–4
RBCs (/HPF)	0–5	1–2
Bact	0	+

Case Questions

I.   Understanding the Diagnosis and Pathophysiology

1.   Outline the metabolic changes that occur during starvation that could result in weight loss.

2.   Identify current definitions of malnutrition in the United States using the current ICD codes.

3.   Current definitions of malnutrition use biochemical markers as a component of the diagnostic criteria. Explain the effect of inflammation on visceral proteins and how that may impact the clinician's ability to diagnose malnutrition. What laboratory values will confirm the presence of inflammation?

4.   What does the AND evidence analysis indicate regarding the correlation of albumin/prealbumin with visceral protein status and risk of malnutrition during periods of prolonged protein-energy restriction?

5.   Read the article: Jensen et al. Adult starvation and disease-related malnutrition: A proposal for etiology-based diagnosis in the clinical practice setting from the International Consensus Guideline Committee. *Clinical Nutrition* 29 (2010):151–153. Explain the differences between malnutrition associated with chronic disease and malnutrition associated with acute illness and inflammation.

II.   Understanding the Nutrition Therapy

6.   Mr. Campbell was ordered a mechanical soft diet when he was admitted to the hospital. Describe the modifications for this diet order.

7.   What is Ensure Plus? Determine additional options for Mr. Campbell that would be appropriate for a high-calorie, high-protein beverage supplement.

III.   Nutrition Assessment

8.   Assess Mr. Campbell's height and weight. Calculate his BMI and % usual body weight.

9.   After reading the physician's history and physical, identify any signs or symptoms that support the diagnosis of malnutrition.

10.   Evaluate Mr. Campbell's initial nursing assessment. What important factors noted in his nutrition assessment may support the diagnosis of malnutrition?

11. What is a Braden score? Assess Mr. Campbell's score. How does this relate to his nutritional status?

12. Identify any signs or symptoms from the physician's history and physical and from the nursing assessment that are consistent with dehydration.

13. Determine Mr. Campbell's energy and protein requirements. Explain the rationale for the method you used to calculate these requirements.

14. Determine Mr. Campbell's fluid requirements. Compare this with the information on the intake/output report.

15. From the nutrition history, assess Mr. Campbell's usual dietary intake. How does this compare to the requirements that you calculated for him?

IV.   Nutrition Diagnosis

16. Identify the pertinent nutrition problems and the corresponding nutrition diagnoses and write at least two PES statements.

V.   Nutrition Intervention

17. Determine the appropriate intervention for each nutrition diagnosis.

18. Based on the criteria established in Jensen et al.'s article as well as the consensus statement from AND and ASPEN, what type of malnutrition is Mr. Campbell experiencing? Provide the specific criteria that support your diagnosis.

VI.   Nutrition Monitoring and Evaluation

19. Identify the steps you would take to monitor Mr. Campbell's nutritional status while he is hospitalized. How would this differ if you were providing follow-up care through his physician's office?

20. Write your ADIME note for this initial nutrition assessment for Mr. Campbell.

Bibliography

Jensen GL, Mirtallo J, Compher C, et al. Adult starvation and disease related malnutrition: A proposal for etiology-based diagnosis in the clinical practice setting from the International Consensus Guideline Committee. *Clinical Nutrition*. 2010;29:151–153.

Jensen GL, Wheeler D. A new approach to defining and diagnosing malnutrition in adult critical illness. *Curr Opin Crit Care*. 2012;18:206–211.

Lim SL, Ong KC, Chan YH, Loke WC, Ferguson M, Daniels L. Malnutrition and its impact on cost of hospitalization, length of stay, readmission and 3-year mortality. *Clinical Nutrition*. 2011; doi:10.1016/j.clnu.2011.11.001 epub ahead of print.

Meijers J, van Bokhorst-de van der Schueren MA, Schols JM, et al. Defining malnutrition: Mission or Mission Impossible? *Nutrition*. 2010;26:432–440.

Mueller C, Compher C, Ellen DM; American Society for Parenteral and Enteral Nutrition (A.S.P.E.N.) Board of Directors. A.S.P.E.N. clinical guidelines: Nutrition screening, assessment, and intervention in adults. *J Parenter Enteral Nutr*. 2011;35(1):16–24.

Nahikian-Nelms ML. Metabolic stress and the critically ill. In: Nelms M, Sucher K, Lacey K, Long S. *Nutrition Therapy and Pathophysiology*, 2nd ed. Belmont, CA: Wadsworth, Cengage Learning; 2011:682–701.

Nahikian-Nelms ML., Habash D. Nutrition assessment: Foundation of the nutrition care process. In: Nelms M, Sucher K, Lacey K, Long S. *Nutrition Therapy and Pathophysiology*. 2nd ed. Belmont, CA: Wadsworth, Cengage Learning; 2011:34–65.

Pirlich M, Schutz T, Norman K, et al. The German hospital malnutrition study. *Clinical Nutrition*. 2006;25:563–572.

White JV, Guenter P, Jensen G, Malone A, Schofield M, Academy of Nutrition and Dietetics Malnutrition Work Group, A.S.P.E.N. Malnutrition Task Force, A.S.P.E.N. Board of Directors. Consensus statement of the Academy of Nutrition and Dietetics/American Society of Parenteral and Enteral Nutrition: Characteristics recommended for the identification and documentation of adult malnutrition (undernutrition). *J Acad Nutr Diet*. 2012;112:730–8.

Internet Resources

ADA Evidence Analysis Library: http://www.adaevidencelibrary.com

Braden Scale: http://www.bradenscale.com

ICD-9 Codes: http://www.icd9data.com/2012/Volume1/default.htm

Nutrition Care Manual: http://www.nutritioncaremanual.org

USDA SuperTracker: http://www.choosemyplate.gov/supertracker-tools/supertracker.html

Unit Two

NUTRITION THERAPY FOR CARDIOVASCULAR DISORDERS

Cardiovascular disease is the leading cause of death in the United States. Risk factors for cardiovascular disease include dyslipidemia, smoking, diabetes mellitus, high blood pressure, obesity, and physical inactivity. Researchers estimate that more than 80 million Americans have one or more forms of cardiovascular disease; as a result, many patients that the health care team encounters will have conditions related to cardiovascular disease.

This section includes three of the most common diagnoses: hypertension (HTN), myocardial infarction (MI), and heart failure (HF). All these diagnoses require a significant medical nutrition therapy component for their care.

Over 74 million people in the United States have hypertension. Hypertension is defined as a systolic blood pressure of 140 mm Hg or higher and a diastolic pressure of 90 mm Hg or higher. Essential hypertension, which is the most common form of hypertension, is of unknown etiology. Case 4 focuses on lifestyle modifications as the first step in treatment of hypertension accompanied by dyslipidemia in a female patient. This case incorporates the pharmacological treatment of hypertension, and you will use the most recent information from *Dietary Approaches to Stop Hypertension* (DASH) as the center of the medical nutrition therapy intervention. Because cardiovascular disease is a complex, multifactorial condition, Case 4 provides the opportunity to evaluate these multiple risk factors through all facets of nutrition assessment. We specifically emphasize interpretation of laboratory indices for dyslipidemia. In this case, you will also determine the clinical classification and treatment of abnormal serum lipids, explore the use of drug therapy to treat dyslipidemias, and develop appropriate nutrition interventions using the Therapeutic Lifestyle Changes (TLC) recommendations as the framework for these diagnoses.

Case 5 focuses on the acute care of an individual suffering a myocardial infarction (MI). Ischemia of the vessels within the heart results in death of the affected heart tissue. This case lets you evaluate pertinent assessment measures for the individual suffering an MI and then develop an appropriate nutrition care plan that complements the medical care for prevention of further cardiac deterioration.

Case 6 addresses the long-term consequences of cardiovascular disease in a patient suffering from heart failure (HF). In HF, the heart cannot pump effectively, and the lack of oxygen and nutrients affects the body's tissues. HF is a major public health problem in the United States, and its incidence is increasing. Without a heart transplant, long-term prognosis is poor. This advanced case requires you to integrate understanding of the physiology of several body systems as you address heart failure's metabolic effects. Additionally, this case allows you to explore the role of the health care team in palliative care.

Hypertension and Cardiovascular Disease

Objectives

After completing this case, the student will be able to:

1. Describe the physiology of blood pressure regulation.
2. Apply knowledge of the pathophysiology of hypertension and dyslipidemias to identify and explain common nutritional problems associated with these diseases.
3. Explain the role of nutrition therapy as an adjunct to the pharmacotherapy, surgical, and other medical treatment of cardiovascular disease.
4. Interpret laboratory parameters for nutritional implications and significance.
5. Analyze nutrition assessment data to evaluate nutritional status and identify specific nutrition problems.

6. Determine nutrition diagnoses and write appropriate PES statements.
7. Develop a nutrition care plan—with appropriate measurable goals, interventions, and strategies for monitoring and evaluation—that addresses the nutrition diagnoses of this case.

Mrs. Cookie Sanders is a 54-year-old housewife. For the past year, she has treated her newly diagnosed hypertension with lifestyle changes including diet, smoking cessation, and exercise. She is in to see her physician for further evaluation and treatment for essential hypertension. Blood drawn 2 weeks prior to this appointment shows an abnormal lipid profile.

Sanders, Cookie, Female, 54 y.o.
Allergies: No known allergies   **Code:** FULL   **Isolation:** None
Pt. Location: University Clinic   **Physician:** A. Thornton   **Appointment Date:** 6/25

Patient Summary: 54-year-old female here for evaluation and treatment for essential hypertension and hyperlipidemia

History:
Onset of disease: Mrs. Sanders is a 54-yo female who is not employed outside the home. She was diagnosed 1 year ago with Stage 2 (essential) HTN. Treatment thus far has been focused on non-pharmacological measures. She began a walking program resulting in a 10-pound weight loss she has been able to maintain during the past year. She walks 30 minutes 4–5 times per week, though she sometimes misses on bingo nights. She was given a nutrition information pamphlet in the MD office outlining a lower-Na diet. Mrs. Sanders was a 2-pack-a-day smoker but quit ("cold turkey") when her HTN was diagnosed last year. No c/o of any symptoms related to HTN. Pt denies chest pain, SOB, syncope, palpitations, or myocardial infarction.
Medical history: Not significant before Dx of HTN
Surgical history: None
Medications at home: None
Tobacco use: No—quit 1 year ago
Alcohol use: 2–4 beers/wk
Family history: What? HTN. Who? Mother died of MI related to uncontrolled HTN.

Demographics:
Marital status: Married; *Spouse name:* Steve Sanders, 60 yo
Number of children: Children are grown and do not live at home.
Years education: High school
Language: English
Occupation: No employment outside of home
Hours of work: Varies
Household members: 2
Ethnicity: African-American
Religious affiliation: Roman Catholic

MD Progress Note:
Review of Systems
Constitutional: Negative
Skin: Negative
Cardiovascular: No carotid bruits
Respiratory: Negative
Gastrointestinal: Negative
Neurological: Negative
Psychiatric: Negative

Physical Exam
General appearance: Healthy, middle-aged female who looks her age
Heart: Regular rate and rhythm, normal heart sounds—no clicks, murmurs, or gallops

Sanders, Cookie, Female, 54 y.o.
Allergies: No known allergies   **Code:** FULL   **Isolation:** None
Pt. Location: University Clinic   **Physician:** A. Thornton   **Appointment Date:** 6/25

HEENT: Eyes: No retinopathy, PERRLA
Genitalia: Normal female
Neurologic: Alert and oriented × 3
Extremities: Noncontributory
Skin: Smooth, warm, dry, excellent turgor, no edema
Chest/lungs: Lungs clear
Peripheral vascular: Pulse 4+ bilaterally, warm, no edema
Abdomen: Nontender, no guarding, normal bowel sounds

Vital Signs:	Temp: 98.6	Pulse: 80	Resp rate: 15	
	BP: 160/100	Height: 5'6"	Weight: 160 lbs	BMI: 25.8

Assessment and Plan:
54-year-old female with Stage 2 HTN here for initiation of pharmacologic therapy with thiazide diuretics and reinforcement of lifestyle modifications. Rule out metabolic syndrome.

Dx: Stage 2 HTN, heart disease, early COPD

Medical Tx plan: Urinalysis; hematocrit; blood chemistry to include plasma glucose, potassium, BUN, creatinine, fasting lipid profile, triglycerides, calcium, uric acid; chest X-ray; nutrition consult; 25 mg hydrochlorothiazide daily; evaluate for initiation of HMGCoA reductase inhibitor therapy; pt to be reassessed in 3 months

.... A. Thornton, MD

Nutrition:
General: Mrs. Sanders describes her appetite as "very good." She does the majority of grocery shopping and cooking, although Mr. Sanders cooks breakfast on the weekends. She usually eats three meals each day, but on bingo nights, she usually skips dinner and just snacks while playing bingo. When she does this, she is really hungry when she gets home in the late evening, so she often eats a bowl of ice cream before going to bed. The Sanders usually eat out on Friday and Saturday evenings at pizza restaurants or steakhouses (Mrs. Sanders usually has 2 regular beers with these meals). She mentions that last year when her HTN was diagnosed, a nurse at the MD's office gave her a sheet of paper with a list of foods to avoid for a lower-salt diet. She and her husband tried to comply with the diet guidelines, but they found foods bland and tasteless, and they soon abandoned the effort.

24-hour recall:
AM: 1 c coffee (black)
            Oatmeal (1 instant packet with 1 tsp margarine and 2 tsp sugar)
            ½ c low-fat (2%) milk
            1 c orange juice

Sanders, Cookie, Female, 54 y.o.
Allergies: No known allergies   **Code:** FULL   **Isolation:** None
Pt. Location: University Clinic   **Physician:** A. Thornton   **Appointment Date:** 6/25

Snack: 2 c coffee (black)
              1 glazed donut
Lunch: 1 can Campbell's® tomato bisque soup prepared with milk
              10 saltines
              1 can diet cola
PM: 6 oz baked chicken (white meat, no skin; seasoned with salt, pepper, garlic)
              1 large baked potato with 1 tbsp butter, salt, and pepper
              1 c glazed carrots (1 tsp sugar, 1 tsp butter)
              Dinner salad with ranch-style dressing (3 tbsp)—lettuce, spinach, croutons, sliced
              cucumber
              2 regular beers
HS snack: 2 c butter pecan ice cream

Food allergies/intolerances/aversions: None
Previous nutrition therapy? Yes. If yes, when: 1 year ago. Where? MD's office
Food purchase/preparation: Self and husband
Vit/min intake: Multivitamin/mineral daily

Laboratory Results

	Ref. Range	6/25	12/25 (6 mos. later)	3/25 (9 mos. later)
Chemistry				
Sodium (mEq/L)	136–145	137	138	139
Potassium (mEq/L)	3.5–5.5	4.5	3.6	3.9
Chloride (mEq/L)	95–105	102	100	101
Carbon dioxide (CO_2, mEq/L)	23–30	30	29	29
BUN (mg/dL)	8–18	20 !↑	18	22 !↑
Creatinine serum (mg/dL)	0.6–1.2	0.9	1.1	1.1
BUN/Crea ratio	10:1–20:1	22.2:1 !↑	16.4:1	20:1
Uric acid (mg/dL)	2.8–8.8 F 4.0–9.0 M	6.8	7.0	7.2
Glucose (mg/dL)	70–110	115 !↑	90	96
Phosphate, inorganic (mg/dL)	2.3–4.7	4.1	3.5	3.5
Magnesium (mg/dL)	1.8–3	2.1	2.3	2.3
Calcium (mg/dL)	9–11	9.2	9.0	9.1
Osmolality (mmol/kg/H_2O)	285–295	294	293	295
Bilirubin, direct (mg/dL)	<0.3	1.1 !↑	0.8 !↑	0.9 !↑
Protein, total (g/dL)	6–8	7	7	6.8

Sanders, Cookie, Female, 54 y.o.
Allergies: No known allergies
Pt. Location: University Clinic

Code: FULL
Physician: A. Thornton

Isolation: None
Appointment Date: 6/25

Laboratory Results *(Continued)*

	Ref. Range	6/25	12/25 (6 mos. later)	3/25 (9 mos. later)
Albumin (g/dL)	3.5–5	4.6	4.3	4.4
Prealbumin (mg/dL)	16–35	32	31	31
Ammonia (NH$_3$, μmol/L)	9–33	19	18	22
Alkaline phosphatase (U/L)	30–120	120	100	109
ALT (U/L)	4–36	30	35	28
AST (U/L)	0–35	34	31	30
CPK (U/L)	30–135 F 55–170 M	100	125	130
Lactate dehydrogenase (U/L)	208–378	314	323	350
Cholesterol (mg/dL)	120–199	270 !↑	230 !↑	210 !↑
HDL-C (mg/dL)	>55 F, >45 M	30 !↓	35 !↓	38 !↓
LDL (mg/dL)	<130	210 !↑	169 !↑	147 !↑
LDL/HDL ratio	<3.22 F <3.55 M	7.0 !↑	4.8 !↑	3.9 !↑
Apo A (mg/dL)	101–199 F 94–178 M	75 !↓	100 !↓	110
Apo B (mg/dL)	60–126 F 63–133 M	140 !↑	120	115
Triglycerides (mg/dL)	35–135 F 40–160 M	150 !↑	130	125
T$_4$ (μg/dL)	4–12	8.6	8.5	7.8
T$_3$ (μg/dL)	75–98	95	92	93
HbA$_{1C}$ (%)	3.9–5.2	4.9	5.0	5.1
Coagulation (Coag)				
PT (sec)	12.4–14.4	13.5	13.7	13.2
Hematology				
WBC ($\times 10^3$/mm^3)	4.8–11.8	6.1	5.7	5.5
RBC ($\times 10^6$/mm^3)	4.2–5.4 F 4.5–6.2 M	5.5 !↑	5.0	5.25
Hemoglobin (Hgb, g/dL)	12–15 F 14–17 M	14.2	14.0	13.9
Hematocrit (Hct, %)	37–47 F 40–54 M	45	44	44

(Continued)

Sanders, Cookie, Female, 54 y.o.
Allergies: No known allergies   **Code:** FULL   **Isolation:** None
Pt. Location: University Clinic   **Physician:** A. Thornton   **Appointment Date:** 6/25

Laboratory Results *(Continued)*

	Ref. Range	6/25	12/25 (6 mos. later)	3/25 (9 mos. later)
Mean cell volume (μm^3)	80–96	88	90	85
Platelet count ($\times 10^3/mm^3$)	140–440	430	350	366
Ferritin (mg/mL)	20–120 F 20–300 M	115	116	112
Vitamin B_{12} (ng/dL)	24.4–100	90	70.5	63.3
Folate (ng/dL)	5–25	5	24	8
Urinalysis				
Collection method	—	rand. spec.	rand. spec.	rand. spec.
Color	—	pale yellow	pale yellow	pale yellow
Appearance		clear	clear	clear
Specific gravity	1.003–1.030	1.025	1.021	1.024
pH	5–7	7.0	5.0	6.0
Protein (mg/dL)	Neg	Neg	Neg	Neg
Glucose (mg/dL)	Neg	Neg	Neg	Neg
Ketones	Neg	Neg	Neg	Neg
Blood	Neg	Neg	Neg	Neg
Bilirubin	Neg	Neg	Neg	Neg
Nitrites	Neg	Neg	Neg	Neg
Urobilinogen (EU/dL)	<1.1	0.02	0.01	0.09
Leukocyte esterase	Neg	Neg	Neg	Neg
Protchk	Neg	Neg	Neg	Neg
WBCs (/HPF)	0–5	0	0	0
RBCs (/HPF)	0–5	0	0	0
Bact	0	0	0	0
Mucus	0	0	0	0
Crys	0	0	0	0
Casts (/LPF)	0	0	0	0
Yeast	0	0	0	0

Case Questions

I.   Understanding the Disease and Pathophysiology

1.   Define blood pressure and explain how it is measured.

2.   How is blood pressure normally regulated in the body?

3.   What causes essential hypertension?

4.   What are the symptoms of hypertension?

5.   How is hypertension diagnosed?

6.   List the risk factors for developing hypertension.

7.   What risk factors does Mrs. Sanders currently have?

8.   Hypertension is classified in stages based on the risk of developing CVD. Complete the following table of hypertension classifications.

Category	Systolic BP	Blood Pressure mmHg	Diastolic BP
Normal		and	
Prehypertension		or	
Stage 1 Hypertension		or	
Stage 2 Hypertension		or	

9.   How is hypertension treated?

10. Dr. Thornton indicated in his note that he will "rule out metabolic syndrome." What is metabolic syndrome?

11. What factors found in the medical and social history are pertinent for determining Mrs. Sanders's CHD risk category?

12. What progression of her disease might Mrs. Sanders experience?

II.   Understanding the Nutrition Therapy

13. Briefly describe the DASH eating plan.

14. Using the EAL, describe the association between sodium intake and blood pressure.

15. Lifestyle modifications reduce blood pressure, enhance the efficacy of antihypertensive medications, and decrease cardiovascular risk. List lifestyle modifications that have been shown to lower blood pressure.

III.   Nutrition Assessment

16. What are the health implications of Mrs. Sanders's body mass index (BMI)?

17. Calculate Mrs. Sanders's resting and total energy needs.

18. What nutrients in Mrs. Sanders's diet are of major concern to you?

19. From the information gathered within the intake domain, list possible nutrition problems using the diagnostic terms.

20. Dr. Thornton ordered the following labs: fasting glucose, cholesterol, triglycerides, creatinine, and uric acid. He also ordered an EKG. In the following table, outline the indication for these tests (tests provide information related to a disease or condition).

Parameter	Normal Value	Pt's Value	Reason for Abnormality	Nutrition Implication
Glucose	70–110 mg/dL			
BUN	8–18 mg/dL			
Creatinine	0.6–1.2 mg/dL			
Total cholesterol	120–199 mg/dL			
HDL-cholesterol	>55 mg/dL F >45 mg/dL M			
LDL-cholesterol	<130 mg/dL			
Apo A	101–199 mg/dL F 94–178 mg/dL M			
Apo B	60–126 mg/dL F 63–133 mg/dL M			
Triglycerides	35–135 mg/dL F 40–160 mg/dL M			

21. Interpret Mrs. Sanders's risk of CAD based on her lipid profile.

22. What is the significance of apolipoprotein A and apolipoprotein B in determining a person's risk of CAD?

23. Indicate the pharmacological differences among the antihypertensive agents listed below.

Medications	Mechanism of Action	Nutritional Side Effects and Contraindications
Diuretics		
Beta-blockers		
Calcium-channel blockers		
ACE inhibitors		
Angiotensin II receptor blockers		
Alpha-adrenergic blockers		

24. What are the most common nutritional implications of taking hydrochlorothiazide?

25. Mrs. Sanders's physician has decided to prescribe an ACE inhibitor and an HMGCoA reductase inhibitor (Zocor). What changes can be expected in her lipid profile as a result of taking these medications?

26. How does an ACE inhibitor lower blood pressure?

27. How does an HMGCoA reductase inhibitor lower serum lipid?

28. What other classes of medications can be used to treat hypercholesterolemia?

29. What are the pertinent drug–nutrient interactions and medical side effects for ACE inhibitors and HMGCoA reductase inhibitors?

30. From the information gathered within the clinical domain, list possible nutrition problems using the diagnostic terms.

31. What are some possible barriers to compliance?

IV. **Nutrition Diagnosis**
32. Select two nutrition problems and complete the PES statement for each.

V. **Nutrition Intervention**
33. When you ask Mrs. Sanders how much weight she would like to lose, she tells you she would like to weigh 125, which is what she weighed most of her adult life. Is this reasonable? What would you suggest as a goal for weight loss for Mrs. Sanders?

34. How quickly should Mrs. Sanders lose this weight?

35. For each of the PES statements that you have written, establish an ideal goal (based on the signs and symptoms) and an appropriate intervention (based on the etiology).

36. Identify the major sources of sodium, saturated fat, and cholesterol in Mrs. Sanders's diet. What suggestions would you make for substitutions and/or other changes that would help Mrs. Sanders reach her medical nutrition therapy goals?

37. What would you want to reevaluate in three to four weeks at a follow-up appointment?

38. Evaluate Mrs. Sanders's labs at six months and then at nine months. Describe the changes that have occurred.

Bibliography

Academy of Nutrition and Dietetics Evidence Based Library. Hypertension (HTN): Classification of Blood Pressure. ADA Evidence Analysis Library. http://www.adaevidencelibrary.com/template .cfm?key=1945&cms_preview=1. Accessed 01/03/12.

Academy of Nutrition and Dietetics Evidence Based Library. Hypertension Evidence-Based Nutrition Practice Guideline. ADA Evidence Analysis Library. https://www.adaevidencelibrary.com/topic .cfm?cat=3260. Accessed 01/11/12.

Academy of Nutrition and Dietetics Nutrition Care Manual. BMI and Weight Range Calculator. http://nutritioncaremanual.org/calculators .cfm?calculator_type=CalcBMI. Accessed 2/5/12.

Academy of Nutrition and Dietetics Nutrition Care Manual. Hypertension. http://nutritioncaremanual .org/topic.cfm?ncm_heading=Diseases%2FCondition s&ncm_toc_id=8480. Accessed 2/5/12.

Appel LJ. American Society of Hypertension position paper: Dietary approaches to lower blood pressure. *J Clin Hypertens*. 2009;11:358–368.

Graudal NA, Hubeck-Graudal T, Jurgens G. Effects of low-sodium diet vs. high-sodium diet on blood pressure, renin, aldosterone, catacholamines, cholesterol, and triglyderide. *Am J Hypertens*. 2012;25(1):1–15.

Grundy SM, Brewer B, Cleeman JI, Smith SC, Lenfant C. Definition of metabolic syndrome: Report of the National Heart, Lung, and Blood Institute/American Heart Association Conference on scientific issues related to definition. *Circulation*. 2004;109:433–438.

Lacey K. The nutrition care process. In: Nelms M, Sucher K, Lacey K, Roth SL. *Nutrition Therapy and Pathophysiology*. 2nd ed. Belmont, CA: Wadsworth, Cengage Learning; 2011:14–33.

National Institutes of Health, National Heart, Lung, and Blood Institutes. *DASH Eating Plan: Lower Your Blood Pressure*. US Dept. of Health and Human Services, NIH Publication No. 06-4082. Originally printed 1998, revised April 2006. http://www.nhlbi.nih.gov/health/public /heart/hbp/dash/new_dash.pdf. Accessed 2/5/12.

National Institutes of Health, National Heart, Lung, and Blood Institutes. *What is metabolic syndrome?* US Dept. of Health and Human Services, revised November 2011. http://www.nhlbi.nih.gov/health /health-topics/topics/ms/. Accessed 2/11/12.

Nelms MN. Nutrition assessment: Foundation of the nutrition care process. In: Nelms M, Sucher K, Lacey K, Roth SL. *Nutrition Therapy and Pathophysiology*. 2nd ed. Belmont, CA: Wadsworth, Cengage Learning; 2011:34–65.

Pronsky ZM, Crowe JP. *Food and Medication Interaction*, 16th ed. Birchrunville, PA: Food–Medication Interactions; 2010.

Pujol TJ, Tucker JE. Diseases of the cardiovascular system. In: Nelms M, Sucher K, Lacey K, Roth SL. *Nutrition Therapy and Pathophysiology*. 2nd ed. Belmont, CA: Wadsworth, Cengage Learning; 2011:283–328.

Seventh Report of the Joint National Committee on Detection, Evaluation, and Treatment of High Blood Pressure (JNC-VI). http://www.nhlbi.nih .gov/guidelines/hypertension/. Accessed 2/5/12.

Third Report of the Expert Panel on Detection, Evaluation, and Treatment of High Blood Cholesterol in Adults (Adult Treatment Panel III). Bethesda, MD: National Institutes of Health (2002). http://www.nhlbi.nih.gov/guidelines /cholesterol/index.htm. Accessed 2/11/12.

Internet Resources:

American Society of Hypertension: http://www.ash-us.org/

Calculate Your Body Mass Index: http://www.nhlbisupport .com/bmi/

Joint National Committee on Prevention, Detection, Evaluation, and Treatment of High Blood Pressure (JNC7): http://www.nhlbi.nih.gov/guidelines /hypertension/

National Heart Lung Blood Institute: What is the DASH eating plan?: http://www.nhlbi.nih.gov/health /health-topics/topics/dash/

National Institutes of Health. Heart Diseases: http://www .nlm.nih.gov/medlineplus/heartdiseases.html

USDA Nutrient Data Laboratory: http://www.ars.usda.gov /main/site_main.htm?modecode=12-35-45-00

Myocardial Infarction

Objectives

After completing this case, the student will be able to:

1. Describe the progression of atherosclerosis and its role in the etiology of a myocardial infarction.
2. Identify and explain common nutritional problems associated with a myocardial infarction.
3. Interpret laboratory parameters for nutritional implications and significance.
4. Analyze nutrition assessment data to evaluate nutritional status and identify specific nutrition problems.
5. Determine nutrition diagnoses and write appropriate PES statements.
6. Develop a nutrition care plan—with appropriate measurable goals, interventions, and strategies for monitoring and evaluation—that addresses the nutrition diagnoses of this case.

Mr. Klosterman, a 61-year-old man, is admitted through the emergency department of University Hospital after experiencing a sudden onset of severe precordial pain on the way home from work. Mr. Klosterman is found to have suffered a myocardial infarction and is treated with an emergency angioplasty of the infarct-related artery.

Klosterman, James, Male, 61 y.o.

Allergies: NKA	**Code:** FULL	**Isolation:** None
Pt. Location: RM 704	**Physician:** R.H. Smith	**Admit Date:** 12/1

Patient Summary: James Klosterman is a 61-year-old male admitted through the emergency department for an emergency coronary angiography with angioplasty of the infarct-related artery.

History:

Onset of disease: 61-yo male who noted the sudden onset of severe precordial pain on the way home from work. The pain is described as pressure-like pain radiating to the jaw and left arm. The patient has noted an episode of emesis and nausea. He denies palpitations or syncope. He denies prior history of pain. He admits to smoking cigarettes (1 pack/day for 40 years). He denies hypertension, diabetes, or high cholesterol. He denies SOB.

Medical history: Not significant before Dx of MI

Surgical history: Surgery; cholecystectomy 10 years ago, appendectomy 30 years ago

Medications at home: None

Allergies: Sulfa drugs

Tobacco use: 1 ppd for 40 years

Alcohol use: 1 glass of wine per day

Family history: What? CAD. Who? Father—MI age 59

Demographics:

Marital status: Married, *Spouse name*: Sally Klosterman, 59 yo

Number of children: Children are grown and do not live at home.

Years education: BS degree

Language: English only

Occupation: Lutheran minister

Hours of work: 40/wk

Household members: 2

Ethnicity: German

Religious affiliation: Lutheran

MD Progress Note:

Review of Systems

Constitutional: Negative

Skin: Negative

Cardiovascular: No carotid bruits

Respiratory: Negative

Gastrointestinal: Negative

Neurological: Negative

Psychiatric: Negative

Physical Exam

General appearance: Mildly overweight male in acute distress from chest pain

Heart: PMI 5 ICS MCL focal. S1 normal intensity. S2 normal intensity and split. S4 gallop at the apex. No murmurs, clicks, or rubs.

Klosterman, James, Male, 61 y.o.
Allergies: NKA   **Code:** FULL   **Isolation:** None
Pt. Location: RM 704   **Physician:** R.H. Smith   **Admit Date:** 12/1

HEENT: Head: Normocephalic
Eyes: EOMI, fundoscopic exam WNL. No evidence of atherosclerosis, diabetic retinopathy, or early hypertensive changes.
Ears: TM normal bilaterally
Nose: WNL
Throat: Tonsils not infected, uvula midline, gag normal
Genitalia: Grossly physiologic
Neurologic: No focal localizing abnormalities; DTR symmetric bilaterally
Extremities: No C, C, E
Skin: Diaphoretic and pale
Chest/lungs: Clear to auscultation and percussion
Peripheral vascular: PPP
Abdomen: RLQ scar and midline suprapubic scar. BS WNL. No hepatomegaly, splenomegaly, masses, inguinal lymph nodes, or abdominal bruits

Vital Signs: Temp: 98.4   Pulse: 92   Resp rate: 20
                   BP: 118/78   Height: 5'10"   Weight: 185 lbs   BMI: 26.6

Nursing Assessment	12/1
Abdominal appearance (concave, flat, rounded, obese, distended)	flat
Palpation of abdomen (soft, rigid, firm, masses, tense)	soft
Bowel function (continent, incontinent, flatulence, no stool)	continent
Bowel sounds (P=present, AB=absent, hypo, hyper)	
RUQ	P
LUQ	P
RLQ	P
LLQ	P
Stool color	light brown
Stool consistency	
Tubes/ostomies	NA
Genitourinary	
Urinary continence	catheter
Urine source	catheter
Appearance (clear, cloudy, yellow, amber, fluorescent, hematuria, orange, blue, tea)	clear, yellow
Integumentary	
Skin color	pale

(Continued)

Klosterman, James, Male, 61 y.o.
Allergies: NKA   **Code:** FULL   **Isolation:** None
Pt. Location: RM 704   **Physician:** R.H. Smith   **Admit Date:** 12/1

Nursing Assessment *(Continued)*

Nursing Assessment	12/1
Skin temperature (DI=diaphoretic, W=warm, dry, CL=cool, CLM=clammy, CD+=cold, M=moist, H=hot)	D, M
Skin turgor (good, fair, poor, TENT=tenting)	TENT
Skin condition (intact, EC=ecchymosis, A=abrasions, P=petechiae, R=rash, W=weeping, S=sloughing, D=dryness, EX=excoriated, T=tears, SE=subcutaneous emphysema, B=blisters, V=vesicles, N=necrosis)	intact
Mucous membranes (intact, EC=ecchymosis, A=abrasions, P=petechiae, R=rash, W=weeping, S=sloughing, D=dryness, EX=excoriated, T=tears, SE=subcutaneous emphysema, B=blisters, V=vesicles, N=necrosis)	intact
Other components of Braden score: special bed, sensory pressure, moisture, activity, friction/shear (>18 = no risk, 15–16 = low risk, 13–14 = moderate risk, ≤12 = high risk)	activity; 22

Orders:

IV heparin—5000 units bolus followed by 1000 unit/hour continuous infusion with a PTT at 2 X control
Chewable aspirin 160 mg PO and continued every day
Lopressor 50 mg twice daily
Lidocaine prn
NPO until procedure completed
Type and cross for 6 units of packed cells

Nutrition:

Meal type: Clear liquids, no caffeine
History: Appetite good. Has been trying to change some things in his diet. Wife indicates that she has been using "corn oil" instead of butter and has tried not to fry foods as often.

24-hour recall:
Breakfast: None
Midmorning snack: 1 large cinnamon raisin bagel with 1 tbsp fat-free cream cheese, 8 oz orange juice, coffee
Lunch: 1 c canned vegetable beef soup, sandwich with 4 oz roast beef, lettuce, tomato, dill pickles, 2 tsp mayonnaise, 1 small apple, 8 oz 2% milk
Dinner: 2 lean pork chops (3 oz each), 1 large baked potato, 2 tsp margarine, ½ c green beans, ½ c coleslaw (cabbage with 1 tbsp salad dressing), 1 slice apple pie
Snack: 8 oz 2% milk, 1 oz pretzels

Food allergies/intolerances/aversions: None
Previous nutrition therapy? Yes. If yes, when: Last year. Where? Community dietitian.

Klosterman, James, Male, 61 y.o.
Allergies: NKA
Pt. Location: RM 704

Code: FULL
Physician: R.H. Smith

Isolation: None
Admit Date: 12/1

Food purchase/preparation: Spouse
Vit/min intake: None

Laboratory Results

	Ref. Range	12/1 1957	12/2 0630	12/3 0645
Chemistry				
Sodium (mEq/L)	136–145	141	142	138
Potassium (mEq/L)	3.5–5.5	4.2	4.1	3.9
Chloride (mEq/L)	95–105	103	102	100
Carbon dioxide (CO_2, mEq/L)	23–30	20 !↓	24	26
BUN (mg/dL)	8–18	14	15	16
Creatinine serum (mg/dL)	0.6–1.2	1.1	1.1	1.1
Glucose (mg/dL)	70–110	136 !↑	106	104
Phosphate, inorganic (mg/dL)	2.3–4.7	3.1	3.2	3.0
Magnesium (mg/dL)	1.8–3	2.0	2.3	2.0
Calcium (mg/dL)	9–11	9.4	9.4	9.4
Osmolality (mmol/kg/H_2O)	285–295	292	290	291
Bilirubin, direct (mg/dL)	<0.3	0.1	0.1	0.2
Protein, total (g/dL)	6–8	6.0	5.9 !↓	6.1
Albumin (g/dL)	3.5–5	4.2	4.3	4.2
Prealbumin (mg/dL)	16–35	30	32	31
Ammonia (NH_3, µmol/L)	9–33	26	22	25
Alkaline phosphatase (U/L)	30–120	75	70	68
ALT (U/L)	4–36	30	215 !↑	185 !↑
AST (U/L)	0–35	25	245 !↑	175 !↑
CPK (U/L)	30–135 F 55–170 M	75	500 !↑	335 !↑
CPK-MB (U/L)	0	0	75 !↑	55 !↑
Lactate dehydrogenase (U/L)	208–378	325	685 !↑	365
Troponin I (ng/dL)	<0.2	2.4 !↑	2.8 !↑	
Troponin T (ng/dL)	<0.03	2.1 !↑	2.7 !↑	
Cholesterol (mg/dL)	120–199	235 !↑	226 !↑	214 !↑
HDL-C (mg/dL)	>55 F,>45 M	30 !↓	32 !↓	33 !↓
LDL (mg/dL)	<130	160 !↑	150 !↑	141 !↑

(Continued)

Klosterman, James, Male, 61 y.o.
Allergies: NKA
Pt. Location: RM 704

Code: FULL
Physician: R.H. Smith

Isolation: None
Admit Date: 12/1

Laboratory Results *(Continued)*

	Ref. Range	12/1 1957	12/2 0630	12/3 0645
LDL/HDL ratio	<3.22 F <3.55 M	5.3 !↑	4.7 !↑	4.3 !↑
Apo A (mg/dL)	101–199 F 94–178 M	72 !↓	80 !↓	98
Apo B (mg/dL)	60–126 F 63–133 M	115	110	105
Triglycerides (mg/dL)	35–135 F 40–160 M	150	140	130
Coagulation (Coag)				
PT (sec)	12.4–14.4	12.6	12.6	12.4
Hematology				
WBC (×10³/mm³)	4.8–11.8	11.0	9.32	8.8
RBC (×10⁶/mm³)	4.2–5.4 F 4.5–6.2 M	4.7	4.75	4.68
Hemoglobin (Hgb, g/dL)	12–15 F 14–17 M	15	14.8	14.4
Hematocrit (Hct, %)	37–47 F 40–54 M	45	45	44
Mean cell volume (μm³)	80–96	91	92	90
Mean cell Hgb (pg)	26–32	30	31	30
Mean cell Hgb content (g/dL)	31.5–36	33	32	33
RBC distribution (%)	11.6–16.5	13.2	12.8	13.0
Platelet count (×10³/mm³)	140–440	320	295	280
Hematology, Manual Diff				
Neutrophil (%)	50–70	55	58	62
Lymphocyte (%)	15–45	17	23	35
Monocyte(%)	3–10	4	4	7
Eosinophil(%)	0–6	0	0	0
Basophil(%)	0–2	0	0	0
Blasts (%)	3–10	3	3	4
Segs(%)	0–60	45	47	52
Bands (%)	0–10	15 !↑	17 !↑	8

Klosterman, James, Male, 61 y.o.
Allergies: NKA
Pt. Location: RM 704

Code: FULL
Physician: R.H. Smith

Isolation: None
Admit Date: 12/1

Laboratory Results *(Continued)*

	Ref. Range	12/1 1957	12/2 0630	12/3 0645
Urinalysis				
Color	—	pale yellow	pale yellow	pale yellow
Appearance	—	clear	clear	clear
Specific gravity	1.003–1.030	1.020	1.015	1.018
pH	5–7	5.8	5.0	6
Protein (mg/dL)	Neg	Neg	Neg	Neg
Glucose (mg/dL)	Neg	Neg	Neg	Neg
Ketones	Neg	Trace !↑	Neg	Neg
Blood	Neg	Neg	Neg	Neg
Urobilinogen (EU/dL)	<1.1	Neg	Neg	Neg
Leukocyte esterase	Neg	Neg	Neg	Neg
Prot chk	Neg	Neg	Neg	Neg
WBCs (/HPF)	0–5	0	0	0
RBCs (/HPF)	0–5	0	0	0
Bact	0	0	0	0

Case Questions

I.   Understanding the Disease and Pathophysiology

1.   Mr. Klosterman had a myocardial infarction. Explain what happened to his heart.

2.   Mr. Klosterman's chest pain resolved after two sublingual NTG at 3-minute intervals and 2 mgm of IV morphine. In the cath lab he was found to have a totally occluded distal right coronary artery and a 70% occlusion in the left circumflex coronary artery. The left anterior descending was patent. Angioplasty of the distal right coronary artery resulted in a patent infarct-related artery with near-normal flow. A stent was left in place to stabilize the patient and limit infarct size. Left ventricular ejection fraction was normal at 42%, and a postero-basilar scar was present with hypokinesis. Explain angioplasty and stent placement. What is the purpose of this medical procedure?

3.   Mr. Klosterman and his wife are concerned about the future of his heart health. What role does cardiac rehabilitation play in his return to normal activities and in determining his future heart health?

II.   Understanding the Nutrition Therapy

4.   What risk factors indicated in his medical record can be addressed through nutrition therapy?

5.   What are the current recommendations for nutritional intake during a hospitalization following a myocardial infarction?

III.   Nutrition Assessment

6.   What is the healthy weight range for an individual of Mr. Klosterman's height?

7.   This patient is a Lutheran minister. He does get some exercise daily. He walks his dog outside for about 15 minutes at a leisurely pace. Calculate his energy and protein requirements.

8.   Using Mr. Klosterman's 24-hour recall, calculate the total number of calories he consumed as well as the energy distribution of calories for protein, carbohydrate, and fat using the exchange system.

9. Examine the chemistry results for Mr. Klosterman. Which labs are consistent with the MI diagnosis? Explain. Why were the levels higher on day 2?

10. What is abnormal about his lipid profile? Indicate the abnormal values.

11. Mr. Klosterman was prescribed the following medications on discharge. What are the food–medication interactions for this list of medications?

Medication	Possible Food–Medication Interactions
Lopressor 50 mg daily	
Lisinopril 10 mg daily	
Nitro-Bid 9.0 mg twice daily	
NTG 0.4 mg sl prn chest pain	
ASA 81 mg daily	

12. You talk with Mr. Klosterman and his wife, a math teacher at the local high school. They are friendly and seem cooperative. They are both anxious to learn what they can do to prevent another heart attack. What questions will you ask them to assess how to best help them?

13. What other issues might you consider to support successful lifestyle changes for Mr. Klosterman?

14. From the information gathered within the assessment, list possible nutrition problems using the correct diagnostic terms.

IV. Nutrition Diagnosis
15. Select two of the identified nutrition problems and complete the PES statement for each.

V. Nutrition Intervention
16. For each of the PES statements you have written, establish an ideal goal (based on the signs and symptoms) and an appropriate intervention (based on the etiology).

17. Mr. Klosterman and his wife ask about supplements. "My roommate here in the hospital told me I should be taking fish oil pills." What does the research say about omega-3-fatty acid supplementation for this patient?

VI.   Nutrition Monitoring and Evaluation

18. What would you want to assess in three to four weeks when he and his wife return for additional counseling?

Bibliography

Academy of Nutrition and Dietetics Evidence Based Library. Disorders of lipid metabolism (DLM) and nutrition monitoring and evaluation, ADA Analysis Library. http://www.adaevidencelibrary.com/template.cfm?template=guide_summary&key=2999. Accessed 02/18/12.

Academy of Nutrition and Dietetics Evidence Based Library. Disorders of lipid metabolism (DLM) and referral to a registered dietitian for medical nutrition therapy. ADA Analysis Library. http://www.adaevidencelibrary.com/template.cfm?template=guide_summary&key=2875. Accessed 02/18/12.

Academy of Nutrition and Dietetics Evidence Based Library. Is waist-to-hip ratio an independent predictor of CHD? ADA Analysis Library. http://www.adaevidencelibrary.com/conclusion.cfm?conclusion_statement_id=298. Accessed 02/18/12.

Academy of Nutrition and Dietetics Evidence Based Library. Recommendation summary: disorders of lipid metabolism (DLM) and nutrition assessment. ADA Evidence Analysis Library. http://www.adaevidencelibrary.com/template.cfm?template=guide_summary&key=2990. Accessed 02/18/12.

Academy of Nutrition and Dietetics Nutrition Care Manual. BMI and Weight Range Calculator. http://nutritioncaremanual.org/calculators.cfm?calculator_type=CalcBMI. Accessed 2/5/12.

Artinian NT, Fletcher GJ, Mozaffarian D, et al.; on behalf of the American Heart Association Prevention Committee of the Council on Cardiovascular Nursing. Interventions to promote physical activity and dietary lifestyle changes for cardiovascular risk factor reduction in adults: a scientific statement from the American Heart Association. *Circulation.* 2010;122:406–441.

Hemilä H, Miller ER. Evidenced-based medicine and vitamin E supplementation. *American J Clin Nutr.* 2007;86:261–262.

Kris-Etherton PM, Harris WS, Appel LJ; American Heart Association Nutrition Committee. Fish consumption, fish oil, omega-3 fatty acids, and cardiovascular disease. *Circulation.* 2003 Jan 28;107(3):512.

Leaf A, Albert CM, Josephson M, et al. Prevention of fatal arrhythmias in high-risk subjects by fish oil n-3 fatty acid intake. *Circulation.* 2005 Nov 1;112(18):2762–8.

Available from http://www.ncbi.nlm.nih.gov/sites/entrez?Db=pubmed&Cmd=Search&Term=%22Fatty%20Acid%20Antiarrhythmia%20Trial%20Investigators%22%5BCorporate%20Author%5D&itool=EntrezSystem2.PEntrez.Pubmed.Pubmed_ResultsPanel.Pubmed_RVAbstract.

Lichtenstein AH, Ausman LM, SJalbert SM, et al. Efficacy of a Therapeutic Lifestyle Change/Step 2 diet in moderately hypercholesterolemic middle-aged and elderly female and male subjects. *J. Lipid Res.* 2002;43:264–273.

Miller ER, Pastor-Barriuso R, Dalal D, Riemersma RA, Appel LJ, Guallar E. Meta-analysis: High-dosage vitamin E supplementation may increase all-cause mortality. *Ann Intern Med.* 2005;142:37–46.

Nelms MN. Nutrition assessment: Foundation of the nutrition care process. In: Nelms M, Sucher K, Lacey K, Roth SL. *Nutrition Therapy and Pathophysiology.* 2nd ed. Belmont, CA: Wadsworth, Cengage Learning; 2011:34–65.

Pronsky Z M, Crowe JP. *Food and Medication Interaction.* 16th ed. Birchrunville, PA: Food–Medication Interactions; 2010.

Pujol TJ, Tucker JE, Barnes JT. Diseases of the cardiovascular system. In: Nelms M, Sucher K, Lacey K, Roth SL. *Nutrition Therapy and Pathophysiology.* 2nd ed. Belmont, CA: Wadsworth, Cengage Learning; 2011:283–339.

Roger VL, Go AS, Lloyd-Jones DM, et al.; on behalf of the American Heart Association Statistics Committee and Stroke Statistics Subcommittee. Heart disease and stroke statistics—2012 update: a report from the American Heart Association. *Circulation.* 2012;125:e2–e220. http://circ.ahajournals.org/content/125/1/e2. Accessed 2/25/12.

Third Report of the Expert Panel on Detection, Evaluation, and Treatment of High Blood Cholesterol in Adults (Adult Treatment Panel III). Bethesda, MD: National Institutes of Health (2002). http://www.nhlbi.nih.gov/guidelines/cholesterol/index.htm. Accessed 2/11/12.

U.S. National Library of Medicine. MedlinePlus. Fish oil. http://www.nlm.nih.gov/medlineplus/druginfo/natural/993.html. Accessed 4/10/2012.

Wang C, Harris WS, Chung M, et al. n-3 Fatty acids from fish or fish-oil supplements, but not alpha-linolenic acid, benefit cardiovascular disease outcomes in primary- and secondary-prevention studies: a systematic review. *Am J Clin Nutr.* 2006 Jul;84(1):5–17.

Internet Resources

ADA Evidence Analysis Library: http://www.adaevidencelibrary.com

The Cleveland Clinic: http://clevelandclinicmeded.com/medicalpubs/diseasemanagement/cardiology/acute-myocardial-infarction/

Nutrition Care Manual: http://www.nutritioncaremanual.org

USDA Nutrient Data Laboratory: www.nal.usda.gov/fnic/foodcomp

Case 6

Heart Failure with Resulting Cardiac Cachexia

Objectives

After completing this case, the student will be able to:

1. Use nutrition assessment information to determine baseline nutritional status.
2. Correlate a patient's signs and symptoms with the pathophysiology of heart failure.
3. Evaluate laboratory indices for nutritional implications and significance.
4. Demonstrate understanding of nutrition support options for heart failure.
5. Identify the roles of pharmacologic intervention and drug–nutrient interactions.

6. Determine appropriate nutritional interventions for the patient with heart failure and cardiac cachexia.

Dr. Charles Peterman, an 85-year-old retired physician, is admitted with acute symptoms related to his heart failure. Dr. Peterman has a long history of cardiac disease, including a previous myocardial infarction and mitral valve disease.

Peterman, Charles, Male, 85 y.o.
Allergies: NKA   **Code:** DNR   **Isolation:** None
Pt. Location: RM 1952   **Physician:** DA Schmidt   **Admit Date:** 2/14

Patient Summary: Charles Peterman has a history of being treated for CAD, HTN, and HF.

History:

Onset of disease: HF × 2 yrs
Medical history: Long-standing history of CAD, HTN, mitral valve insufficiency, previous anterior MI
Surgical history: No surgeries
Medications at home: Lanoxin 0.125 mg once daily, Lasix 80 mg twice daily, Aldactone 25 mg once daily, lisinopril 30 mg po once daily, Lopressor 25 mg once daily, Zocor 20 mg once daily, Metamucil 1 tbsp twice daily, calcium carbonate 500 mg twice daily, Centrum 2 tablets once daily
Tobacco use: No
Alcohol use: No
Family history: What? HTN, CAD. Who? Parents.

Demographics:

Marital status: Married—lives with wife; *Spouse name:* Jean
Number of children: 0
Years education: Postgraduate
Language: English only
Occupation: Retired physician
Hours of work: N/A
Ethnicity: Caucasian
Religious affiliation: Presbyterian

Admitting History/Physical:

Chief complaint: Patient collapsed at home and was brought to the emergency room by ambulance.
General appearance: Elderly male in acute distress

Vital Signs: Temp: 98   Pulse: 110   Resp rate: 24
                          BP: 90/70   Height: 5'10"   Weight: 165 lbs

Heart: Diffuse PMI in AAL in LLD; Grade II holosystolic murmur at the apex radiating to the left sternal border; first heart sound diminished and second heart sound preserved; third heart sound present
HEENT: Head: Temporal wasting
              Eyes: Ophthalmoscopic exam reveals AV crossing changes and arteriolar spasm
              Ears: WNL
              Nose: WNL
              Throat: Jugular venous distension in sitting position with a positive hepatojugular reflux
Genitalia: WNL
Neurologic: WNL
Extremities: 4 + pedal edema; weak hand grip

Peterman, Charles, Male, 85 y.o.
Allergies: NKA
Pt. Location: RM 1952

Code: DNR
Physician: DA Schmidt

Isolation: None
Admit Date: 2/14

Skin: Gray, moist
Chest/lungs: Rales in both bases posteriorly
Peripheral vascular: WNL
Abdomen: Ascites, no masses, liver tender to A&P

Nursing Assessment	2/14
Abdominal appearance (concave, flat, rounded, obese, distended)	distended
Palpation of abdomen (soft, rigid, firm, masses, tense)	firm
Bowel function (continent, incontinent, flatulence, no stool)	continent
Bowel sounds (P=present, AB=absent, hypo, hyper)	
RUQ	P
LUQ	P
RLQ	P
LLQ	P
Stool color	light brown
Stool consistency	formed
Tubes/ostomies	NA
Genitourinary	
Urinary continence	catheter
Urine source	catheter
Appearance (clear, cloudy, yellow, amber, fluorescent, hematuria, orange, blue, tea)	clear, yellow
Integumentary	
Skin color	gray
Skin temperature (DI=diaphoretic, W=warm, dry, CL=cool, CLM=clammy, CD+=cold, M=moist, H=hot)	M
Skin turgor (good, fair, poor, TENT=tenting)	TENT
Skin condition (intact, EC=ecchymosis, A=abrasions, P=petechiae, R=rash, W=weeping, S=sloughing, D=dryness, EX=excoriated, T=tears, SE=subcutaneous emphysema, B=blisters, V=vesicles, N=necrosis)	intact
Mucous membranes (intact, EC=ecchymosis, A=abrasions, P=petechiae, R=rash, W=weeping, S=sloughing, D=dryness, EX=excoriated, T=tears, SE=subcutaneous emphysema, B=blisters, V=vesicles, N=necrosis)	intact
Other components of Braden score: special bed, sensory pressure, moisture, activity, friction/shear (>18 = no risk, 15–16 = low risk, 13–14 = moderate risk, ≤12 = high risk)	activity, 15

Peterman, Charles, Male, 85 y.o.
Allergies: NKA   **Code:** DNR   **Isolation:** None
Pt. Location: RM 1952   **Physician:** DA Schmidt   **Admit Date:** 2/14

Orders:
Admit to CCU
Parenteral dopamine and IV diuretics
100 mg thiamin IV
Telemetry
Vitals every 1 hr × 8, every 2 hrs × 8 for first 24 hours
Daily ECG and chest X-rays
Echocardiogram
Chem 24
Urinalysis
Strict I&Os

Nutrition:
Meal type: 2 g Na$^+$
Intake % of meals: <5%, sips of liquids for past 24 hrs
Fluid requirement: 1,500 mL
History: Wife reports that Dr. Peterman's appetite has been poor for the last 6 months, with no real weight loss that she can determine. "It's very hard to know the difference between his real weight and any fluid he is retaining." She describes difficulty eating due to SOB and nausea.
Usual dietary intake: Generally likes all foods but has recently been eating only soft foods, esp. ice cream. Tries to drink 2 cans Ensure Plus each day.
Food allergies/intolerances/aversions: Shellfish
Previous nutrition therapy? Not specifically, but has monitored salt intake for the past 2 years as well as followed a low-fat, low-cholesterol diet for at least the previous 10 years
Food purchase/preparation: Spouse
Vit/min intake: Centrum Silver 2×/day, calcium supplement 1000 mg/day

Laboratory Results

	Ref. Range	2/14 1952	2/16 0645	2/20 0630
Chemistry				
Sodium (mEq/L)	136–145	132 !↓	133 !↓	133 !↓
Potassium (mEq/L)	3.5–5.5	3.7	3.6	3.8
Chloride (mEq/L)	95–105	98	100	99
Carbon dioxide (CO_2, mEq/L)	23–30	26	24	25
BUN (mg/dL)	8–18	32 !↑	34 !↑	30 !↑
Creatinine serum (mg/dL)	0.6–1.2	1.6 !↑	1.7 !↑	1.5 !↑
Glucose (mg/dL)	70–110	110	106	102
Phosphate, inorganic (mg/dL)	2.3–4.7	4.0	3.8	3.6
Magnesium (mg/dL)	1.8–3	2.0	1.9	1.8
Calcium (mg/dL)	9–11	9.0	8.8	8.9

Peterman, Charles, Male, 85 y.o.
Allergies: NKA
Pt. Location: RM 1952

Code: DNR
Physician: DA Schmidt

Isolation: None
Admit Date: 2/14

Laboratory Results *(Continued)*

	Ref. Range	2/14 1952	2/16 0645	2/20 0630
Bilirubin, direct (mg/dL)	<0.3	1.0 !↑	1.1 !↑	0.9 !↑
Protein, total (g/dL)	6–8	5.8 !↓	5.6 !↓	5.5 !↓
Albumin (g/dL)	3.5–5	2.8 !↓	2.7 !↓	2.6 !↓
Prealbumin (mg/dL)	16–35	15 !↓	11 !↓	10 !↓
Ammonia (NH$_3$, µmol/L)	9–33	32	30	33
Alkaline phosphatase (U/L)	30–120	112	115	118
ALT (U/L)	4–36	100 !↑	120 !↑	115 !↑
AST (U/L)	0–35	70 !↑	80 !↑	85 !↑
CPK (U/L)	30–135 F 55–170 M	180 !↑	200 !↑	205 !↑
Lactate dehydrogenase (U/L)	208–378	350	450 !↑	556 !↑
Troponin I (ng/L)	<0.2	0.026	0.028	0.027
Troponin T (ng/L)	<0.03	0.035 !↑	0.037 !↑	0.036 !↑
Cholesterol (mg/dL)	120–199	150	162	149
HDL-C (mg/dL)	>55 F, >45 M	30 !↓	31 !↓	30 !↓
LDL (mg/dL)	<130	180 !↑	160 !↑	152 !↑
LDL/HDL ratio	<3.22 F <3.55 M	5 !↑	5.23 !↑	4.97 !↑
Apo A (mg/dL)	101–199 F 94–178 M	60 !↓	65 !↓	70 !↓
Apo B (mg/dL)	60–126 F 63–133 M	140 !↑	138 !↑	136 !↑
Triglycerides (mg/dL)	35–135 F 40–160 M	150	145	140
Coagulation (Coag)				
PT (sec)	12.4–14.4	13.2	13.3	13.3
Hematology				
WBC (×10^3/mm^3)	4.8–11.8	12 !↑	12 !↑	10.5
RBC (×10^6/mm^3)	4.2–5.4 F 4.5–6.2 M	5.5	6.5 !↑	6.4 !↑
Hemoglobin (Hgb, g/dL)	12–15 F 14–17 M	14	14.3	14.5
Hematocrit (Hct, %)	37–47 F 40–54 M	41	42	42
Mean cell volume (µm^3)	80–96	90	89	91

(Continued)

Peterman, Charles, Male, 85 y.o.
Allergies: NKA
Pt. Location: RM 1952

Code: DNR
Physician: DA Schmidt

Isolation: None
Admit Date: 2/14

Laboratory Results *(Continued)*

	Ref. Range	2/14 1952	2/16 0645	2/20 0630
Mean cell Hgb (pg)	26–32	31	31	30
Mean cell Hgb content (g/dL)	31.5–36	33	34	32
Platelet count ($\times 10^3$/mm^3)	140–440	300	290	310
Transferrin (mg/dL)	250–380 F 215–365 M	350	355	352
Hematology, Manual Diff				
Lymphocyte (%)	15–45	20	17	26
Monocyte (%)	3–10	4	1 !↓	2 !↓
Eosinophil (%)	0–6	4	1	2
Segs (%)	0–60	65 !↑	73 !↑	66 !↑

Intake/Output

Date		2/14 0701–2/15 0700			
Time		0701–1500	1501–2300	2301–0700	Daily total
IN	P.O.	**10**	**15**	**0**	**25**
	I.V. (mL/kg/hr)	**336** (0.56)	**336** (0.56)	**336** (0.56)	**1008** (0.56)
	I.V. piggyback				
	TPN				
	Total intake (mL/kg)	**346** (4.6)	**351** (4.7)	**336** (4.5)	**1033** (13.8)
OUT	Urine (mL/kg/hr)	**200** (0.33)	**175** (0.29)	**250** (0.42)	**625** (0.35)
	Emesis output				
	Other				
	Stool				
	Total output (mL/kg)	**200** (2.7)	**175** (2.3)	**250** (3.3)	**625** (8.3)
Net I/O		**+146**	**+176**	**+86**	**+408**
Net since admission (2/14)		**+146**	**+322**	**+408**	**+408**

Case Questions

I.   Understanding the Disease and Pathophysiology

1.   Outline the typical pathophysiology of heart failure. Onset of heart failure usually can be traced to damage from an MI and atherosclerosis. Is this consistent with Dr. Peterman's history?

2.   Identify specific signs and symptoms in the patient's physical examination that are consistent with heart failure. For any three of these signs and symptoms, write a brief discussion that connects them to physiological changes that you described in question #1.

3.   Heart failure is often described as R-sided failure or L-sided failure. What is the difference? How are the clinical manifestations different?

4.   Dr. Peterman's admitting diagnosis was cardiac cachexia. What is cardiac cachexia? What are the characteristic symptoms? Explain the role of the underlying heart disease in development of cardiac cachexia.

II.   Understanding the Nutrition Therapy

5.   Dr. Peterman's wife states that they have monitored their salt intake for several years. What is the role of sodium restriction in the treatment of heart failure? What level of sodium restriction is recommended for the outpatient with heart failure?

6.   Should he be placed on a fluid restriction? If so, how would this assist with the treatment of his heart failure? What specific foods are typically "counted" as a fluid?

7.   Identify any common nutrient deficiencies found in patients with heart failure.

III.   Nutrition Assessment

8.   Identify factors that would affect interpretation of Dr. Peterson's weight and body composition. Look at the I/O record. What will likely happen to Dr. Peterson's weight if this trend continues?

9.   Calculate Dr. Peterman's energy and protein requirements. Explain your rationale for the weight you have used in your calculation.

10. Dr. Peterman was started on an enteral feeding when he was admitted to the hospital. Outline a nutrition therapy regimen for him that includes formula choice, total volume, and goal rate.

11. Identify any abnormal biochemical values and assess them using the following table:

Parameter	Normal Value	Pt's Value	Reason for Abnormality	Nutrition Implication

12. The following chart lists drugs/supplements that were prescribed for Dr. Peterman. Give the rationale for the use of each. In addition, describe any nutrition implications for these medications.

Medication	Rationale for Use	Nutrition Implications
Lanoxin		
Lasix		
Dopamine		
Thiamin		

IV. Nutrition Diagnosis
13. Select two nutrition problems and complete a PES statement for each.

V.   Nutrition Intervention

14. Dr. Peterman was not able to tolerate the enteral feeding because of diarrhea. What recommendations could be made to improve tolerance to the tube feeding?

15. The tube feeding was discontinued because of continued intolerance. Parenteral nutrition was not initiated. What recommendations could you make to optimize Dr. Peterman's oral intake?

16. An echocardiogram indicated severe cardiomegaly secondary to end-stage heart failure. Mr. Peterson had a living will that stated he wanted no extraordinary measures taken to prolong his life. He was able to express his wishes verbally and requested oral feedings and palliative care only. Mr. Peterman expired after a two-week hospitalization. What is a living will? What is palliative care?

17. During his final days of life, Dr. Peterman was not receiving parenteral or enteral nutritional support. What is the role of the registered dietitian during palliative care?

Bibliography

Academy of Nutrition and Dietetics Evidence Based Library. Heart failure evidence-based nutrition practice guidelines. ADA Analysis Library. http://www.adaevidencelibrary.com/topic.cfm?cat=2800. Accessed 03/04/12.

Academy of Nutrition and Dietetics Nutrition Care Manual. Heart failure. http://nutritioncaremanual.org/topic.cfm?ncm_heading=Diseases%2FConditions&ncm_toc_id=8585. Accessed 03/04/2012.

Hunt SA, Abraham WT, Chin MH, et al. 2009 Focused update incorporated into the ACC/AHA 2005 guidelines for the diagnosis and management of heart failure in adults: A report of the American College of Cardiology Foundation/American Heart Association Task Force on Practice Guidelines. *Circulation.* 2009;119:e391–e479.

Kociol RD, Pang PS, Gheorghiade M, Fonarow GC, O'Connor CM, Felker GM. Troponin elevation in patients with heart failure. *J Am Coll Cardiol.* 2010; 56:1071–1078.

Pujol TJ, Tucker JE, Barnes JT. Diseases of the cardiovascular system. In: Nelms M, Sucher K, Lacey K, Roth SL. *Nutrition Therapy and Pathophysiology.* 2nd ed. Belmont, CA: Wadsworth, Cengage Learning; 2011:283–339.

Ukleja A, Freeman KL, Gilbert K, et al. Standards for nutrition support. Adult hospitalized patients. *Nutr Clin Pract.* 2010; 25:403–414.

Van Horn L, McCoin M, Kris-Etherton PM, et al. The evidence for dietary prevention and treatment for cardiovascular disease. *J Am Diet Assoc.* 2008; 108:287–331.

Internet Resources

Academy of Nutrition and Dietetics: Nutrition Care Manual (by subscription). www.nutritioncaremanual.org.

American Heart Association: http://www.heart.org/HEARTORG/Conditions/HeartFailure/Heart-Failure_UCM_002019_SubHomePage.jsp

Heart Failure animation: http://www.medmovie.com/mmdatabase/mediaplayer.aspx?Message=VG9waWNpZD01NjM7Q2xpZW50SUQ9NjU7VmVybmFjdWxhcklEPTE%3D-R2waapTywBA%3D&SrchVisible=True&SrchID=563

Heart Failure Society of America: http://www.hfsa.org/

San Diego Cardiac Center–Heart Failure Online: www.heartfailure.org

National Library of Medicine–National Institutes of Health: www.nlm.nih.gov/medlineplus/ency/article/000158.htm

WebMD-Cardiac Cachexia and Heart Failure: http://www.webmd.com/heart-disease/heart-failure/cardiac-cachexia-and-heart-failure

Unit Three

NUTRITION THERAPY FOR UPPER GASTROINTESTINAL DISORDERS

The five cases presented in Units Three and Four cover a wide array of diagnoses that ultimately affect normal digestion and absorption. These conditions use medical nutrition therapy as a cornerstone of their treatment.

In some disorders, such as celiac disease, medical nutrition therapy is the *only* treatment. With other GI problems, it is important to understand that, because of the symptoms the patient experiences, nutritional status is often in jeopardy. Nausea, vomiting, diarrhea, constipation, and malabsorption are common with these disorders. Interventions in these cases are focused on treating such symptoms in order to restore nutritional health.

Case 7 targets gastroesophageal reflux disease (GERD). More than 20 million Americans suffer from symptoms of gastroesophageal reflux daily, and more than 100 million suffer occasional symptoms. Gastroesophageal reflux disease most frequently results from lower esophageal sphincter (LES) incompetence. Factors that influence LES competence include both physical and lifestyle factors. This case identifies the common symptoms of GERD

and challenges you to develop and analyze both nutritional and medical care for this patient.

Case 8 focuses on peptic ulcer disease (PUD) treated pharmacologically and surgically. Peptic ulcer disease involves ulcerations that penetrate the submucosa, usually in the antrum of the stomach or in the duodenum. Erosion may proceed to other levels of tissue and can eventually result in perforation. The breakdown in tissue allows continued insult by the highly acidic environment of the stomach. *Helicobacter pylori* is established as a major cause of chronic gastritis and peptic ulcer disease. Nutrition therapy for peptic ulcer disease is highly individualized. Treatment plans should avoid foods that increase gastric acid secretions and restrict any particular food or beverage that the patient does not tolerate. This case describes the complications of PUD resulting in hemorrhage and perforation that require surgical intervention. Nutritional complications, such as dumping syndrome and malabsorption, often accompany gastric surgery. This case also introduces the transition from enteral nutrition support to the appropriate oral diet for postoperative use.

Gastroesophageal Reflux Disease

Objectives

After completing this case, the student will be able to:

1. Apply knowledge of the pathophysiology of gastroesophageal reflux disease (GERD) in order to identify and explain common nutritional problems associated with this disease.
2. Describe basic principles of drug action required for medical treatment of GERD.
3. Discuss the rationale for nutrition recommendations to minimize adverse symptoms of GERD.
4. Interpret pertinent laboratory parameters for nutritional implications and significance.
5. Analyze nutrition assessment data to evaluate nutritional status and identify specific nutrition problems.
6. Determine nutrition diagnoses and write appropriate PES statements.
7. Develop a nutrition care plan—with appropriate measurable goals, interventions, and strategies for monitoring and evaluation—that addresses the nutrition diagnoses of this case.

Jack Nelson, a 48-year-old male, visits his physician for evaluation of increasing complaints of severe indigestion. Intraesophageal pH monitoring and a barium esophagram support a diagnosis of gastroesophageal reflux disease.

Nelson, Jack, Male, 48 y.o.
Allergies: NKA
Pt. Location: RM 1952

Code: FULL
Physician: P Phelps

Isolation: None
Admit Date: 9/22

Patient Summary: 48-yo male here for evaluation and treatment for increased indigestion

History:
Onset of disease: Patient has been experiencing increased indigestion over last year. Previously it was only at night but now he experiences indigestion almost constantly. He has been taking Tums several times daily. Mr. Nelson has gained almost 35 lbs since his knee surgery, which he attributes to a decrease in his ability to run and not being able to find a consistent replacement for exercise. Patient states he plays with his children on the weekends, but that is the extent of his physical activity. He states he probably has been eating and drinking more over the last year, which he attributes to stress. He is worried about his family history of heart disease, which is why he takes an aspirin each day. He has not really followed any diet restrictions.
Medical history: Essential HTN—Dx 1 year ago
Surgical history: s/p R knee arthroplasty 5 years ago
Medications at home: Atenolol 50 mg daily; 325 mg aspirin daily; multivitamin daily; 500 mg ibuprofen twice daily for last month
Tobacco use: No
Alcohol use: Yes; 1–2 beers 3–4 times/week
Family history: What? CAD. Who? Father

Demographics:
Marital status: Married—lives with wife and 2 sons
Spouse name: Mary
Number of children: 2
Years education: BA
Language: English only
Occupation: Retail manager of local department store
Hours of work: M–F, works consistently in evenings and on weekends as well
Ethnicity: Caucasian
Religious affiliation: Protestant

MD Progress Note:
Review of Systems
Constitutional: Negative
Skin: Negative
Cardiovascular: No carotid bruits
Respiratory: Negative
Gastrointestinal: Heme + stool
Neurological: Negative
Psychiatric: Negative

Nelson, Jack, Male, 48 y.o.

Allergies: NKA	**Code:** FULL	**Isolation:** None
Pt. Location: RM 1952	**Physician:** P Phelps	**Admit Date:** 9/22

Physical Exam

General appearance: Mildly obese 48-year-old white male in mild distress

Heart: Noncontributory

HEENT: Noncontributory

Genitalia: WNL

Neurologic: Oriented × 4

Extremities: No edema; normal strength, sensations, and DTR

Skin: Warm, dry

Chest/lungs: Lungs clear to auscultation and percussion

Peripheral vascular: Pulses full—no bruits

Abdomen: No distention. BS present in all regions. Liver percusses approx 8 cm at the midclavicular line, one fingerbreadth below the right costal margin. Epigastric tenderness without rebound or guarding.

Vital Signs:	Temp: 98.6	Pulse: 90	Resp rate: 16
	BP: 119/75	Height: 5'9"	Weight: 215 lbs

Assessment and Plan:

Rule out GERD, decrease aspirin to 75 mg daily

Dx: Gastroesophageal reflux disease, HTN

Medical Tx plan: Hematology, Chem 24, Ambulatory 48-hour pH monitoring with Bravo™ pH Monitoring System, Barium esophagram—request radiologist to attempt to demonstrate reflux using abdominal pressure and positional changes; Endoscopy with biopsy to r/o *H. pylori* infection; Begin omeprazole 30 mg every am; Decrease aspirin to 75 mg daily; D/C self-medication of ibuprofen daily; Nutrition consult

P. Phelps, MD

Nutrition:

History: Patient relates he has gained almost 35 lbs since his knee surgery. He attributes this to a decrease in his ability to run, and he has not found a consistent replacement for exercise. He plays with his children on weekends, but that is the extent of his physical activity. He states he probably has been eating and drinking more over the last year, which he attributes to stress. He is worried about his family history of heart disease, which is why he takes an aspirin each day. He has not really followed any diet restrictions.

Usual dietary intake:

AM:	1½–2 c dry cereal (Cheerios, bran flakes, Crispix), ½–¾ c skim milk, 16–32 oz orange juice
Lunch:	1½ oz ham on ww bagel, 1 apple or other fruit, 1 c chips, diet soda

Nelson, Jack, Male, 48 y.o.
Allergies: NKA   **Code:** FULL   **Isolation:** None
Pt. Location: RM 1952   **Physician:** P Phelps   **Admit Date:** 9/22

Snack when he comes home: Handful of crackers, cookies, or chips, 1–2 16-oz beers
PM: 6–9 oz of meat (grilled, baked usually), pasta, rice, or potatoes,1–2 c fresh fruit, salad or other vegetable, bread, iced tea
Late PM: Ice cream, popcorn, or crackers. Drinks 5–6 12-oz diet sodas daily as well as iced tea. Relates that his family's schedule has been increasingly busy, so they order pizza or stop for fast food 1–2 times per week instead of cooking.

24-hr recall:
(at home PTA): Crispix—2 c, 1 c skim milk, 16 oz orange juice
At work: 3 12-oz Diet Pepsis
Lunch: Fried chicken sandwich from McDonald's, small French fries, 32-oz iced tea
Late afternoon: 2 c chips, 1 beer
Dinner: 1 breast, fried, from Kentucky Fried Chicken, 1½ c potato salad, ¼ c green bean casserole, ½ c fruit salad, 1 c baked beans, iced tea
Bedtime: 2 c ice cream mixed with 1 c skim milk for milkshake

Food allergies/intolerances/aversions: Fried foods seem to make the indigestion worse
Previous nutrition therapy? No
Food purchase/preparation: Wife or eats out
Vit/min intake: One-A-Day for Men multivitamin daily

Laboratory Results

	Ref. Range	9/22 1952
Chemistry		
Sodium (mEq/L)	136–145	144
Potassium (mEq/L)	3.5–5.5	4.5
Chloride (mEq/L)	95–105	102
Carbon dioxide (CO_2, mEq/L)	23–30	28
BUN (mg/dL)	8–18	9
Creatinine serum (mg/dL)	0.6–1.2	0.7
BUN/Crea ratio	10.0–20.0	12.8
Glucose (mg/dL)	70–110	110
Phosphate, inorganic (mg/dL)	2.3–4.7	3.8
Magnesium (mg/dL)	1.8–3	2.0
Calcium (mg/dL)	9–11	9.1
Bilirubin total (mg/dL)	≤1.5	1.4
Bilirubin, direct (mg/dL)	<0.3	0.2

Nelson, Jack, Male, 48 y.o.
Allergies: NKA
Pt. Location: RM 1952

Code: FULL
Physician: P Phelps

Isolation: None
Admit Date: 9/22

Laboratory Results *(Continued)*

	Ref. Range	9/22 1952
Protein, total (g/dL)	6–8	6.8
Albumin (g/dL)	3.5–5	4.9
Prealbumin (mg/dL)	16–35	33
Ammonia (NH$_3$, μmol/L)	9–33	20
Alkaline phosphatase (U/L)	30–120	80
ALT (U/L)	4–36	30
AST (U/L)	0–35	22
CPK (U/L)	30–135 F 55–170 M	100
Lactate dehydrogenase (U/L)	208–378	219
Cholesterol (mg/dL)	120–199	220 !↑
HDL-C (mg/dL)	>55 F, >45 M	20 !↓
LDL (mg/dL)	<130	165 !↑
LDL/HDL ratio	<3.22 F <3.55 M	8.25 !↑
Triglycerides (mg/dL)	35–135 F 40–160 M	178 !↑
Coagulation (Coag)		
PT (sec)	12.4–14.4	13.8
INR	0.9–1.1	1.0
PTT (sec)	24–34	28
Hematology		
WBC (×10^3/mm^3)	4.8–11.8	5.6
RBC (×10^6/mm^3)	4.2–5.4 F 4.5–6.2 M	5.2
Hemoglobin (Hgb, g/dL)	12–15 F 14–17 M	14.0
Hematocrit (Hct, %)	37–47 F 40–54 M	40
Mean cell volume (μm^3)	80–96	85
Mean cell Hgb (pg)	26–32	28
Mean cell Hgb content (g/dL)	31.5–36	32
RBC distribution (%)	11.6–16.5	15.5
Platelet count (×10^3/mm^3)	140–440	345

(Continued)

Nelson, Jack, Male, 48 y.o.
Allergies: NKA
Pt. Location: RM 1952

Code: FULL
Physician: P Phelps

Isolation: None
Admit Date: 9/22

Laboratory Results *(Continued)*

	Ref. Range	9/22 1952
Hematology, Manual Diff		
Neutrophil (%)	50–70	55
Lymphocyte (%)	15–45	28
Monocyte (%)	3–10	6
Eosinophil (%)	0–6	0
Basophil (%)	0–2	0
Blasts (%)	3–10	3

Case Questions

I.   **Understanding the Disease and Pathophysiology**

1.   How is acid produced and controlled within the gastrointestinal tract?

2.   What role does lower esophageal sphincter (LES) pressure play in the etiology of gastro-esophageal reflux disease? What factors affect LES pressure?

3.   What are the complications of gastroesophageal reflux disease?

4.   The physician biopsied for *H. pylori*. What is this?

5.   Identify the patient's signs and symptoms that could suggest the diagnosis of gastroesopha-geal reflux disease.

6.   Describe the diagnostic tests performed for this patient.

7.   What risk factors does the patient present with that might contribute to his diagnosis? (Be sure to consider lifestyle, medical, and nutritional factors.)

8.   The MD has decreased the patient's dose of daily aspirin and recommended dis-continuing his ibuprofen. Why? How do aspirin and NSAIDs affect gastroesophageal disease?

9.   The MD has prescribed omeprazole. What class of medication is this? What is the basic mechanism of the drug? What other drugs are available in this class? What other groups of medications are used to treat GERD?

II.   Understanding the Nutrition Therapy

10. Summarize the current recommendations for nutrition therapy for GERD.

III.   Nutrition Assessment

11. Calculate the patient's %UBW and BMI. What does this assessment of weight tell you? In what ways may this contribute to his diagnosis?

12. Calculate energy and protein requirements for Mr. Nelson. How would this recommendation be modified to support a gradual weight loss?

13. Complete a computerized nutrient analysis for this patient's usual intake and 24-hour recall. How does his caloric intake compare to your calculated requirements?

14. Are there any other abnormal labs that should be addressed to improve Mr. Nelson's overall health? Explain.

15. What other components of lifestyle modification would you address in order to help in treating his disorder?

IV.   Nutrition Diagnosis

16. Identify pertinent nutrition problems and corresponding nutrition diagnoses and write at least two PES statements for them.

V.   Nutrition Intervention

17. Determine the appropriate intervention for each nutritional diagnosis.

18. Using Mr. Nelson's 24-hour recall, outline necessary modifications you could use as a teaching tool.

Food Item	Modification	Rationale
Crispix		
Skim milk		
Orange juice		
Diet Pepsi		
Fried chicken sandwich		
French fries		
Iced tea		
Chips		
Beer		
Fried chicken		
Potato salad		
Green bean casserole		
Fruit salad		
Baked beans		
Milkshake		

Bibliography

Academy of Nutrition and Dietetics Nutrition Care Manual. http://nutritioncaremanual.org/topic .cfm?ncm_heading=Diseases%2FConditions&ncm _toc_id=19379. Accessed 07/14/12.

Bredenoord AJ. Mechanisms of reflux perception in gastroesophageal reflux disease: A review. *Am J Gastroenterol.* 2012;107:8–15.

National Digestive Diseases Information Clearinghouse. Bethesda, MD: National Digestive Diseases Information Clearinghouse; May 2007; last updated April 30, 2012. Heartburn, gastroesophageal reflux (GER), and gastroesophageal reflux disease (GERD); NIH Publication No. 07-0882: http://digestive.niddk .nih.gov/ddiseases/pubs/gerd/index.aspx. Accessed 07/14/12.

Nelms MN. Nutrition assessment: Foundation of the nutrition care process. In: Nelms M, Sucher K, Lacey K, Roth SL. *Nutrition Therapy and Pathophysiology.* 2nd ed. Belmont, CA: Wadsworth, Cengage Learning; 2011:34–65.

Nelms MN. Diseases of the upper gastrointestinal tract. In: Nelms M, Sucher K, Lacey K, Roth SL. *Nutrition Therapy and Pathophysiology.* 2nd ed. Belmont, CA: Wadsworth, Cengage Learning; 2011:340–375.

Patrick L. Gastroesophageal reflux disease (GERD): A review of conventional and alternative treatments. *Altern Med Rev.* 2011;16(2):116–133.

Pronsky Z. M. Food-Medication Interactions. 18th ed. Birchrunville, PA: Food-Medication Interactions; 2012.

Singh M, Lee J, Gupta N, et al. Weight loss can lead to resolution of gastroesophageal reflux disease of symptoms: a prospective intervention trial. *Obesity,* accepted as online preview article 25 June 2012. http://www.nature.com/oby/journal/vaop/naam/abs /oby2012180a.html. Accessed 07/14/12.

Internet Resources

American College of Gastroenterology: http://patients.gi.org/topics/acid-reflux/

National Digestive Diseases Information Clearinghouse: http://digestive.niddk.nih.gov

National Library of Medicine and National Institutes of Health: http://www.nlm.nih.gov/medlineplus /gerd.html

USDA Nutrient Data Laboratory: http://www.ars.usda.gov /main/site_main.htm?modecode=12-35-45-00

Ulcer Disease: Medical and Surgical Treatment

Objectives

After completing this case, the student will be able to:

1. Discuss the etiology and risk factors for development of ulcer disease.
2. Identify classes of medications used to treat ulcer disease and determine possible drug–nutrient interactions.
3. Describe surgical procedures used to treat refractory ulcer disease and explain common nutritional problems associated with this treatment.
4. Apply knowledge of nutrition therapy guidelines for ulcer disease and gastric surgery.
5. Analyze nutrition assessment data to evaluate nutritional status and identify specific nutrition problems.
6. Determine nutrition diagnoses and write appropriate PES statements.
7. Calculate enteral nutrition prescriptions.
8. Evaluate a standard enteral nutritional regimen.
9. Develop a nutrition care plan—with appropriate measurable goals, interventions, and strategies for monitoring and evaluation—that addresses the nutrition diagnoses of this case.

Maria Rodriguez is a 38-year-old female who has been treated as an outpatient for her gastroesophageal reflux disease. Her increasing symptoms of hematemesis, vomiting, and diarrhea lead her to be admitted for further gastrointestinal workup. She undergoes a gastrojejunostomy to treat her perforated duodenal ulcer.

Rodriguez, Maria, Female, 38 y.o.
Allergies: Codeine causes N/V
Pt. Location: RM 1145

Code: FULL
Physician: A. Gustat

Isolation: None
Admit Date: 8/30

Patient Summary: Maria Rodriguez is a 38-yo female admitted through ER for a surgical consult for possible perforated duodenal ulcer.

History:

Onset of disease: Diagnosed with GERD approx. 11 months ago; diagnosed with duodenal ulcer 2 weeks ago

Medical history: Gravida 2 para 2. No other significant history except history of GERD.

Surgical history: Two weeks ago as an outpatient, she is s/p endoscopy that revealed 2-cm duodenal ulcer with generalized gastritis with a positive biopsy for *Helicobacter pylori*.

Medications at home: She has completed 10 days of a 14-day course of bismuth subsalicylate 525 mg 4 × daily, metronidazole 250 mg 4 × daily, tetracycline 500 mg 4 × daily, and omeprazole 20 mg 2 × daily, prescribed for total of 28 days

Tobacco use: Yes

Alcohol use: No

Family history: What? DM, PUD. Who? DM: maternal grandmother; PUD: father and grandfather.

Demographics:

Marital status: Widowed—lives with 2 daughters ages 12 and 14

Number of children: 2

Years education: Associate's degree

Language: English and Spanish

Occupation: Computer programmer

Hours of work: M–F 9–5

Ethnicity: Hispanic

Religious affiliation: Catholic

Physical Exam

General appearance: 38-year-old Hispanic female—thin, pale, and in acute distress

Heart: Regular rate and rhythm, heart sounds normal

HEENT: Noncontributory

Genitalia: WNL

Neurologic: Oriented X4

Extremities: Noncontributory

Skin: Warm and dry to touch

Chest/lungs: Rapid breath sounds, lungs clear

Peripheral vascular: Pulses full—no bruits

Abdomen: Tender with guarding, absent bowel sounds

Vital Signs: Temp: 102   Pulse: 68   Resp rate: 32
                    BP: 78/60   Height: 5'2"   Weight: 110 lbs

Rodriguez, Maria, Female, 38 y.o.
Allergies: Codeine causes N/V
Pt. Location: RM 1145

Code: FULL
Physician: A. Gustat

Isolation: None
Admit Date: 8/30

Assessment and Plan:

On 8/31, a gastrojejunostomy (Billroth II) was completed. Patient is now s/p gastrojejunostomy secondary to perforated duodenal ulcer.

Dx: Perforated duodenal ulcer

Medical Tx plan: Feeding jejunostomy was placed during surgery and patient is receiving Peptamen AF@ 25 mL/hr via continuous drip. Nutrition consult; orders have been left to advance the enteral feeding to 50 mL/hr. She is receiving only ice chips by mouth.

.... A. Gustat, MD

Nutrition:

History: Patient relates that she understands about the feeding she is receiving through her tube. She explains that she has eaten very little since her ulcer was diagnosed and wonders how long it will be before she can eat again. Her physicians have told her they might like her to try something by mouth in the next few days.

Usual dietary intake (prior to current illness)*:*
AM: Coffee, 1 slice dry toast. On weekends, cooked large breakfasts for family, which included omelets, rice or grits, or pancakes, waffles, fruit.
Lunch: Sandwich from home (2 oz turkey on whole-wheat bread with mustard); 1 piece of raw fruit, cookies (2–3 Chips Ahoy)
Dinner: 2 c rice, some type of meat (2–3 oz chicken), fresh vegetables (steamed tomatoes, peppers, and onions—1 c), coffee
Usual intake includes 8–10 c coffee and 1–2 soft drinks (12-oz cans) daily

24-hr recall: Has been NPO since admission.
Food allergies/intolerances/aversions: See nutrition history.
Previous nutrition therapy? No
Food purchase/preparation: Self and daughters
Vit/min intake: None

Rodriguez, Maria, Female, 38 y.o.
Allergies: Codeine causes N/V
Pt. Location: RM 1145

Code: FULL
Physician: A. Gustat

Isolation: None
Admit Date: 8/30

Intake/Output

Date		9/3/2012 0701 – 9/4/2012 0700			
Time		0701–1500	1501–2300	2301–0700	Daily total
IN	Tube feeding: Formula	150	100	200	450
	Tube feeding: Flush	50	50	50	150
	(mL/kg/hr)	(0.5)	(0.38)	(0.63)	(0.5)
	I.V.	**400**	**400**	**380**	**1180**
	(mL/kg/hr)	(1)	(1)	(0.95)	(0.98)
	I.V. piggyback				
	TPN				
	Total intake	**600**	**550**	**630**	**1780**
	(mL/kg)	(12)	(11)	(12.6)	(35.6)
OUT	Urine	**550**	**200**	**480**	**1230**
	(mL/kg/hr)	(1.38)	(0.5)	(1.2)	(1.03)
	Emesis output				
	Other: Drains	**275**	**320**	**220**	**815**
	Stool	**200**		**128**	**328**
	Total output	**1025**	**520**	**828**	**2373**
	(mL/kg)	(20.5)	(10.4)	(16.56)	(47.46)
Net I/O		**−425**	**+30**	**−198**	**−593**
Net since admission (8/30/2012)		**−425**	**−395**	**−593**	**−593**

Laboratory Results

	Ref. Range	8/30 0800	9/3 0600
Chemistry			
Sodium (mEq/L)	136–145	141	140
Potassium (mEq/L)	3.5–5.5	4.5	4.2
Chloride (mEq/L)	95–105	103	101
Carbon dioxide (CO_2, mEq/L)	23–30	26	24
BUN (mg/dL)	8–18	24 !↑	15
Creatinine serum (mg/dL)	0.6–1.2	1.1	0.9
BUN/Crea ratio	10.0–20.0	22 !↑	17
Glucose (mg/dL)	70–110	80	128 !↑
Phosphate, inorganic (mg/dL)	2.3–4.7	3.7	3.5
Magnesium (mg/dL)	1.8–3	1.9	1.7 !↓

Rodriguez, Maria, Female, 38 y.o.
Allergies: Codeine causes N/V
Pt. Location: RM 1145

Code: FULL
Physician: A. Gustat

Isolation: None
Admit Date: 8/30

Laboratory Results *(Continued)*

	Ref. Range	8/30 0800	9/3 0600
Calcium (mg/dL)	9–11	9.0	8.7 !↓
Bilirubin total (mg/dL)	≤1.5	1.7 !↑	1.0
Bilirubin, direct (mg/dL)	<0.3	1.3 !↑	0.6 !↑
Protein, total (g/dL)	6–8	5.7 !↓	5.8 !↓
Albumin (g/dL)	3.5–5	3.0 !↓	3.3 !↓
Prealbumin (mg/dL)	16–35	15 !↓	14 !↓
Ammonia (NH$_3$, μmol/L)	9–33	11	10
Alkaline phosphatase (U/L)	30–120	98	90
ALT (U/L)	4–36	30	24
AST (U/L)	0–35	31	17
Cholesterol (mg/dL)	120–199	121	122
Triglycerides (mg/dL)	35–135 F 40–160 M	100	101
Coagulation (Coag)			
PT (sec)	12.4–14.4	12.5	12.8
INR	0.9–1.1	1.0	0.9
PTT (sec)	24–34	27	29
Hematology			
WBC (× 10^3/mm^3)	4.8–11.8	16.3 !↑	12.5 !↑
RBC (× 10^6/mm^3)	4.2–5.4 F 4.5–6.2 M	4.9	5.0
Hemoglobin (Hgb, g/dL)	12–15 F 14–17 M	11.2 !↓	10.2 !↓
Hematocrit (Hct, %)	37–47 F 40–54 M	33 !↓	31 !↓
Mean cell volume (μm^3)	80–96	91	86
Mean cell Hgb (pg)	26–32	25.9 !↓	25.5 !↓
Mean cell Hgb content (g/dL)	31.5–36	31 !↓	28.5 !↓
Platelet count (× 10^3/mm^3)	140–440	345	356
Hematology, Manual Diff			
Neutrophil (%)	50–70		
Lymphocyte (%)	15–45	12 !↓	22
Monocyte (%)	3–10	5	4
Eosinophil (%)	0–6	2	3
Segs (%)	0–60	87 !↑	78 !↑

Case Questions

I.   Understanding the Disease and Pathophysiology

1. Identify the patient's risk factors for ulcer disease.

2. How is smoking related to ulcer disease?

3. What role does *H. pylori* play in ulcer disease?

4. Four different medications were prescribed for treatment of this patient's *H. pylori* infection. Identify the drug functions/mechanisms. (Use table below.)

Drug	Action
Metronidazole	
Tetracycline	
Bismuth subsalicylate	
Omeprazole	

5. What are the possible drug–nutrient side effects from Mrs. Rodriguez's prescribed regimen? (See table above.) Which drug–nutrient side effects are most pertinent to her current nutritional status?

6. Explain the surgical procedure the patient received.

7. How may the normal digestive process change with this procedure?

II.   Understanding the Nutrition Therapy

8. The most common physical side effects from this surgery are development of early or late dumping syndrome. Describe each of these syndromes, including symptoms the patient might experience, etiology of the symptoms, and standard interventions for preventing/treating the symptoms.

9. What other potential nutritional deficiencies may occur after this surgical procedure? Why might Mrs. Rodriguez be at risk for iron-deficiency anemia, pernicious anemia, and/or megaloblastic anemia?

10. Should Mrs. Rodriguez be on any type of vitamin/mineral supplementation at home when she is discharged? Would you make any recommendations for specific types? Explain.

III.   Nutrition Assessment

11. Prior to being diagnosed with GERD, Mrs. Rodriguez weighed 145 lbs. Calculate %UBW and BMI. Which of these is the most pertinent in identifying the patient's nutrition risk? Why?

12. What other anthropometric measures could be used to further confirm her nutritional status?

13. Calculate energy and protein requirements for Mrs. Rodriguez.

14. This patient was started on an enteral feeding postoperatively. What type of enteral formula is Peptamen AF? Using the current guidelines for initiation of nutrition support, state whether you agree with this choice and provide a rationale for your response.

15. Why was the enteral formula started at 25 mL/hr?

16. Is the current enteral prescription meeting this patient's nutritional needs? Compare her energy and protein requirements to what is provided by the formula. If her needs are not being met, what should be the goal for her enteral support?

17. What would the RD assess to monitor tolerance to the enteral feeding?

18. Using the intake/output record for postoperative day 3, how much enteral nutrition did the patient receive? How does this compare to what was prescribed?

19. As the patient is advanced to solid food, what modifications in diet would the RD address? Why? What would be a typical first meal for this patient?

20. What other advice would you give to Mrs. Rodriguez to maximize her tolerance of solid food?

21. Mrs. Rodriguez asks to speak with you because she is concerned about having to follow a special diet forever. What might you tell her?

22. Using her admission chemistry and hematology values, which biochemical measures are abnormal? Explain.

    a. Which values can be used to further assess her nutritional status? Explain.

    b. Which laboratory measures (see lab results, pages 84–85) are related to her diagnosis of a duodenal ulcer? Why would they be abnormal?

23. Do you think this patient is malnourished? If so, what criteria can be used to support a diagnosis of malnutrition? Using the guidelines proposed by ASPEN and AND, what type of malnutrition can be suggested as the diagnosis for this patient?

IV.   Nutrition Diagnosis
24. Select two nutrition problems and complete the PES statement for each.

V.   Nutrition Intervention
25. For each of the PES statements that you have written, establish an ideal goal (based on the signs and symptoms) and an appropriate intervention (based on the etiology).

26. What nutrition education should this patient receive prior to discharge?

27. Do any lifestyle issues need to be addressed with this patient? Explain.

Bibliography

Academy of Nutrition and Dietetics Nutrition Care Manual. Accessed 07/20/12 from http://nutritioncare-manual.org/topic.cfm?ncm_toc_id=20009.

Nelms MN. Diseases of the upper gastrointestinal tract. In: Nelms M, Sucher K, Lacey K, Roth SL. *Nutrition Therapy and Pathophysiology*. 2nd ed. Belmont, CA: Wadsworth, Cengage Learning; 2011:340–375.

Nelms MN. Enteral and parenteral nutrition support. In Nelms M, Sucher K, Lacey K, Roth SL. *Nutrition Therapy and Pathophysiology*. 2nd ed. Belmont, CA: Wadsworth, Cengage Learning; 2011:80–105.

Nelms MN. Nutrition assessment: Foundation of the nutrition care process. In: Nelms M, Sucher K, Lacey K, Roth SL. *Nutrition Therapy and Pathophysiology*. 2nd ed. Belmont, CA: Wadsworth, Cengage Learning; 2011:34–65.

Pronsky ZM. Food-Medication Interactions, 18th ed. Birchrunville, PA: Food-Medication Interactions; 2012.

White JV, Guenter P, Jensen G, Malone A, Schofield M, Academy of Nutrition and Dietetics Malnutrition Work Group, A.S.P.E.N. Malnutrition Task Force, A.S.P.E.N. Board of Directors. Consensus statement of the Academy of Nutrition and Dietetics/American Society of Parenteral and Enteral Nutrition: characteristics recommended for the identification and documentation of adult malnutrition (undernutrition). J Acad Nutr Diet. 2012;112: 730-8.

Internet Resources

Medscape Reference: http://emedicine.medscape.com/article/181753-overview

National Digestive Diseases Information Clearinghouse–NIDDK/National Institutes of Health: http://digestive.niddk.nih.gov/ddiseases/pubs/pepticulcers_ez/

National Library of Medicine and National Institutes of Health MedlinePlus: http://www.nlm.nih.gov/medlineplus/pepticulcer.html#cat5

NUTRITION THERAPY FOR LOWER GASTROINTESTINAL DISORDERS

The next three cases target conditions affecting the small and large intestines. These conditions, whose etiologies are all different, involve the symptoms of diarrhea, constipation, and sometimes malabsorption. In all three cases, nutrition therapy is one of the major modes of treatment.

Celiac disease, explored in Case 9, is an autoimmune disease triggered by exposure to gliadin, a protein found in the gluten portion of wheat, rye, and barley. This case explores an atypical presentation of celiac disease, new diagnostic procedures for celiac disease, secondary malabsorption syndromes, and the use of medical nutrition therapy.

Case 10 examines irritable bowel syndrome, which is the most common gastrointestinal complaint in the United States and Canada. The improved recognition and understanding of IBS in recent years has allowed for development of additional treatments.

The final case in this section targets inflammatory bowel disease. Crohn's disease and ulcerative colitis are two conditions that fall under the diagnosis of inflammatory bowel disease. Both of these conditions dramatically affect nutritional status and often require nutritional support during periods of exacerbation. This case involves the effects of Crohn's disease on digestion and absorption, the diagnosis of malnutrition, and parenteral nutrition support.

Celiac Disease

Objectives

After completing this case, the student will be able to:

1. Apply knowledge of the pathophysiology of celiac disease to identify and explain common nutritional problems associated with the disease.
2. Apply knowledge of nutrition therapy for celiac disease.
3. Analyze nutrition assessment data to evaluate nutritional status and identify specific nutrition problems.
4. Determine nutrition diagnoses and write appropriate PES statements.
5. Develop a nutrition care plan—with appropriate measurable goals, interventions, and strategies for monitoring and evaluation—that addresses the nutrition diagnoses of this case.

After experiencing joint pain, tingling in her legs and feet, mouth sores, itchy skin rash, muscle cramps, and depression over the past year, Mrs. Melissa Gaines visits her physician for evaluation.

Gaines, Melissa, Female, 46 y.o.
Allergies: NKA
Pt. Location: RM 926

Code: FULL
Physician: R. Smith

Isolation: None
Admit Date: 11/12

Patient Summary: 46-year-old female here for evaluation of symptoms of neuropathy; mouth sores; blistering, itchy skin rash over elbows, knees, and buttocks; muscle cramps; and depression

History:

Onset of disease: Patient relates having joint pain and neuropathy for the past year. She also reports having muscle cramps on a regular basis; an itchy skin rash over her torso, scalp, and buttocks; and canker sores inside her mouth almost constantly. These symptoms have made it hard for her to work, and she generally feels uncomfortable, which, she believes, is why she feels depressed.

Medical history: 3 pregnancies—2 live births, 1 miscarriage at 22 weeks. No other significant medical history.

Surgical history: N/A

Medications at home: Vitamins

Tobacco use: Yes

Alcohol use: No

Family history: What? CAD. Who? Father.

Demographics:

Marital status: Married—lives with husband and 2 sons; *Spouse name:* Michael

Number of children: 2

Years education: Bachelor's degree

Language: English only

Occupation: Secretary for hospital administrator

Hours of work: 8–4:30

Ethnicity: Caucasian

Religious affiliation: None

Physical Exam

General appearance: Well-nourished, pale woman who complains of fatigue, itchy rash, joint pain, and neuropathy

HEENT: Eyes: PERRLA sclera pale; fundi benign
           Throat: Pharynx clear without postnasal drainage

Genitalia: Deferred

Neurologic: Intact; alert and oriented

Extremities: No edema, strength 4/5

Skin: Pale, dermatitis herpetiformis on torso, scalp, and buttocks

Chest/lungs: Lungs clear to percussion and auscultation

Peripheral vascular: Pulses full—no bruits

Abdomen: Not distended; bowel sounds present

Vital Signs: Temp: 98.2   Pulse: 78   Resp rate: 17
                   BP: 108/72   Height: 5'3"   Weight: 125 lbs

Gaines, Melissa, Female, 46 y.o.
Allergies: NKA
Pt. Location: RM 926

Code: FULL
Physician: R. Smith

Isolation: None
Admit Date: 11/12

Assessment and Plan:

46-year-old female with fatigue, itchy rash, joint pain, and neuropathy here for workup. Rule out celiac disease.

Dx: R/O Celiac disease and anemia

Medical Tx plan: IgA-tTG, total serum IgA, IgA-EMA, IgG-tTG, Chem 24, hematology with differential
Dapsone, 50 mg/d po
Gluten-free diet
Nutrition consult

.... R. Smith, MD

Nutrition:

History: Pt denies any problems with appetite or foods; denies diarrhea, nausea, or vomiting. States her greatest nonpregnant weight was prior to her last pregnancy, when she weighed 150 lbs. She gained 25 lbs with her pregnancy, and her full-term son weighed 6 lbs 6 oz.

Usual dietary intake:
Likes a variety of foods.

24-hr recall (prior to admission):
AM: Toast—ww 2 slices, 1 tsp butter, hot tea with 2 tsp sugar
Lunch: 1 c chicken noodle soup, peanut butter & jelly sandwich (2 slices ww bread, 2 tbsp peanut butter, 2 tsp grape jelly), 1 c applesauce, 12 oz Sprite
Dinner: 1 c ww pasta, ½ c marinara sauce (no meat), 1 c sautéed green beans, 1 slice garlic bread, ½ c rainbow sherbet

Food allergies/intolerances/aversions: Maybe Nutrasweet?
Previous nutrition therapy? No
Food purchase/preparation: Self
Vit/min intake: Multivitamin/mineral

Laboratory Results

	Ref. Range	11/12 0926
Chemistry		
Sodium (mEq/L)	136–145	138
Potassium (mEq/L)	3.5–5.5	3.7
Chloride (mEq/L)	95–105	101

(Continued)

Gaines, Melissa, Female, 46 y.o.
Allergies: NKA
Pt. Location: RM 926

Code: FULL
Physician: R. Smith

Isolation: None
Admit Date: 11/12

Laboratory Results *(Continued)*

	Ref. Range	11/12 0926
Carbon dioxide (CO$_2$, mEq/L)	23–30	27
BUN (mg/dL)	8–18	9
Creatinine serum (mg/dL)	0.6–1.2	0.7
Glucose (mg/dL)	70–110	72
Phosphate, inorganic (mg/dL)	2.3–4.7	2.8
Magnesium (mg/dL)	1.8–3	1.6 !↓
Calcium (mg/dL)	9–11	9.1
Bilirubin total (mg/dL)	≤1.5	1.0
Bilirubin, direct (mg/dL)	<0.3	0.2
Protein, total (g/dL)	6–8	6.0
Albumin (g/dL)	3.5–5	3.5
Prealbumin (mg/dL)	16–35	16
Ammonia (NH$_3$, μmol/L)	9–33	10
Alkaline phosphatase (U/L)	30–120	125 !↑
ALT (U/L)	4–36	12
AST (U/L)	0–35	8
Cholesterol (mg/dL)	120–199	117 !↓
AGA antibody	0	+ !↑
EMA antibody	0	+ !↑
tTG IgA antibody (units)	<19	41 !↑
Coagulation (Coag)		
PT (sec)	12.4–14.4	13.3
INR	0.9–1.1	1.0
PTT (sec)	24–34	26
Hematology		
WBC (× 10^3/mm^3)	4.8–11.8	5.2
RBC (× 10^6/mm^3)	4.2–5.4 F 4.5–6.2 M	4.9
Hemoglobin (Hgb, g/dL)	12–15 F 14–17 M	9.5 !↓
Hematocrit (Hct, %)	37–47 F 40–54 M	34 !↓
Mean cell volume (μm^3)	80–96	90

Gaines, Melissa, Female, 46 y.o.
Allergies: NKA
Pt. Location: RM 926

Code: FULL
Physician: R. Smith

Isolation: None
Admit Date: 11/12

Laboratory Results *(Continued)*

	Ref. Range	11/12 0926
Mean cell Hgb (pg)	26–32	27
Mean cell Hgb content (g/dL)	31.5–36	30 !↓
RBC distribution (%)	11.6–16.5	11.9
Platelet count ($\times 10^3/mm^3$)	140–440	220
Transferrin (mg/dL)	250–380 F 215–365 M	395 !↑
Ferritin (mg/mL)	20–120 F 20–300 M	18 !↓
Iron (µg/dL)	65–165 F 75–175 M	60 !↓
Total iron binding capacity (µg/dL)	240–450	475 !↑
Iron saturation (%)	15–50 F 10–50 M	12 !↓
Vitamin B_{12} (ng/dL)	24.4–100	21.2 !↓
Folate (ng/dL)	5–25	3 !↓
Hematology, Manual Diff		
Neutrophil (%)	50–70	61
Lymphocyte (%)	15–45	21.4
Monocyte (%)	3–10	5
Eosinophil (%)	0–6	1
Segs (%)	0–60	55

Case Questions

I.   **Understanding the Disease and Pathophysiology**

1.   What is the etiology of celiac disease? Is anything in Mrs. Gaines's history typical of patients with celiac disease? Explain. The prevalence of celiac disease appears to be increasing. What does the current literature suggest as contributors to this change in celiac disease prevalence?

2.   What are AGA and EMA antibodies? Explain the connection between the presence of antibodies and etiology of celiac disease.

3.   AGA, EMA, and tTG antibodies are used in serological testing to diagnose celiac disease. Each test is sensitive and specific for the antibody it measures. The tTG test has a sensitivity of more than 90%. What does this mean? It also has a specificity of more than 95%. What does this mean?

4.   Mrs. Gaines presents with many nongastrointestinal symptoms of celiac disease. List the nongastrointestinal as well as the gastrointestinal clinical manifestations of celiac disease.

5.   Biopsy of the small intestine continues to be the "gold standard" for diagnosis of celiac disease. Briefly describe the procedure.

6.   How does celiac disease damage the small intestine?

II.   **Understanding the Nutrition Therapy**

7.   Gluten restriction is the major component of the medical nutrition therapy for celiac disease. What is gluten? Where is it found?

8.   Can patients on a gluten-free diet tolerate oats?

9.   What sources other than foods might introduce gluten to the patient?

10. Are there any known health benefits of following a gluten-free diet if a person does not have celiac disease?

11. Can patients with celiac disease also be lactose-intolerant?

12. There is a high prevalence of anemia among individuals with celiac disease. How can this be explained? What tests are used for anemia?

III. Nutrition Assessment

13. Calculate this patient's total energy and protein needs.

14. Evaluate Mrs. Gaines's laboratory measures for nutritional significance. Identify all laboratory values that are indicative of a potential nutrition problem.

15. Are the abnormalities identified in question #14 related to the consequences of celiac disease? Explain.

16. Are any symptoms from Mrs. Gaines's physical examination consistent with her laboratory values? Explain.

IV. Nutrition Diagnosis

17. Select two nutrition problems and complete the PES statement for each.

V. Nutrition Intervention

18. For each of the PES statements that you have written, establish an ideal goal (based on the signs and symptoms) and an appropriate intervention (based on the etiology).

19. What type of diet would you initially prescribe, considering the possibility that Mrs. Gaines has suffered intestinal damage?

VI.   Nutrition Monitoring and Evaluation

20. Evaluate the following excerpt from Mrs. Gaines's food diary. Identify the foods that might not be tolerated on a gluten-/gliadin-free diet. For each food identified, provide an appropriate substitute.

Cornflakes _____

Bologna slices _____

Lean cuisine—ginger garlic stir fry with chicken _____

Skim milk _____

Cheddar cheese spread _____

Green bean casserole (mushroom soup, onions, green beans) _____

Coffee _____

Rice crackers _____

Fruit cocktail _____

Sugar _____

Pudding _____

V8 juice _____

Banana _____

Cola _____

Bibliography

Academy of Nutrition and Dietetics Nutrition Care Manual. Accessed 07/14/12 from http://www.nutritioncaremanual.org/topic.cfm?ncm_toc_id=22684.

Guandalini S, Newland C. Differentiating food allergies from food intolerances. *Curr Gastroenterol Rep.* 2011;13:426–434.

Nelms MN, Habash D. Nutrition assessment: Foundation of the nutrition care process. In: Nelms M, Sucher K, Lacey K, Roth SL. *Nutrition Therapy and Pathophysiology.* 2nd ed. Belmont, CA: Wadsworth, Cengage Learning; 2011:34–65.

Nelms MN. Diseases of the lower gastrointestinal tract. In: Nelms M, Sucher K, Lacey K, Roth SL. *Nutrition Therapy and Pathophysiology.* 2nd ed. Belmont, CA: Wadsworth, Cengage Learning; 2011:376–436.

Pronsky ZM. Food-Medication Interactions, 18th ed. Birchrunville, PA: Food-Medication Interactions; 2012.

Riddle MS, Murray JA, Porter CK. The incidence and risk of celiac disease in a healthy US adult population. *Am J Gastroenterol.* 2012;107:1248–1255.

Rubio-Tapia A, Ludvigsson JF, Brantner TL, Murray JA, Everhart JE. The prevalence of celiac disease in the United States. *Am J Gastroenterol,* advance online publication. 31 July 2012.

Westerberg DP, Gill JM, Dave B, DiPrinzio MJ, Quisel A, Foy A. New strategies for diagnosis and management of celiac disease. *JAOA.* 2006;106(3):145–51.

Internet Resources

Celiac Disease Center at Columbia University: http://www.celiacdiseasecenter.org/CF-HOME.htm

Celiac Disease Foundation: http://www.celiac.org/

Celiac Sprue Association: http://www.csaceliacs.info/celiac_disease.jsp

National Digestive Diseases Information Clearinghouse (NDDIC): http://digestive.niddk.nih.gov/ddiseases/pubs/celiac/

National Library of Medicine/National Institutes of Health: http://www.nlm.nih.gov/medlineplus/celiacdisease.html

University of Chicago Celiac Disease Center: http://www.uchospitals.edu/specialties/celiac/

USDA Nutrient Data Laboratory: http://www.ars.usda.gov/nutrientdata

Case 10

Irritable Bowel Syndrome (IBS)

Objectives

After completing this case, the student will be able to:
1. Describe the proposed etiologies of IBS.
2. Use the current medical diagnostic criteria to identify IBS signs and symptoms found within the patient's history and physical exam.
3. Apply the appropriate nutrition assessments for an individual with IBS.

4. Develop a nutrition care plan—with appropriate measurable goals, interventions, and strategies for monitoring and evaluation—that addresses the nutrition diagnoses for this case.

Alicia Clarke is a 42-year-old female who presents to the gastroenterology clinic with stomach and intestinal complaints.

Clarke, Alicia, Female, 42 y.o.
Allergies: NKA
Pt. Location: Gastroenterology Clinic

Code: FULL
Physician: F. Cryan, MD

Isolation: None
Date: 6/30

Patient Summary: Alicia Clarke is a 42-year-old female who presents to outpatient gastroenterology clinic with stomach and intestinal complaints.

History:

Onset of disease: Patient presents upon referral from her family practice physician after experiencing both diarrhea and constipation for many years. Family physician found negative stool cultures. Colonoscopy negative for active disease.

Medical history: Hypothyroidism, gastroesophageal reflux disease, obesity

Surgical history: Caesarean × 2

Medications at home: Omeprazole 50 mg twice daily; levothyroxine 25 mg; vitamin D 600 IU; 800 mg calcium; Lomotil prn

Tobacco use: 1 ppd × 10 years—quit at age 30

Alcohol use: 3-4 × per week

Family history: Father—HTN, atherosclerosis; mother, sister—hypothyroidism, type 2 DM

Demographics:

Marital status: Divorced; 2 children—ages 12 and 14

Years education: 16 years

Language: English only

Occupation: Kindergarten teacher

Hours of work: 8–4 weekdays

Household members: Self, two children, and mother. Husband lives in city and shares custody of children.

Ethnicity: Caucasian

Religious affiliation: Baptist

History/Physical:

Chief complaint: "I am here for a workup of my stomach and intestinal problems. I have always had a 'funny' stomach, I think. As far back as I can remember, I have had times when I had diarrhea and others when I would go for days without going to the bathroom. The diarrhea is much worse now and I have had several accidents when I didn't make it to the bathroom. This is really interfering with my daily life." Patient describes ongoing abdominal pain almost every day with alternating constipation with diarrhea. Diarrhea has been more predominant lately with several episodes per day.

General appearance: Obese, anxious-appearing female

Vital Signs: Temp: 98.6   Pulse: 100   Resp rate: 16
                    BP: 128/72   Height: 5'5"   Weight: 191 lbs

Heart: Regular rate and rhythm

Clarke, Alicia, Female, 42 y.o.
Allergies: NKA
Pt. Location: Gastroenterology Clinic

Code: FULL
Physician: F. Cryan, MD

Isolation: None
Date: 6/30

HEENT: Head: WNL
          Eyes: PERRLA
          Ears: Clear
          Nose: Clear
          Throat: Dry mucous membranes without exudates or lesions
Genitalia: Deferred
Neurologic: Alert and oriented × 3
Extremities: WNL
Skin: Warm and dry
Chest/lungs: WNL—clear to auscultation and percussion
Peripheral vascular: Pulse 4 + bilaterally, warm, no edema
Abdomen: Hyperactive bowel sounds × 4; no organomegaly or masses—lower abdominal tenderness

Assessment/Plan:

Patient meets Rome III criteria for IBS-D. Will begin Elavil 25 mg daily. Initiate Metamucil 1 tbsp in 8 oz of liquid twice daily. Schedule laboratory for hydrogen breath test, anti-tTG. Nutrition consult. Patient return to clinic in three weeks.

.... F. Cryan, MD

Nutrition:

History: Patient states that her appetite is good even with her abdominal pain and diarrhea. She has steadily gained weight since her pregnancies and in the last 5 years has gained over 20 lbs. She feels that there is not any one food that causes diarrhea more than others. She likes most foods— likes to cook and prepares most meals at home. Patient states that she has been trying to follow a high-fiber diet so that it might help with her gastrointestinal symptoms. Patient also asks if her yogurt intake will provide enough probiotics as she has heard she needs to balance the bacteria in her intestines. She does admit that she isn't quite sure what that means.

Usual intake (for past several months):

AM:	Homemade yogurt smoothie with 1 c fresh fruit (peaches and cherries) + 8 oz yogurt, or dry cereal with dried fruit, nuts mixed with yogurt; 2–3 c coffee with half and half and artificial sweetener
Midmorning:	Diet Pepsi, ½ c dried fruit and nuts
Lunch:	Salad with kidney beans or lentils, cheese, tomatoes, carrots, asparagus; wheat crackers—approx. 12–15, Diet Pepsi
PM:	Some type of meat—varies, but mostly chicken; pasta or potatoes, variety of vegetables, some type of bread or roll with butter
Snacks:	Ice cream, cake, or cookies, usually each night but lately has been trying to eat sugar-free candies to help with weight loss. Wine or beer—2–3 times per week.

Clarke, Alicia, Female, 42 y.o.
Allergies: NKA   **Code:** FULL   **Isolation:** None
Pt. Location: Gastroenterology Clinic   **Physician:** F. Cryan, MD   **Date:** 6/30

FODMAP Assessment

Food	In an average week, how often do you:	Daily or several × per week	Some-times (1×/week)	Rarely (1×/month)	Never
Meals	1. Eat meals from sit-down or take-out restaurants?		X		
Grains	2. Eat wheat or white breads?	X			
	3. Eat wheat pasta and/or noodles?	X			
	4. Eat wheat-based breakfast cereals?	X			
	5. Eat wheat-based cookies, cakes, and/or crackers?	X			
Fruit	6. Eat apples, pears, guava, honeydew melon, mango, papaya, quince, star fruit, and/or watermelon?		X		
	7. Eat stone fruits such as apricots, peaches, cherries, plums, and/or nectarines?		X		
	8. Eat grapes, persimmons, and/or lychee?		X		
	9. Eat dried fruits?	X			
	10. Drink fruit juice?	X			
Vegetables	11. Eat onion, leeks, asparagus, artichokes, cabbage, Brussels sprouts, and/or green beans?	X			
Dairy	12. Drink milk (whole, 1%, 2%, or skim)?	X			
	13. Drink coconut milk?				X
	14. Eat ice cream, yogurt, and/or cream-based products?		X		
Protein	15. Eat legumes such as baked beans, kidney beans, lentils, black-eyed peas, chickpeas, and/or butter beans?	X			

Clarke, Alicia, Female, 42 y.o.
Allergies: NKA
Pt. Location: Gastroenterology Clinic

Code: FULL
Physician: F. Cryan, MD

Isolation: None
Date: 6/30

FODMAP Assessment *(Continued)*

Food	In an average week, how often do you:	Daily or several × per week	Some-times (1×/week)	Rarely (1×/ month)	Never
Sweets	16. Eat products with high-fructose corn syrup such as fruit drinks, carbonated sugar drinks, pancake syrup, jams, and/or jellies?	X			
	17. Add fructose to your food?				X
	18. Add honey to your food or beverages?	X			
	19. Use foods or medicines with artificial sweeteners such as sorbitol, mannitol, isomalt, and/or xylitol?	X			
Beverages	20. Drink carbonated beverages?	X			
	21. Drink fortified wines such as port wines and/or sherry?				X
	22. Drink chicory-based coffee substitute?				X
Condiments	23. Eat catsup, tomato paste, chutney, pickle relish, plum sauce, sweet and sour sauce, and/or barbecue sauce?		X		

Laboratory Results

	Ref. Range	6/30 1000
Chemistry		
Sodium (mEq/L)	136–145	141
Potassium (mEq/L)	3.5–5.5	3.7
Chloride (mEq/L)	95–105	101
Carbon dioxide (CO_2, mEq/L)	23–30	25
BUN (mg/dL)	8–18	11
Creatinine serum (mg/dL)	0.6–1.2	0.7

(Continued)

Clarke, Alicia, Female, 42 y.o.
Allergies: NKA
Pt. Location: Gastroenterology Clinic

Code: FULL
Physician: F. Cryan, MD

Isolation: None
Date: 6/30

Laboratory Results *(Continued)*

	Ref. Range	6/30 1000
Glucose (mg/dL)	70–110	115 ↑!
Phosphate, inorganic (mg/dL)	2.3–4.7	3.1
Magnesium (mg/dL)	1.8–3	1.8
Calcium (mg/dL)	9–11	10.1
Osmolality (mmol/kg/H_2O)	285–295	292
Bilirubin total (mg/dL)	≤1.5	0.2
Bilirubin, direct (mg/dL)	<0.3	0.1
Protein, total (g/dL)	6–8	6.2
Albumin (g/dL)	3.5–5	4.1
Prealbumin (mg/dL)	16–35	22
Ammonia (NH_3, μmol/L)	9–33	11
Alkaline phosphatase (U/L)	30–120	35
ALT (U/L)	4–36	6
AST (U/L)	0–35	1
CPK (U/L)	30–135 F 55–170 M	32
Lactate dehydrogenase (U/L)	208–378	209
Cholesterol (mg/dL)	120–199	201 ↑!
Triglycerides (mg/dL)	35–135 F 40–160 M	171 ↑!
T_4 (μg/dL)	4–12	4.2
T_3 (μg/dL)	75–98	78
HbA_{1C} (%)	3.9–5.2	6.1 ↑!
Hematology		
WBC (×10^3/mm^3)	4.8–11.8	5.5
RBC (×10^6/mm^3)	4.2–5.4 F 4.5–6.2 M	5.1
Hemoglobin (Hgb, g/dL)	12–15 F 14–17 M	12.1
Hematocrit (Hct, %)	37–47 F 40–54 M	37
Hematology, Manual Diff		
Neutrophil (%)	50–70	55
Lymphocyte (%)	15–45	16

Clarke, Alicia, Female, 42 y.o.
Allergies: NKA
Pt. Location: Gastroenterology Clinic

Code: FULL
Physician: F. Cryan, MD

Isolation: None
Date: 6/30

Laboratory Results *(Continued)*

	Ref. Range	6/30 1000
Monocyte (%)	3–10	3
Eosinophil (%)	0–6	0
Basophil (%)	0–2	0
Blasts (%)	3–10	3
Urinalysis		
Collection method	—	clean catch
Color	—	yellow
Appearance	—	clear
Specific gravity	1.003–1.030	1.004
pH	5–7	5.1
Protein (mg/dL)	Neg	Neg
Glucose (mg/dL)	Neg	Neg
Ketones	Neg	Neg
Blood	Neg	Neg
Bilirubin	Neg	Neg
Urobilinogen (EU/dL)	<1.1	Ncg
Leukocyte esterase	Neg	Neg
Prot chk	Neg	Neg
WBCs (/HPF)	0–5	0

Case Questions

I. **Understanding the Diagnosis and Pathophysiology**

1. IBS is considered to be a functional disorder. What does this mean? How does this relate to Mrs. Clarke's history of having a colonoscopy and her physician's order for a hydrogen breath test and measurements of anti-tTG?

2. What are the ACG and the Rome III criteria? Using the information from Mrs. Clarke's history and physical, determine how Dr. Cryan made her diagnosis of IBS-D.

3. Discuss the primary factors that may be involved in IBS etiology. You must include in your discussion the possible roles of genetics, infection, and serotonin.

4. Mrs. Clarke's physician prescribed two medications for her IBS. What are they and what is the proposed mechanism of each? She discusses the potential use of Lotronex if these medications do not help. What is this medication and what is its mechanism? Identify any potential drug–nutrient interactions for these medications.

II. **Understanding the Nutrition Therapy**

5. For each of the following foods, outline the possible effect on IBS symptoms.

    a. lactose

    b. fructose

    c. sugar alcohols

    d. high-fat foods

6. What is FODMAP? What does the current literature tell us about this intervention?

7. Define the terms *prebiotic* and *probiotic*. What does the current research indicate regarding their use for treatment of IBS?

III. Nutrition Assessment

8. Assess Mrs. Clarke's weight and BMI. What is her desirable weight?

9. Identify any abnormal laboratory values measured at this clinic visit and explain their significance for the patient with IBS.

10. List Mrs. Clarke's other medications and identify the rationale for each prescription. Are there any drug–nutrient interactions you should discuss with Mrs. Clarke?

11. Determine Mrs. Clarke's energy and protein requirements. Be sure to explain what standards you used to make this estimation.

12. Assess Mrs. Clarke's recent diet history. How does this compare to her estimated energy and protein needs? Identify foods that may potentially aggravate her IBS symptoms.

IV. Nutrition Diagnosis

13. Prioritize two nutrition problems and complete the PES statement for each.

V. Nutrition Intervention

14. The RD that counsels Mrs. Clarke discusses the use of an elimination diet. How may this be used to treat Mrs. Clarke's IBS?

15. The RD discusses the use of the FODMAP assessment to identify potential trigger foods. Describe the use of this approach for Mrs. Clarke. How might a food diary help her determine which foods she should avoid?

16. Should the RD recommend a probiotic supplement? If so, what standards might the RD use to make this recommendation?

17. Mrs. Clarke is interested in trying other types of treatment for IBS including acupuncture, herbal supplements, and hypnotherapy. What would you tell her about the use of each of these in IBS? What is the role of the RD in discussing complementary and alternative therapies?

VI.   Nutrition Monitoring and Evaluation

18. Write an ADIME note for your initial nutrition assessment with your plans for education and follow-up.

Bibliography

American College of Gastroenterology Task Force on Irritable Bowel Syndrome, Brandt LJ, Chey WD, et al. An evidenced-based systematic review on the management of irritable bowel syndrome. *Am J Gastroenterol.* 2009;104:S1–S35.

Barrett JS, Gibson PR. Development and validation of a comprehensive semi-quantitative food frequency questionnaire that includes FODMAP and glycemic index. *J Am Diet Assoc.* 2010;110:1469–70.

Barrett JS, Gibson PR. Fermentable oligosaccharides, disaccharides, monosaccharides and polyols (FODMAPs) and nonallergic food intolerance: FODMAPs or food chemicals? *Therap Adv Gastroenterol.* 2012;5:261–8.

Brenner DM, Moeller MJ, Chey WD, Schoenfield PS. The utility of probiotics in the treatment of irritable bowel syndrome: A systematic review. *Am J Gastroenterol.* 2009;104:1033–1049.

Cabre E. Irritable bowel syndrome: Can nutrient manipulation help? *Curr Op Clin Nutr Metab Care.* 2012;13:581–587.

Camilleri M. Pharmacology of the new treatments for lower gastrointestinal motility disorders and irritable bowel syndrome. *Nature.* 2012;91:44–59.

Chouinard LE. The role of psyllium fibre supplementation in treating irritable bowel syndrome. *Can J Diet Pract Res.* 2011;72:e107–e114.

Clarke G, Cryan JF, Dinan TG, Quigley EM. Review article: Probiotics for the treatment of irritable bowel syndrome—Focus on lactic acid bacteria. *Aliment Pharmacol Ther.* 2012;35:403–413.

Crowell MD. Role of serotonin in the pathophysiology of the irritable bowel syndrome. *Br J Pharmacol.* 2004;141(8):1285–1293.

Drisko J, Bischoff B, Hall M, McCallum R. Treating irritable bowel syndrome with a food elimination diet followed by food challenge and probiotics. *J Am Coll Nutr.* 2006;25(6):514–522.

Drossman DA. The functional gastrointestinal disorders and the Rome III process. *Gastroenterology.* 2006;130:1377–1390.

Douglas LC, Sanders ME. Probiotics and prebiotics in dietetics practice. *J Am Diet Assoc.* 2008;108:510–521.

DuPont AW. Postinfectious irritable bowel syndrome [review]. *Clin Infect Dis.* 2008;46:594–599.

El-Salhy M, Ostgaard H, Gundersen D, Hatlebakk JG, Hausen T. The role of diet in the pathogenesis and management of irritable bowel syndrome (Review). *Int J Mol Med.* 2012;29:723–31.

Floch MH. Use of diet and probiotic therapy in the irritable bowel syndrome: Analysis of the literature. *J Clin Gastroenterol.* 2005;39(4 Suppl 3):S243–S246.

Gibson PR, Shepherd SJ. Food choice as a key management strategy for functional gastrointestinal symptoms. *Am J Gastroenterol.* 2012;107:657–66.

Harris LA, Heitkemper MM. Practical considerations for recognizing and managing severe irritable bowel syndrome. *Gastroenterology Nursing.* 2012;35:12–21.

Heitkemper M, Cain KC, Schulman R, Burr R, Poppe A, Jarrett M. Subtypes of irritable bowel syndrome based on abdominal pain/discomfort severity and bowel pattern. *Dig Dis Sci.* 2011;56:2050–2058.

Heizer WD, Southern S, McGovern S. The role of diet in symptoms of irritable bowel syndrome: A narrative review. *J Am Diet Assoc.* 2009;109:1204–14.

Lim B, Manheimer E, Lao L, et al. Acupuncture for treatment of irritable bowel syndrome. *Cochrane Database Syst Rev.* 2006;(4):CD005111.

Liu JP, Yang M, Liu YX, Wei ML, Grimsgaard S. Herbal medicines for treatment of irritable bowel syndrome. *Cochrane Database Syst Rev.* 2006;(4):CD005111.

Longstreth GF, Thompson WG, Chey WD, Houghton LA, Mearin F, Spiller RC. Functional bowel disorders. *Gastroenterology.* 2006;130:1480–1491.

Mohammed I, Cherkas LF, Riley SA, Spector TD, Trudgill NJ. Genetic influences in irritable bowel syndrome: A twin study. *Am J Gastroenterol.* 2005;100(6):1340–1344.

Nahikian-Nelms M. Diseases of the lower gastrointestinal tract. In: Nelms M, Sucher K, Lacey K, Roth SL. *Nutrition Therapy and Pathophysiology.* 2nd ed. Belmont, CA: Wadsworth, Cengage Learning; 2011:376–436.

Ong DK, Mitchell SB, Barrett JS, et al. Manipulation of dietary short chain carbohydrates alters the pattern of gas production and genesis of symptoms in irritable bowel syndrome. *J Gastroenterology and Hepatology.* 2010;25:1366–1373.

Park MI, Camilleri M. Genetics and genotypes in irritable bowel syndrome: Implications for diagnosis and treatment. *Gastroenterol Clin North Am.* 2005;34(2):305–317.

Ruepert L, Quartero AO, de Wit NJ, van der Heijden GJ, Muris JW. Bulking agents, antispasmodics and antidepressants for the treatment of irritable bowel syndrome. *Cochrane Database Syst Rev.* 2011;10:CD003460.

Quigley EM, Abdel-Hamid H, Barbara G. et al. A global perspective on irritable bowel syndrome. A consensus statement of the World Gastroenterology Organisation Summit Task Force on irritable bowel syndrome. *J Clin Gastroenterol.* 2012;46:356–366.

Sanders ME, Douglas LC. Probiotics and Prebiotics in Dietetics Practice. *Am J Dietetics.* August 2008. Available at http://www.journals.elsevierhealth.com/periodicals

/yjada/article/S0002-8223(07)02210-9/abstract. Accessed on July 10, 2012.

Scarlata K. Successful low-FODMAP living—Experts discuss meal-planning strategies to help IBS clients better control GI distress. *Today's Dietitian.* 2012;14:36–38.

Shepherd SJ, Parker FC, Muir JG, Gibson PR. Dietary triggers of abdominal symptoms in patients with irritable bowel syndrome: Randomized placebo controlled evidence. *Clin Gastroenterol Hepatol.* 2008;6:765–771.

Shepherd SJ, Gibson PR. Fructose malabsorption and symptoms of irritable bowel syndrome: Guidelines for effective dietary management. *J Am Diet Assoc.* 2006;106:1631–1639.

Staudacher HM, Whelan K, Irving PM, Lomer MC. Comparison of symptom response following advice for a diet low in fermentable carbohydrates (FODMAPs) versus standard dietary advice in patients with irritable bowel syndrome. *J Hum Nutr Diet.* 2011;24:487–95.

Thomas JR, Nanda R, Shu LH. A FODMAP Diet Update: Craze or Credible? *Practical Gastroenterology.* 2012; 112:37–46.

Webb AN, Kukuruzovic RH, Catto-Smith AG, Sawyer SM. Hypnotherapy for treatment of irritable bowel syndrome. *Cochrane Database Syst Rev.* 2006;(1):CD004116.

Williams EA, Nai X, Corfe BM. Dietary intakes of people with irritable bowel syndrome. *BMC Gastroenterology.* 2011;11:9–16.

Internet Resources

ADA Evidence Analysis Library: http://www .adaevidencelibrary.com

International Foundation for Functional Gastrointestinal Disorders: http://www.aboutibs.org/

Lotronex: https://www.lotronex.com/

Medicinenet: http://www.medicinenet.com

Nutrition Care Manual: http://www.nutritioncaremanual.org

Rome Foundation: http://www.romecriteria.org/criteria/

WebMD: http://www.webmd.com/ibs/alternative-therapies

Inflammatory Bowel Disease: Crohn's Disease

Objectives

After completing this case, the student will be able to:

1. Apply knowledge of the pathophysiology of Crohn's disease to identify and explain common nutritional problems associated with this disease.
   a. Describe physiological changes resulting from Crohn's disease.
   b. Identify nutritional consequences of Crohn's disease.
   c. Identify nutritional consequences of surgical resection of the small intestine.
2. Describe current medical care for Crohn's disease.
3. Identify potential drug–nutrient interactions.

4. Analyze nutrition assessment data to evaluate nutritional status and identify specific nutrition problems.
5. Determine nutrition diagnoses and write appropriate PES statements.
6. Calculate parenteral nutrition formulations.
7. Evaluate a parenteral nutrition regimen.
8. Develop a nutrition care plan—with appropriate measurable goals, interventions, and strategies for monitoring and evaluation—that addresses the nutrition diagnoses of this case.

Matt Sims was diagnosed with Crohn's disease 2½ years ago. He is now admitted with an acute exacerbation of that disease.

Sims, Matt, Male, 35 y.o.
Allergies: Milk
Pt. Location: RM 1952

Code: Full
Physician: D Tucker

Isolation: None
Admit Date: 12/15

Patient Summary: Matt Sims was hospitalized this past September with an abscess and acute exacerbation of Crohn's disease.

History:
Onset of disease: Dx. Crohn's disease 2½ years ago
Medical history: Initial diagnostic workup indicated acute disease within last 5–7 cm of jejunum and first 5 cm of ileum. Regimens have included corticosteroids, mesalamine. Plans are to initiate Humira.
Surgical history: No surgeries
Medications at home: 6-mercaptopurine
Tobacco use: No
Alcohol use: No
Family history: Noncontributory

Demographics:
Marital status: Married—lives with wife, Mary and son, age 5; both are well.
Years education: Bachelor's degree
Language: English only
Occupation: High school math teacher
Hours of work: 8–4:30, some after-school meetings and responsibilities as advisor for school clubs
Ethnicity: Caucasian
Religious affiliation: Episcopalian

Admitting History/Physical:
Chief complaint: "I was diagnosed with inflammatory bowel disease almost 3 years ago. At first they thought I had ulcerative colitis but six months later it was identified as Crohn's disease. I was really sick at that time and was in the hospital for more than two weeks. I have done OK until school started this fall. I've noticed more diarrhea and abdominal pain, but I tried to keep going since school just started. I was here in September and we switched medicine. I was a little better and I went back to work. Now my abdominal pain is unbearable—I seem to have diarrhea constantly, and now I am running a fever."
General appearance: Thin, 36-year-old white male in apparent distress

Vital Signs: Temp: 101.5°F   Pulse: 81   Resp rate: 18
                   BP: 125/82   Height: 5'9"   Weight: 140 lbs

Heart: RRR without murmurs or gallops
HEENT: Eyes: PERRLA, normal fundi
          Ears: Noncontributory

Sims, Matt, Male, 35 y.o.
Allergies: Milk
Pt. Location: RM 1952

Code: Full
Physician: D Tucker

Isolation: None
Admit Date: 12/15

Nose: Noncontributory
Throat: Pharynx clear
Rectal: No evidence of perianal disease
Neurologic: Oriented × 4
Extremities: No edema; pulses full; no bruits; normal strength, sensation, and DT
Skin: Warm, dry
Chest/lungs: Lungs clear to auscultation and percussion
Peripheral vascular: WNL
Abdomen: Distension, extreme tenderness with rebound and guarding; minimal bowel sounds

Nursing Assessment	12/15
Abdominal appearance (concave, flat, rounded, obese, distended)	rounded
Palpation of abdomen (soft, rigid, firm, masses, tense)	soft
Bowel function (continent, incontinent, flatulence, no stool)	continent
Bowel sounds (P=present, AB=absent, hypo, hyper)	
RUQ	P
LUQ	P
RLQ	P
LLQ	P
Stool color	light brown
Stool consistency	soft to liquid
Tubes/ostomies	N/A
Genitourinary	
Urinary continence	catheter
Urine source	catheter
Appearance (clear, cloudy, yellow, amber, fluorescent, hematuria, orange, blue, tea)	clear, yellow
Integumentary	
Skin color	pale
Skin temperature (DI=diaphoretic, W=warm, dry, CL=cool, CLM=clammy, CD += cold, M=moist, H=hot)	W
Skin turgor (good, fair, poor, TENT=tenting)	good
Skin condition (intact, EC=ecchymosis, A=abrasions, P=petechiae, R=rash, W=weeping, S=sloughing, D=dryness, EX=excoriated, T=tears, SE=subcutaneous emphysema, B=blisters, V=vesicles, N=necrosis)	intact

(Continued)

Sims, Matt, Male, 35 y.o.
Allergies: Milk
Pt. Location: RM 1952

Code: Full
Physician: D Tucker

Isolation: None
Admit Date: 12/15

Nursing Assessment *(Continued)*

Nursing Assessment	12/15
Mucous membranes (intact, EC=ecchymosis, A=abrasions, P=petechiae, R=rash, W=weeping, S=sloughing, D=dryness, EX=excoriated, T=tears, SE=subcutaneous emphysema, B=blisters, V=vesicles, N=necrosis)	intact
Other components of Braden score: special bed, sensory pressure, moisture, activity, friction/shear (>18=no risk, 15–16=low risk, 13–14=moderate risk, ≤12=high risk)	activity, 17

Orders:

R/O acute exacerbation of Crohn's disease vs. infection vs. small bowel obstruction.
CBC/Chem 24
ASCA
CT scan of abdomen and possible esophagogastroduodenoscopy
D5W NS @ 75 cc/hr; Clear liquids
Surgical consult
Nutrition support consult

Nutrition:

General: Patient states he has been eating fairly normally for the last year. After hospitalization at initial diagnosis, he had lost almost 25 lbs, which he regained. He initially ate a low-fiber diet and worked hard to regain the weight he had lost. He drank Boost between meals for several months. His usual weight before his illness was 166–168 lbs. He was at his highest weight (168 lbs) about 6 months ago but now states he has lost most of what he regained and has even lost more since his last hospitalization when he was at 140 lbs.

Recent dietary intake:
AM: Cereal, small amount of skim milk, toast or bagel; juice
AM snack: Cola—sometimes crackers or pastry
Lunch: Sandwich (ham or turkey) from home, fruit, chips, cola
Dinner: Meat, pasta or rice, some type of bread; rarely eats vegetables
Bedtime snack: Cheese and crackers, cookies, cola

24-hr recall: Clear liquids for past 24 hours since admission

Food allergies/intolerances/aversions: Previous nutrition therapy? Yes. If yes, when: Last hospitalization.

Sims, Matt, Male, 35 y.o.
Allergies: Milk
Pt. Location: RM 1952

Code: Full
Physician: D Tucker

Isolation: None
Admit Date: 12/15

What? "The dietitian talked to me about ways to decrease my diarrhea—ways to keep from being dehydrated—and then we worked out a plan to help me regain weight. I know what to do—it is just that the pain and diarrhea make my appetite so bad. It is really hard for me to eat."
Food purchase/preparation: Self and spouse
Vit/min intake: Multivitamin daily

Intake/Output

Date		12/20 0701 – 2/21 0700			
Time		0701–1500	1501–2300	2301–0700	Daily total
IN	P.O.	**NPO**	**NPO**	**NPO**	
	I.V.	**680**	**680**	**680**	**2040**
	(mL/kg/hr)	(1.34)	(1.34)	(1.34)	(1.34)
	I.V. piggyback				
	TPN				
	Total intake	**680**	**680**	**680**	**2040**
	(mL/kg)	(10.7)	(10.7)	(10.7)	(32.1)
OUT	Urine	**500**	**570**	**520**	**1590**
	(mL/kg/hr)	(0.98)	(1.12)	(1.02)	(1.04)
	Emesis output				
	Other				
	Stool				
	Total output	**500**	**570**	**520**	**1590**
	(mL/kg)	(7.86)	(8.96)	(8.18)	(25)
Net I/O		+180	+110	+160	+450
Net since admission (12/15)		+180	+290	+450	+450

Laboratory Results

	Ref. Range	2/15 1952
Chemistry		
Sodium (mEq/L)	136–145	136
Potassium (mEq/L)	3.5–5.5	3.7
Chloride (mEq/L)	95–105	101
Carbon dioxide (CO_2, mEq/L)	23–30	26
BUN (mg/dL)	8–18	11
Creatinine serum (mg/dL)	0.6–1.2	0.8
Glucose (mg/dL)	70–110	82
Phosphate, inorganic (mg/dL)	2.3–4.7	2.9

(Continued)

Sims, Matt, Male, 35 y.o.
Allergies: Milk
Pt. Location: RM 1952

Code: Full
Physician: D Tucker

Isolation: None
Admit Date: 12/15

Laboratory Results *(Continued)*

	Ref. Range	2/15 1952
Magnesium (mg/dL)	1.8–3	1.8
Calcium (mg/dL)	9–11	9.1
Bilirubin, direct (mg/dL)	<0.3	0.3
Protein, total (g/dL)	6–8	5.5 !↓
Albumin (g/dL)	3.5–5	3.2 !↓
Prealbumin (mg/dL)	16–35	11 !↓
Ammonia (NH$_3$, µmol/L)	9–33	11
Alkaline phosphatase (U/L)	30–120	120
ALT (U/L)	4–36	35
AST (U/L)	0–35	22
C-reactive protein (mg/dL)	<1.0	2.8 !↑
Cholesterol (mg/dL)	120–199	149
HDL-C (mg/dL)	>55 F, >45 M	38 !↓
LDL (mg/dL)	<130	111
LDL/HDL ratio	<3.22 F <3.55 M	2.92
Triglycerides (mg/dL)	35–135 F 40–160 M	85
T-transglutaminase IgA AB	Neg	Neg
Tissue transglutaminase IgG	Neg	Neg
ASCA	Neg	+ !↑
Coagulation (Coag)		
PT (sec)	12.4–14.4	15 !↑
Hematology		
WBC (×10^3/mm^3)	4.8–11.8	11.1
RBC (×10^6/mm^3)	4.2–5.4 F 4.5–6.2 M	4.9
Hemoglobin (Hgb, g/dL)	12–15 F 14–17 M	12.9 !↓
Hematocrit (Hct, %)	37–47 F 40–54 M	38 !↓
Mean cell volume (µm^3)	80–96	87
Mean cell Hgb (pg)	26–32	30
Mean cell Hgb content (g/dL)	31.5–36	33

Sims, Matt, Male, 35 y.o.
Allergies: Milk
Pt. Location: RM 1952

Code: Full
Physician: D Tucker

Isolation: None
Admit Date: 12/15

Laboratory Results *(Continued)*

	Ref. Range	2/15 1952
Platelet count ($\times 10^3$/mm^3)	140–440	422
Transferrin (mg/dL)	250–380 F 215–365 M	180 !↓
Ferritin (mg/mL)	20–120 F 20–300 M	16 !↓
ZPP (μmol/mol)	30–80	85 !↑
Vitamin B$_{12}$ (ng/dL)	24.4–100	75
Folate (ng/dL)	5–25	6
Zinc, serum (mcg/mL)	0.60–1.2	0.64
Vitamin D 25 hydroxy (ng/mL)	30–100	22.7 !↓
Free retinol (vitamin A; μg/dL)	20–80	17.2 !↓
Ascorbic acid (mg/dL)	0.2–2.0	<0.1 !↓
Selenium (ng/mL)	70–150	123

Case Questions

I. **Understanding the Disease and Pathophysiology**

1. What is inflammatory bowel disease? What does current medical literature indicate regarding its etiology?

2. Mr. Sims was initially diagnosed with ulcerative colitis and then diagnosed with Crohn's. How could this happen? What are the similarities and differences between Crohn's disease and ulcerative colitis?

3. A CT scan indicated bowel obstruction and the Crohn's disease was classified as severe-fulminant disease. CDAI score of 400. What does a CDAI score of 400 indicate? What does a classification of severe-fulminant disease indicate?

4. What did you find in Mr. Sims' history and physical that is consistent with his diagnosis of Crohn's? Explain.

5. Crohn's patients often have extraintestinal symptoms of the disease. What are some examples of these symptoms? Is there evidence of these in his history and physical?

6. Mr. Sims has been treated previously with corticosteroids and mesalamine. His physician had planned to start Humira prior to this admission. Explain the mechanism for each of these medications in the treatment of Crohn's.

7. Which laboratory values are consistent with an exacerbation of his Crohn's disease? Identify and explain these values.

8. Mr. Sims is currently on several vitamin and mineral supplements. Explain why he may be at risk for vitamin and mineral deficiencies.

9. Is Mr. Sims a likely candidate for short bowel syndrome? Define *short bowel syndrome*, and provide a rationale for your answer.

10. What type of adaptation can the small intestine make after resection?

11. For what classic symptoms of short bowel syndrome should Mr. Sims' health care team monitor?

12. Mr. Sims is being evaluated for participation in a clinical trial using high-dose immunosuppression and autologous peripheral blood stem cell transplantation (autoPBSCT). How might this treatment help Mr. Sims?

II.   Understanding the Nutrition Therapy

13. What are the potential nutritional consequences of Crohn's disease?

14. Mr. Sims underwent resection of 200 cm of jejunum and proximal ileum with placement of jejunostomy. The ileocecal valve was preserved. Mr. Sims did not have an ileostomy, and his entire colon remains intact. How long is the small intestine, and how significant is this resection?

15. What nutrients are normally digested and absorbed in the portion of the small intestine that has been resected?

III.   Nutrition Assessment

16. Evaluate Mr. Sims' % UBW and BMI.

17. Calculate Mr. Sims' energy requirements.

18. What would you estimate Mr. Sims' protein requirements to be?

19. Identify any significant and/or abnormal laboratory measurements from both his hematology and his chemistry labs.

IV.   Nutrition Diagnosis

20. Select two nutrition problems and complete the PES statement for each.

V.   Nutrition Intervention

21. The surgeon notes Mr. Sims probably will not resume eating by mouth for at least 7–10 days. What information would the nutrition support team evaluate in deciding the route for nutrition support?

22. The members of the nutrition support team note his serum phosphorus and serum magnesium are at the low end of the normal range. Why might that be of concern?

23. What is refeeding syndrome? Is Mr. Sims at risk for this syndrome? How can it be prevented?

24. Mr. Sims was placed on parenteral nutrition support immediately postoperatively, and a nutrition support consult was ordered. Initially, he was prescribed to receive 200 g dextrose/L, 42.5 g amino acids/L, and 30 g lipid/L. His parenteral nutrition was initiated at 50 cc/hr with a goal rate of 85 cc/hr. Do you agree with the team's decision to initiate parenteral nutrition? Will this meet his estimated nutritional needs? Explain. Calculate: pro (g); CHO (g); lipid (g); and total kcal from his PN.

25. For each of the PES statements you have written, establish an ideal goal (based on the signs and symptoms) and an appropriate intervention (based on the etiology).

VI.   Nutrition Monitoring and Evaluation

26. Indirect calorimetry revealed the following information:

Measure	Mr. Sims' data
Oxygen consumption (mL/min)	295
CO_2 production (mL/min)	261
RQ	0.88
RMR	2022

What does this information tell you about Mr. Sims?

27. Would you make any changes to his prescribed nutrition support? What should be monitored to ensure adequacy of his nutrition support? Explain.

28. What should the nutrition support team monitor daily? What should be monitored weekly? Explain your answers.

29. Mr. Sims' serum glucose increased to 145 mg/dL. Why do you think this level is now abnormal? What should be done about it?

30. Evaluate the following 24-hour urine data: 24-hour urinary nitrogen for 12/20: 18.4 grams. By using the daily input/output record for 12/20 that records the amount of PN received, calculate Mr. Sims' nitrogen balance on postoperative day 4. How would you interpret this information? Should you be concerned? Are there problems with the accuracy of nitrogen balance studies? Explain.

31. On post-op day 10, Mr. Sims' team notes he has had bowel sounds for the previous 48 hours and had his first bowel movement. The nutrition support team recommends consideration of an oral diet. What should Mr. Sims be allowed to try first? What would you monitor for tolerance? If successful, when can the parenteral nutrition be weaned?

32. What would be the primary nutrition concerns as Mr. Sims prepares for rehabilitation after his discharge? Be sure to address his need for supplementation of any vitamins and minerals. Identify two nutritional outcomes with specific measures for evaluation.

Bibliography

American Nutrition and Dietetics. *Nutrition Care Manual.* http://www.nutritioncaremanual.org/topic.cfm?ncm_heading = Diseases%2FConditions&ncm_toc_id = 19449. Accessed 9/30/12.

Campbell J, Borody TJ, Leis S. The many faces of Crohn's disease: Latest concepts in etiology. *OJIM.* 2012;2: 107–115. Published online June 2012 (http://www.SciRP.org/journal/ojim/).

Hasselblatt P, Drognitz K, Potthoff K, et al. Remission of refractory Crohn's disease by high-dose cyclophosphamide and autologous peripheral blood stem cell transplantation. Alimentary Pharmacology and Therapeutics. 2012;36(8):725–735.

Jensen GL, Mirtallo J, Compher C, et al. Adult starvation and disease related malnutrition: A proposal for etiology-based diagnosis in the clinical practice setting from the International Consensus Guideline Committee. *Clinical Nutrition.* 2010;29:151–153.

Lashner BA. Inflammatory bowel disease. The Cleveland Clinic Disease Management Project. Published August 1, 2010. Available at URL http://www.clevelandclinicmeded.com/medicalpubs/diseasemanagement/gastroenterology/inflammatory-bowel-disease/. Accessed 9/30/12.

Nelms MN. Diseases of the lower gastrointestinal tract. In: Nelms M, Sucher K, Lacey K, Roth SL. *Nutrition Therapy and Pathophysiology.* 2nd ed. Belmont, CA: Wadsworth, Cengage Learning; 2011:376–436.

Nelms MN. Enteral and parenteral nutrition support. In: Nelms M, Sucher K, Lacey K, Roth SL. *Nutrition Therapy and Pathophysiology.* 2nd ed. Belmont, CA: Wadsworth, Cengage Learning; 2011:80–105.

Nelms MN, Habash D. Nutrition assessment: Foundations of the nutrition care process. In: Nelms M, Sucher K, Lacey K, Roth SL. Nutrition Therapy and Pathophysiology. 2nd ed. Belmont, CA: Wadsworth, Cengage Learning; 2011:34–65.

Nelms MN. Metabolic stress and the critically ill. In: Nelms M, Sucher K, Lacey K, Roth SL. *Nutrition Therapy and Pathophysiology.* 2nd ed. Belmont, CA: Wadsworth, Cengage Learning; 2011:682–701.

Semrad CE. Use of parenteral nutrition in patients with inflammatory bowel disease. *Gastroenterol Hepatol.* 2012;8(6):393–395.

Smith MA, Smith T, Trebble TM. Nutritional management of adults with inflammatory bowel disease: Practical lessons from the available evidence. *Frontline Gastroenterology.* 2012;3:172–179.

Internet Resources

Academy of Nutrition and Dietetics Nutrition Care Manual: http://www.nutritioncaremanual.org

Crohn's & Colitis Foundation of America: http://www.ccfa.org

Mayo Clinic: http://www.mayoclinic.com/health/crohns-disease/DS00104

National Digestive Diseases Information Clearinghouse (NDDIC): http://digestive.niddk.nih.gov/ddiseases/pubs/crohns/#treatment

Unit Five

NUTRITION THERAPY FOR HEPATOBILIARY AND PANCREATIC DISORDERS

The liver and pancreas are often called ancillary organs of digestion; however, the term *ancillary* does little to describe their importance in digestion, absorption, and metabolism of carbohydrate, protein, and lipid. The cases in this section portray common conditions affecting these organs and outline their effects on nutritional status. The incidence of hepatobiliary disease has significantly increased over the last several decades, with cirrhosis being the most frequent diagnosis. The most common cause of cirrhosis is chronic alcohol ingestion; the second most common cause is viral hepatitis. The first case in this section focuses on those etiologies.

The incidence of malnutrition is very high in these disease states. Generalized symptoms of these diseases center around interruption of normal metabolism within these organs. Jaundice, anorexia, fatigue, abdominal pain, steatorrhea, and malabsorption are signs or symptoms of hepatobiliary disease.

These symptoms may be responsive to nutrition therapy but also interfere with maintenance of an adequate nutritional status.

In Case 12 the complications of end-stage cirrhosis are explored. Treatment of cirrhosis is primarily supportive. The only cure is a liver transplant. Therefore, nutrition therapy is crucial for preventing protein–calorie malnutrition, minimizing the symptoms of the disease, and maintaining quality of life. This case also introduces infection with the hepatitis C virus (HCV), which is the most common hepatic viral infection in the United States and a common cause of liver failure.

Case 13 focuses on the diagnosis of acute pancreatitis associated with chronic alcohol use. The use of nutrition support for acute pancreatitis has changed dramatically over the last decade, and this case allows for the application of new evidenced-based guidelines when planning the care for this patient.

Case 12

Cirrhosis of the Liver

Objectives

After completing this case, the student will be able to:

1. Integrate knowledge of the pathophysiology of cirrhosis with development of a nutrition care plan.
2. Research and discuss the current role of nutrition in the development and treatment of hepatic encephalopathy.
3. Identify and apply pertinent nutrition assessment indices for the patient with cirrhosis.
4. Develop nutrition diagnoses for the patient.
5. Identify appropriate nutrition therapy goals.
6. Determine key components of nutrition education for the patient with cirrhosis.

Teresa Wilcox is admitted to University Hospital with increasing symptoms of liver disease 3½ years after being diagnosed with acute hepatitis. A liver biopsy and CT scan confirm her diagnosis of cirrhosis of the liver secondary to chronic hepatitis C infection.

Wilcox, Teresa (Terri), Female, 26 y.o.
Allergies: NKA   **Code:** Full   **Isolation:** None
Pt. Location: RM 1012   **Physician:** P. Horowitz   **Admit Date:** 12/19

Patient Summary: Terri Wilcox is a 26-year-old architecture doctoral student who was in relatively good health until 3 years ago when she was diagnosed with hepatitis C.

History:

Onset of disease: Hepatitis C Dx 3 years ago—previously treated with alpha-interferon and ribavirin; seasonal allergies treated with antihistamines
Medical history: Currently, she c/o fatigue, anorexia, N/V, and weakness. She has lost 10# since her last office visit, which was 6 months ago. She also reports that she has been experiencing bruising of her skin that did not happen previously and does not appear to be related to injury.
Surgical history: No surgeries
Medications at home: Allegra 60 mg po qd
Tobacco use: No
Alcohol use: Yes
Family history: Mother (living)—HTN, diverticulitis, cholecystitis, carpal tunnel syndrome; father (deceased)—diabetes mellitus, peptic ulcer disease; maternal grandmother—cholecystitis, bilateral breast cancer; maternal grandfather—leukemia; paternal grandfather—cirrhosis; paternal grandmother—amyotrophic lateral sclerosis

Demographics:

Marital status: Single—lives with roommate who is a law student; *Roommate's name:* Kevin Gustat
Number of children: 0
Years education: Postgraduate
Language: English only
Occupation: Doctoral graduate student in architecture
Hours of work: Teaches late morning and late afternoon; takes classes and conducts research during most evenings
Ethnicity: European American
Religious affiliation: Unitarian

Admitting History/Physical:

Chief complaint: "It just seems as if I can't get enough rest. I feel so weak. Sometimes I'm so tired, I can't go to campus to teach my classes. Does my skin look yellow to you?"
General appearance: Tired-looking young female

Vital Signs: Temp: 96.9   Pulse: 72   Resp rate: 19
                   BP: 102/65   Height: 5'8"   Weight: 125 lbs

Heart: Regular rate and rhythm, no gallops or rubs, point of maximal impulse at the fifth intercostal space in the midclavicular line
HEENT: Head: Normocephalic
         Eyes: Wears contact lenses to correct myopia; PERRLA
         Ears: Tympanic membranes w/out lesions
         Nose: Dry mucous membranes w/out lesions
         Throat: Enlarged esophageal veins

Wilcox, Teresa (Terri), Female, 26 y.o.
Allergies: NKA
Pt. Location: RM 1012

Code: Full
Physician: P. Horowitz

Isolation: None
Admit Date: 12/19

Genitalia: Normal female
Neurologic: Alert and oriented × 3
Extremities: Normal muscular tone, normal ROM; no edema; no asterixis noted
Skin: Warm and dry; bruising noted on lower arms and legs; telangiectasias noted on chest
Chest/lungs: Respirations normal; no crackles, rhonchi, wheezes, or rubs noted
Peripheral vascular: Pulse 3+ bilaterally
Abdomen: Pierced umbilicus, upper right abdomen; mild distension, hepatomegaly; no ascites

Nursing Assessment	12/19
Abdominal appearance (concave, flat, rounded, obese, distended)	distended
Palpation of abdomen (soft, rigid, firm, masses, tense)	soft
Bowel function (continent, incontinent, flatulence, no stool)	continent
Bowel sounds (P=present, AB=absent, hypo, hyper)	
RUQ	P
LUQ	P
RLQ	P
LLQ	P
Stool color	light brown
Stool consistency	formed
Tubes/ostomies	NA
Genitourinary	
Urinary continence	catheter
Urine source	catheter
Appearance (clear, cloudy, yellow, amber, fluorescent, hematuria, orange, blue, tea)	amber, cloudy
Integumentary	
Skin color	pale
Skin temperature (DI=diaphoretic, W=warm, dry, CL=cool, CLM=clammy, CD +=cold, M=moist, H=hot)	W
Skin turgor (good, fair, poor, TENT=tenting)	good
Skin condition (intact, EC=ecchymosis, A=abrasions, P=petechiae, R=rash, W=weeping, S=sloughing, D=dryness, EX=excoriated, T=tears, SE=subcutaneous emphysema, B=blisters, V=vesicles, N=necrosis)	intact, P
Mucous membranes (intact, EC=ecchymosis, A=abrasions, P=petechiae, R=rash, W=weeping, S=sloughing, D = dryness, EX=excoriated, T=tears, SE=subcutaneous emphysema, B=blisters, V=vesicles, N=necrosis)	intact

(Continued)

Wilcox, Teresa (Terri), Female, 26 y.o.

Allergies: NKA	**Code:** Full	**Isolation:** None
Pt. Location: RM 1012	**Physician:** P. Horowitz	**Admit Date:** 12/19

Nursing Assessment *(Continued)*

Nursing Assessment	12/19
Other components of Braden score: special bed, sensory pressure, moisture, activity, friction/shear (>18 = no risk, 15–16 = low risk, 13–14 = moderate risk, ≤12 = high risk)	23

Orders:

CT scan of liver and biopsy
Test stool for occult blood
Daily I/O
Spironolactone 25 mg qid
Propranolol 40 mg bid
Soft, high-kcal, high-protein diet—small, frequent meals
Multivitamin/mineral supplement
Bed rest

Dx: Cirrhosis—MELD score 23

Nutrition:

Current diet order: Soft, 4 gram sodium, high-kcal
General: Has not had an appetite for the past few weeks. She states that she drinks calcium-fortified orange juice for breakfast most mornings. Lunch is usually soup and crackers with a Diet Coke. Dinner at home, but may be carry-out. If carry-out, it's usually Chinese or Italian food.
Usual dietary intake: Sips of water, juice, and Diet Coke only. Has not eaten for the past 2 days.
Food allergies/intolerances/aversions: Does not like liver or lima beans
Previous nutrition therapy: 3 yrs ago: small, frequent meals, plenty of liquids
Food purchase/preparation: Self and/or significant other
Vit/min intake: 400 mg vitamin E, 600 mg calcium with 400 IU vitamin D; multivitamin/mineral daily; 200 mg milk thistle twice daily; chicory 3 grams daily; 500 mg ginger at least twice daily

Laboratory Results

	Ref. Range	12/19 1012
Chemistry		
Sodium (mEq/L)	136–145	136
Potassium (mEq/L)	3.5–5.5	5.0
Chloride (mEq/L)	95–105	102
Carbon dioxide (CO_2, mEq/L)	23–30	28
BUN (mg/dL)	8–18	21 !↑

Wilcox, Teresa (Terri), Female, 26 y.o.
Allergies: NKA
Pt. Location: RM 1012

Code: Full
Physician: P. Horowitz

Isolation: None
Admit Date: 12/19

Laboratory Results *(Continued)*

	Ref. Range	12/19 1012
Creatinine serum (mg/dL)	0.6–1.2	1.4 !↑
Glucose (mg/dL)	70–110	115 !↑
Phosphate, inorganic (mg/dL)	2.3–4.7	3.6
Magnesium (mg/dL)	1.8–3	2.1
Calcium (mg/dL)	9–11	9.3
Bilirubin, direct (mg/dL)	<0.3	3.7 !↑
Protein, total (g/dL)	6–8	5.4 !↓
Albumin (g/dL)	3.5–5	2.1 !↓
Prealbumin (mg/dL)	16–35	15 !↓
Ammonia (NH_3, µmol/L)	9–33	33
Alkaline phosphatase (U/L)	30–120	275 !↑
ALT (U/L)	4–36	62 !↑
AST (U/L)	0–35	230 !↑
CPK (U/L)	30–135 F 55–170 M	138 !↑
Lactate dehydrogenase (U/L)	208–378	658 !↑
Cholesterol (mg/dL)	120–199	199
HDL-C (mg/dL)	>55 F, >45 M	50 !↓
LDL (mg/dL)	<130	125
Triglycerides (mg/dL)	35–135 F 40–160 M	256 !↑
HbA$_{1C}$ (%)	3.9–5.2	4.9
Coagulation (Coag)		
PT (sec)	12.4–14.4	18.5 !↑
INR	0.9–1.1	2.2 !↑
PTT (sec)	24–34	41 !↑
Hematology		
WBC (×10^3/mm^3)	4.8–11.8	4.8
RBC (×10^6/mm^3)	4.2–5.4 F 4.5–6.2 M	4.1 !↓
Hemoglobin (Hgb, g/dL)	12–15 F 14–17 M	10.9 !↓
Hematocrit (Hct, %)	37–47 F 40–54 M	35.9 !↓

(Continued)

Wilcox, Teresa (Terri), Female, 26 y.o.
Allergies: NKA
Pt. Location: RM 1012

Code: Full
Physician: P. Horowitz

Isolation: None
Admit Date: 12/19

Laboratory Results *(Continued)*

	Ref. Range	12/19 1012
Mean cell volume (μm^3)	80–96	94
Mean cell Hgb (pg)	26–32	29
Mean cell Hgb content (g/dL)	31.5–36	35.4
Platelet count ($\times 10^3/mm^3$)	140–440	342
Ferritin (mg/mL)	20–120 F 20–300 M	120
Vitamin B_{12} (ng/dL)	24.4–100	100
Folate (ng/dL)	5–25	25
Hematology, Manual Diff		
Lymphocyte (%)	15–45	20.6
Monocyte (%)	3–10	4.2
Eosinophil (%)	0–6	2.8
Segs (%)	0–60	51
Urinalysis		
Collection method	—	random specimen
Color	—	dark
Appearance	—	slightly hazy
Specific gravity	1.003–1.030	1.025
pH	5–7	5.9
Protein (mg/dL)	Neg	1+ !↑
Glucose (mg.dL)	Neg	Neg
Ketones	Neg	Neg
Occult blood	Neg	Neg
Bilirubin	Neg	1+ !↑
Nitrites	Neg	Neg
Urobilinogen (EU/dL)	<1.1	1.8 !↑
Leukocyte esterase	Neg	Neg
Prot chk	Neg	Neg
WBCs (/HPF)	0–5	3.8
RBCs (/HPF)	0–5	2.7

Case Questions

I.   Understanding the Disease and Pathophysiology

1. The liver is an extremely complex organ that has a particularly important role in nutrient metabolism. Identify three functions of the liver related to each of the following:

   a. carbohydrate metabolism

   b. protein metabolism

   c. lipid metabolism

   d. vitamin and mineral metabolism

2. The CT scan and liver biopsy confirm the diagnosis of cirrhosis. Explain this diagnosis. The diagnosis also includes a MELD score. What is this, and how does her score relate to the severity of liver failure?

3. The most common cause of cirrhosis is alcohol ingestion. What are other potential causes of cirrhosis? What is the cause of this patient's cirrhosis?

4. Explain the systemic physiological changes that occur as a result of cirrhosis.

5. List the most common signs and symptoms of cirrhosis, and relate each of these to the physiological changes discussed in question #4.

6. After reading this patient's history and physical, identify her signs and symptoms consistent with her diagnosis.

7. Hypoglycemia is a symptom cirrhotic patients may experience. What is the physiological basis for this? How might this affect Ms. Wilcox's nutritional status?

8. What are the current medical treatments for cirrhosis?

9. What is hepatic encephalopathy? Identify the stages of encephalopathy, and outline the major theories regarding the etiology of this condition.

10. Protein-energy malnutrition is commonly associated with cirrhosis. What are potential causes of malnutrition in cirrhosis? Explain each cause.

II. Understanding the Nutrition Therapy

11. Outline the nutrition therapy for stable cirrhosis and the rationale for each modification.

III. Nutrition Assessment

12. Measurements used to assess nutritional status may be affected by the disease process and not necessarily be reflective of nutritional status. Are there any components of nutrition assessment that would be affected by cirrhosis? Explain.

13. Dr. Horowitz notes that Ms. Wilcox has lost 10 lbs since her last exam. Assess and interpret Ms. Wilcox's weight.

14. Calculate the patient's energy and protein needs. Provide the rationale for the standards you used for these calculations.

15. Evaluate the patient's usual nutritional intake using nutrient analysis.

16. Her appetite and intake have been significantly reduced for the past several days. Describe factors that may have contributed to this change in her ability to eat.

17. Why was a soft, 4-g Na, high-kcalorie diet ordered? Should there be any other modifications?

18. This patient takes multiple dietary supplements. Identify the possible rationale for each, and identify any that may pose a risk for someone with cirrhosis.

19. Examine the patient's chemistry values. Which labs support the diagnosis of cirrhosis? Explain their connection to the diagnosis.

20. Examine the patient's hematology values. Which are abnormal, and why? Does she have any physical symptoms consistent with your findings?

21. What signs and/or symptoms would you monitor to determine further liver decompensation?

22. Dr. Horowitz prescribes two medications to assist with the patient's symptoms. What is the rationale for these medications, and what are pertinent nutritional implications of each?

IV.   Nutrition Diagnosis

23. Select two nutrition problems and complete the PES statements for each.

V.   Nutrition Intervention

24. Ms. Wilcox is discharged on a soft, 4-g Na diet with a 2-L fluid restriction. Do you agree with this decision? Are there additional nutrition concerns that you would want to discuss with her?

25. Ms. Wilcox asks if she can use a salt substitute at home. What would you tell her?

26. Using the information from her usual dietary intake, what suggestions might you make to assist with compliance with the fluid and sodium restrictions?

VI.   Evaluation and Monitoring

27. When you see Ms. Wilcox one month later, her weight is now 140 lbs. She is wearing sandals because she says her shoes do not fit. What condition is she most probably experiencing? How could you confirm this?

28. Her diet history is as follows:

Breakfast: 1 slice toast with 2 tbsp peanut butter, 1 c skim milk
Lunch: 2 oz potato chips, grilled cheese sandwich (1 oz American cheese with 2 slices of whole-wheat bread; grilled with 1 tbsp margarine), 1 c skim milk
Supper: 8 barbeque chicken wings, 1 c French fries, 2 c lemonade

What changes might you make to her nutrition therapy? Identify foods that should be eliminated and make suggestions for substitutions.

Bibliography

American Nutrition and Dietetics. *Nutrition Care Manual.* http://nutritioncaremanual.org/topic.cfm?ncm_toc _id = 18609. Accessed 10/17/12.

Cabral CM, Burns DL. Low-protein diets for hepatic encephalopathy debunked. Let them eat steak. *Nutrition in Clinical Practice.* 2011;26(2):155–159.

Hansen L, Sasaki A, Zucker B. End-stage liver disease: Challenges and practice implications. *Nurs Clin N Am.* 2010;45:411–426.

Koretz RL, Avenell A, Lipman TO. Nutritional support for liver disease. *Cochrane Database Syst Rev.* 2012;5: Cochrane AN: CD008344; PMID: 22592729.

Masuda T, Shirabe K, Shohei Y, et al. Nutrition support and infections associated with hepatic resection and liver transplantation in patients with chronic liver disease. *J Parenter Enteral Nutr.* 0148607112456041 Published online August 16, 2012.

MedlinePlus: Cirrhosis. http://www.nlm.nih.gov /medlineplus/cirrhosis.html. Accessed 10/17/12.

O'Brien A, Williams R. Nutrition in end-stage liver disease: Principles and practice. *Gastroenterology.* 2008;134:1729–1740.

Singal AK, Charlton MR. Nutrition in alcoholic liver disease. *Clin Liver Dis.* 2012;16:805–26.

Sucher K, Mattfeldt-Beman M. Diseases of the hepatobiliary: Liver, gallbladder, exocrine pancreas. In: Nelms M, Sucher K, Lacey K, Roth SL. *Nutrition Therapy and Pathophysiology.* 2nd ed. Belmont, CA: Wadsworth, Cengage Learning; 2011:437–470.

Zhao VM, Zeigler TR. Nutrition support in end-stage liver disease. *Crit Care Nurs Clin N Am.* 2010;22:369–380.

Internet Resources

American Liver Foundation: http://www.liverfoundation.org/

Centers for Disease Control and Prevention: http://www .cdc.gov/hepatitis/index.htm

Hepatitis Central: http://www.hepatitis-central.com /cirrhosis/causes-of-cirrhosis.html

Hepatitis Foundation International: http://www.hepfi.org

National Digestive Diseases Information Clearinghouse (NDDIC): http://digestive.niddk.nih.gov/ddiseases /pubs/cirrhosis/index.aspx

Office of Dietary Supplements: http://ods.od.nih.gov/

Case 13

Acute Pancreatitis

Objectives

After completing this case, the student will be able to:

1. Understand the physiological role of the pancreas.
2. Determine the etiology of acute versus chronic pancreatitis.
3. Evaluate the signs and symptoms consistent with pancreatitis.
4. Evaluate the current literature regarding nutrition support for acute pancreatitis.
5. Develop a nutrition care plan—with appropriate measurable goals, interventions, and strategies for monitoring and evaluation—that addresses the nutrition diagnoses for this case.

Mr. Mahon is a 29-year-old graduate student who has been suffering from nausea, vomiting, and acute abdominal pain for several days. He is now admitted after a worsening of symptoms with a diagnosis of acute pancreatitis, which is most likely related to chronic alcohol intake.

Mahon, Jon, Male, 29 y.o.
Allergies: NKA
Pt. Location: MICU Bed 5

Code: FULL
Physician: C. Wexler

Isolation: None
Admit Date: 4/21

Patient Summary: Patient is 29-year-old white male who presented to the ER 48 hours ago with increasing abdominal pain. Patient returned to the ER when abdominal pain continued to get worse and is now admitted to MICU.

History:

Onset of disease: Patient is a graduate student who began experiencing acute abdominal pain, N, V several days ago. Assumed it was a viral illness but when symptoms continued he was brought to the ER by neighbor and friend. When questioned about alcohol intake, patient states that he didn't realize how much he had been drinking. States he was trying to stop his antidepressant medications and guesses he steadily increased his alcohol intake.
Medical history: Depression
Surgical history: s/p appendectomy age 12
Medications at home: None
Tobacco use: None
Alcohol use: 6 pack beer, 4–5 shots bourbon daily; weekends: wine and other mixed drinks
Family history: Mother—breast cancer; father—hypertension

Demographics:

Marital status: Single
Years education: 16+
Language: English only
Occupation: PhD student in English
Hours of work: In school full-time; works as research assistant in department
Household members: Lives with roommate
Ethnicity: Caucasian
Religious affiliation: Jewish

Admitting History/Physical:

Chief complaint: "My stomach pain is so bad—I just can't stand it. I can't seem to quit vomiting and cannot keep anything down."
General appearance: Pale, obese male in obvious distress

Vital Signs: Temp: 101.7   Pulse: 108   Resp rate: 27
                    BP: 132/96   Height: 5'11"   Weight: 245 lbs

Heart: RRR, unremarkable
HEENT: Head: WNL
          Eyes: PERRLA
          Ears: Clear

Mahon, Jon, Male, 29 y.o.
Allergies: NKA
Pt. Location: MICU Bed 5

Code: FULL
Physician: C. Wexler

Isolation: None
Admit Date: 4/21

    Nose: Dry mucous membranes
    Throat: Dry mucous membranes
Genitalia: Deferred
Neurologic: Alert and oriented
Extremities: WNL
Skin: Warm and dry
Chest/lungs: Respirations are rapid but clear to auscultation and percussion
Peripheral vascular: Diminished pulses bilaterally
Abdomen: Hyperactive bowel sounds ×4; extreme tenderness, rebound and guarding

Nursing Assessment	4/22
Abdominal appearance (concave, flat, rounded, obese, distended)	obese
Palpation of abdomen (soft, rigid, firm, masses, tense)	tense
Bowel function (continent, incontinent, flatulence, no stool)	continent
Bowel sounds (P=present, AB=absent, hypo, hyper)	
RUQ	P, hypo
LUQ	P, hypo
RLQ	P, hypo
LLQ	P, hypo
Stool color	none
Stool consistency	
Tubes/ostomies	NA
Genitourinary	
Urinary continence	catheter
Urine source	catheter
Appearance (clear, cloudy, yellow, amber, fluorescent, hematuria, orange, blue, tea)	cloudy, amber
Integumentary	
Skin color	pale
Skin temperature (DI=diaphoretic, W=warm, dry, CL=cool, CLM=clammy, CD+=cold, M=moist, H=hot)	CLM, DI
Skin turgor (good, fair, poor, TENT=tenting)	TENT
Skin condition (intact, EC=ecchymosis, A=abrasions, P=petechiae, R=rash, W=weeping, S=sloughing, D=dryness, EX=excoriated, T=tears, SE=subcutaneous emphysema, B=blisters, V=vesicles, N=necrosis)	D, T

(Continued)

Mahon, Jon, Male, 29 y.o.
Allergies: NKA   **Code:** FULL   **Isolation:** None
Pt. Location: MICU Bed 5   **Physician:** C. Wexler   **Admit Date:** 4/21

Nursing Assessment *(Continued)*

Nursing Assessment	4/22
Mucous membranes (intact, EC=ecchymosis, A=abrasions, P=petechiae, R=rash, W=weeping, S=sloughing, D=dryness, EX=excoriated, T=tears, SE=subcutaneous emphysema, B=blisters, V=vesicles, N=necrosis)	intact, D
Other components of Braden score: special bed, sensory pressure, moisture, activity, friction/shear (>18=no risk, 15–16=low risk, 13–14=moderate risk, ≤12=high risk)	19

Admission Orders:
Laboratory: CBC, CMP, Amylase, Lipase, and Mg
Repeat CBC, Amylase, Lipase in 12 hrs
Repeat Chem 7 every 6 hrs
Radiology:
Abdominal series: pancreatitis, rule out small bowel obstruction
CT abdomen/pelvis: pancreatitis, rule out pseudo cyst
Abdominal U/S: pancreatitis, rule out pseudo cyst, biliary tract disease
Vital Signs: Every 4 hrs
I & O recorded every 8 hrs
NG tube to low intermittent suction (if persistent vomiting, obstruction, severe ileus)
Notify MD for fever >101, WBC >16,000, Calcium <8 mg/dL, unstable vital signs or worsening condition
Diet: NPO
Activity: Bed rest
IVF: D_5W 40 MEq KCl 125 mL/hr

Scheduled Medications:
Imipenen 1000 mg every 6 hrs
Pepcid 20 mg IVP every 12 hrs
Meperidine 50–150 mg IV every 3 hrs prn
Ondansetron 2–4 mg IV every 4–6 hrs prn
Colace (docusate) 100 mg po two times daily prn; if no bowel movement
Milk of Magnesia (MOM) 30 mL po daily prn
Ativan 0.5–1 mg po every 8 hrs prn

Nutrition:
Meal type: NPO
Fluid requirement: 1900–2400 mL
History: Patient states that he has gained weight over the last 5 years—almost 50 lbs. Eats out usually for dinner—drinks coffee at breakfast with a bagel or toast—lunch is usually a sub sandwich

Mahon, Jon, Male, 29 y.o.
Allergies: NKA
Pt. Location: MICU Bed 5

Code: FULL
Physician: C. Wexler

Isolation: None
Admit Date: 4/21

___

or pizza. Patient states that he has eaten very little over the past 3 days because of pain, nausea, and vomiting. When questioned about alcohol intake, patient states that he didn't realize how much he had been drinking but states he was trying to stop his antidepressant medications and guesses he increased his alcohol intake.

Intake/Output

Date		4/21 0701– 4/22 0700			
Time		0701–1500	1501–2300	2301–0700	Daily total
IN	P.O.	**0**	**0**	**0**	**0**
	I.V.	**1,000**	**1,000**	**1,000**	**3,000**
	(mL/kg/hr)	(1.12)	(1.12)	(1.12)	(1.12)
	I.V. piggyback	**500**	**500**	**500**	**1,500**
	TPN				
	Total intake	**1,500**	**1,500**	**1,500**	**4,500**
	(mL/kg)	(13.5)	(13.5)	(13.5)	(40.4)
OUT	Urine	**1,312**	**1,410**	**1,622**	**4,344**
	(mL/kg/hr)	(1.47)	(1.58)	(1.82)	(1.63)
	Emesis output				
	Other	**NG tube to suction−100**	120	115	335
	Stool	**200**	**0**	**0**	**200**
	Total output	**1,612**	**1,530**	**1,737**	**4,879**
	(mL/kg)	(14.5)	(13.7)	(15.6)	(13.8)
Net I/O		−112	−30	−237	−379
Net since admission (4/21)		−112	−142	−379	−379

MD Progress Note:

4/22 0840
Subjective: Jon Mahon's previous 24 hours reviewed
Vitals: Temp: 101.5 Pulse: 82 Resp rate: 25 BP: 122/78
Urine Output: 4344 mL (39 mL/kg)

<u>Physical Exam</u>
General: Well-developed, alert and oriented to person, place, and time. Continues with acute abdominal pain. APACHE Score 4
HEENT: WNL
Neck: WNL
Heart: WNL

Mahon, Jon, Male, 29 y.o.
Allergies: NKA   **Code:** FULL   **Isolation:** None
Pt. Location: MICU Bed 5   **Physician:** C. Wexler   **Admit Date:** 4/21

Lungs: Clear to auscultation
Abdomen: Hypoactive bowel sounds, abdominal tenderness, rebound, guarding, N, V
Assessment/Plan: Results from abdominal ultrasound: inflammation, peripancreatic stranding, and fluid collection. Elevated lipase, amylase, and CRP.

Dx: Acute Pancreatitis

Plan: Continue D5W 40 MEq KCl 125 mL/hr with current medications. Monitor urine output to ensure >40 mL/kg. Place enteral postpyloric feeding tube. Nutrition consult for enteral feeds.

.... C. Wexler, MD

Laboratory Results

	Ref. Range	4/22 1522
Chemistry		
Sodium (mEq/L)	136–145	144
Potassium (mEq/L)	3.5–5.5	3.5
Chloride (mEq/L)	95–105	96
Carbon dioxide (CO_2, mEq/L)	23–30	30
BUN (mg/dL)	8–18	30 !↑
Creatinine serum (mg/dL)	0.6–1.2	1.6 !↑
BUN/Crea ratio	10:1–20:1	18:1
Uric acid (mg/dL)	2.8–8.8 F 4.0–9.0 M	5.2
Glucose (mg/dL)	70–110	78
Phosphate, inorganic (mg/dL)	2.3–4.7	3.4
Magnesium (mg/dL)	1.8–3	1.9
Calcium (mg/dL)	9–11	10.2
Osmolality (mmol/kg/H_2O)	285–295	303 !↑
Bilirubin total (mg/dL)	≤1.5	1.9 !↑
Bilirubin, direct (mg/dL)	<0.3	0.9 !↑
Protein, total (g/dL)	6–8	6.2
Albumin (g/dL)	3.5–5	3.5
Prealbumin (mg/dL)	16–35	18
Ammonia (NH_3, µmol/L)	9–33	15
Alkaline phosphatase (U/L)	30–120	256 !↑
ALT (U/L)	4–36	38 !↑

Mahon, Jon, Male, 29 y.o.
Allergies: NKA
Pt. Location: MICU Bed 5

Code: FULL
Physician: C. Wexler

Isolation: None
Admit Date: 4/21

Laboratory Results *(Continued)*

	Ref. Range	4/22 1522
AST (U/L)	0–35	56 !↑
CPK (U/L)	30–135 F 55–170 M	219 !↑
Lactate dehydrogenase (U/L)	208–378	402 !↑
Lipase (U/L)	0–110	980 !↑
Amylase (U/L)	25–125	543 !↑
CRP (mg/dL)	<1	18 !↑
Cholesterol (mg/dL)	120–199	210 !↑
HDL-C (mg/dL)	>55 F, >45 M	54
VLDL (mg/dL)	7–32	29
LDL (mg/dL)	<130	127
LDL/HDL ratio	<3.22 F <3.55 M	2.35
Triglycerides (mg/dL)	35–135 F 40–160 M	285 !↑
Coagulation (Coag)		
PT (sec)	12.4–14.4	13.2
PTT (sec)	24–34	27
Hematology		
WBC (×10³/mm³)	4.8–11.8	19.8 !↑
RBC (×10⁶/mm³)	4.2–5.4 F 4.5–6.2 M	5.2
Hemoglobin (Hgb, g/dL)	12–15 F 14–17 M	15.8
Hematocrit (Hct, %)	37–47 F 40–54 M	49
Hematology, Manual Diff		
Neutrophil (%)	50–70	90 !↑
Lymphocyte (%)	15–45	32
Monocyte (%)	3–10	5
Eosinophil (%)	0–6	1
Basophil (%)	0–2	2
Blasts (%)	3–10	3

(Continued)

Mahon, Jon, Male, 29 y.o.
Allergies: NKA   **Code:** FULL   **Isolation:** None
Pt. Location: MICU Bed 5   **Physician:** C. Wexler   **Admit Date:** 4/21

Laboratory Results *(Continued)*

	Ref. Range	4/22 1522
Segs (%)	0–60	74 !↑
Bands (%)	0–10	16 !↑
Urinalysis		
Collection method	—	cath
Color	—	dark, amber
Appearance	—	cloudy
Specific gravity	1.003–1.030	1.029
pH	5–7	5.7
Protein (mg/dL)	Neg	+ !↑
Glucose (mg/dL)	Neg	Neg
Ketones	Neg	+ !↑
Blood	Neg	Neg
Bilirubin	Neg	+ !↑
Nitrites	Neg	Neg
Urobilinogen (EU/dL)	<1.1	1.2 !↑
Leukocyte esterase	Neg	Neg
Prot chk	Neg	+ !↑
WBCs (/HPF)	0–5	2
RBCs (/HPF)	0–5	0
Bact	0	0
Mucus	0	0
Crys	0	0
Casts (/LPF)	0	0
Yeast	0	0
Arterial Blood Gases (ABGs)		
pH	7.35–7.45	7.40
pCO_2 (mm Hg)	35–45	40
SO_2 (%)	≥95	97
CO_2 content (mmol/L)	25–30	29
O_2 content (%	15–22	18
pO_2 (mm Hg)	≥80	95
HCO_3^- (mEq/L)	24–28	24

Case Questions

I.   **Understanding the Diagnosis and Pathophysiology**

1.   Describe the normal exocrine and endocrine functions of the pancreas.

2.   Determine the potential etiology of both acute and chronic pancreatitis. What information provided in the physical assessment supports the diagnosis of acute pancreatitis?

3.   What laboratory values or other tests support this diagnosis? List all abnormal values and explain the likely cause for each abnormal value.

4.   The physician lists an APACHE score in his note. What factors are used to determine this score? What does this mean?

5.   What are the potential complications of acute pancreatitis?

II.   **Understanding the Nutrition Therapy**

6.   Historically, the patient with acute pancreatitis was made NPO. Why? This patient has an NG tube placed—why?

7.   The physician has written an order for a nutrition consult to start enteral feedings. Using the most current literature and ASPEN guidelines, explain the role of enteral feeding in acute pancreatitis. Do you agree with the initiation of enteral feeding? Why or why not?

8.   Does this patient's case indicate the use of an immune-modulating formula?

9.   What research supports the use of probiotics in acute pancreatitis? Is there any evidence supporting the use of supplemental glutamine?

III. Nutrition Assessment

10. Assess Mr. Mahon's height and weight. Calculate his BMI and % usual body weight.

11. Evaluate Mr. Mahon's initial nursing assessment. What important factors noted in his nutrition assessment will affect your nutrition recommendations?

12. Determine Mr. Mahon's energy and protein requirements. Explain the rationale for the method you used to calculate these requirements.

13. Determine Mr. Mahon's fluid requirements. Compare this with the information on the intake/output record.

14. From the nutrition history, assess Mr. Mahon's alcohol intake. What is his average caloric intake from alcohol each day using the information that he provided to you?

15. List all medications that Mr. Mahon is receiving. Determine the action of each medication and identify any drug–nutrient interactions that you should monitor.

IV. Nutrition Diagnosis

16. Identify the pertinent nutrition problems and the corresponding nutrition diagnoses.

17. Write your PES statement for each nutrition problem.

V. Nutrition Intervention

18. Determine your enteral feeding recommendations for Mr. Mahon. Provide a formula choice, goal rate, and instructions for initiation and advancement.

19. What recommendations can you make to the patient's critical care team to help improve tolerance to the enteral feeding?

VI.   Nutrition Monitoring and Evaluation

20. List factors that you would monitor to assess tolerance and adequacy of nutrition support.

21. If this patient's acute pancreatitis resolves, what will be the recommendations for him regarding nutrition and his alcohol intake when he is discharged?

22. Write an ADIME note that provides your initial nutrition assessment and enteral feeding recommendations.

Bibliography

Bankhead R, Boullata J, Brantley S, et al. Enteral nutrition practice recommendations. JPEN J *Parenter Enteral Nutr.* 2009;33(2):122–167.

Bengmark S. Bio-ecological control of acute pancreatitis: The role of enteral nutrition, pro- and synbiotics. *Curr Opin Clin Nutr Metab Care.* 2005;8:557–561.

Binnekade JM. Review: Enteral nutrition reduces infections, need for surgical intervention, and length of hospital stay more than parenteral nutrition in acute pancreatitis. *Evid Based Nurs.* 2005;8:19.

Clancy TE, Benoit EP, Ashley SW. Current management of acute pancreatitis. *J Gastrointest Surg.* 2005;9:440–452.

Doley RP, Yadav TD, Wig JD, et al. Enteral nutrition in severe acute pancreatitis. *JOP.* 2009;10:157–162.

Eatock FC, Chong P, Menezes N, et al. A randomized study of early nasogastric versus nasojejunal feeding in severe acute pancreatitis. *Am J Gastroenterol.* 2005;100:440–441.

Flesher ME, Archer KA, Leslie BD, McCollom RA, Martinka GP. Assessing the metabolic and clinical consequences of early enteral feeding in the malnourished patient. *J Parenter Enteral Nutr.* 2005;29:108–117.

Ioannidis O, Lavrentieva A, Botsios D. Review: Nutrition support in acute pancreatitis. JOP. 2008;9:375–390.

Krenitsky J, Makola D, Parrish CR. Pancreatitis part II—Revenge of the cyst: A practical guide to jejunal feeding. *Pract Gastroenterol.* 2007;31:54.

Krenitsky J, Makola D, Parrish CR. Parenteral nutrition in pancreatitis is passe: But are we ready for gastric feeding? *Pract Gastroenterol.* 2007;31:92.

Malone AM, Seres DS, Lord L. Chapter 13: Complications of enteral nutrition. In: *The ASPEN Nutrition Support Core Curriculum: A Case-Based Approach—The Adult Patient.* Silver Spring, MD: American Society for Enteral and Parenteral Nutrition; 2007.

Marik PE, Zaloga GP. Meta-analysis of parenteral nutrition versus enteral nutrition in patients with acute pancreatitis. *BMJ.* 2004;328(7453):1407.

McClave SA, Heyland DK. The physiologic response and associated clinical benefits from provision of early enteral nutrition. *Nutr Clin Prac.* 2009;24:305–315.

McClave SA, Martindale RG, Vanek VW, et al. Guidelines for the provision and assessment of nutrition support therapy in the adult critically ill patient: Society of Critical Care Medicine (SCCM) and American Society for Parenteral and Enteral Nutrition (ASPEN). *J Parenter Enteral Nutr.* 2009;33(3):277–316.

Meier R. Enteral fish oil in acute pancreatitis. *Clin Nutr.* 2005;24(2):169–171.

Mirtallo J. et al. for the International Consensus Guideline Committee for Pancreatitis Task Force. International consensus guidelines for nutrition therapy in pancreatitis. *J Parenter Enteral Nutr.* 2012;36:284–291.

Moraes JM, Felga GE, Chebli LA, et al. A full solid diet as the initial meal in mild acute pancreatitis is safe and results in a shorter length of hospitalization. J *Clin Gastroenterol.* Jan 5, 2010;44(7):517–522.

Nahikian-Nelms M. Metabolic stress and the critically ill. In: Nelms M, Sucher K, Lacey K, Roth SL. *Nutrition Therapy and Pathophysiology.* 2nd ed. Belmont, CA: Wadsworth, Cengage Learning; 2011:682–701.

Nahikian-Nelms M, Habash D. Nutrition assessment: Foundation of the nutrition care process. In: Nelms M, Sucher K, Lacey K, Roth SL. *Nutrition Therapy and Pathophysiology.* 2nd ed. Belmont, CA: Wadsworth, Cengage Learning; 2011:34–65.

Petrov MS, Loveday BPT, Pylypchuk RD, McIlroy K, Phillips ARJ, Windsor JA. Systematic review and meta-analysis of enteral nutrition formulations in acute pancreatitis. *Br J Surg.* 2009;96:1243–1252.

Petrov MS, Pylypchuk RD, Emelyanov NV. Systematic review: Nutritional support in acute pancreatitis. *Aliment Pharmacol Ther.* 2008;28:704–12 .

Sathiaraj E, Murthy S, Mansard MJ, Rao GV, Mahukar S, Reddy DN. Clinical trial: Oral feeding with a soft diet compared with clear liquid diet as initial meal in mild acute pancreatitis. *Aliment Pharmacol Ther.* 2008;28:777–781.

Stanga Z, Giger U, Marx A, DeLegge MH. Effect of jejunal long-term feeding in chronic pancreatitis. *J Parenter Enteral Nutr.* 2005;29(1):12–20.

Sucher K, Mattfeldt-Beman M. Diseases of the liver, gallbladder and exocrine pancreas. In: Nahikian-Nelms M, Sucher K, Lacey K, Roth S. *Nutrition Therapy and Pathophysiology.* 2nd ed. Belmont, CA: Wadsworth, Cengage Learning; 2011:437–470.

Sun S, Yang K, He-Jinjui Tian X, Ma B, Jiang L. Probiotics in patients with severe acute pancreatitis: A meta-analysis. *Langenbecks Arch Surg.* 2009;394:171–177.

Thomson A. Enteral versus parenteral nutritional support in acute pancreatitis: A clinical review. *J Gastroenterol Hepatol.* 2006;21:22–25.

Internet Resources

ADA Evidence Analysis Library: http://www.adaevidencelibrary.com

American Gastroenterology Association: http://www.gastro.org/patient-center /digestive-conditions/pancreatitis

Nutrition Care Manual: http://www .nutritioncaremanual.orgBibliography

NUTRITION THERAPY FOR ENDOCRINE DISORDERS

The most common of all endocrine disorders is diabetes mellitus. Diabetes mellitus is actually a group of diseases characterized by hyperglycemia resulting from cessation of insulin production or impairment in insulin secretion and/or insulin action. Type 1 and type 2 diabetes are the major classifications for this disorder. The four cases in this section provide examples of patients at all stages of treatment and diagnosis. The first two focus on newly diagnosed type 1 diabetes—not only the typical presentation in a pediatric patient but also the atypical presentation in an adult. Cases 16 and 17 target the diagnosis of type 2, but again in both the typical adult case and the atypical presentation in the pediatric patient.

Diabetes mellitus is a chronic disease that has no cure and is the seventh leading cause of death in the United States. New diagnoses of diabetes have tripled in the last 20 years. Over 18.8 million people in the United States have diabetes, and it is estimated that at least 7 million more are undiagnosed (available from: http://www.cdc.gov/diabetes/statistics; accessed December 12, 2012).

Diabetes affects men and women equally, but minorities (especially American Indians and Alaska Natives) are almost twice as likely as non-Hispanic whites to develop diabetes in their lifetime. In addition, diabetes is one of the most costly health problems in the United States. In 2007, health care and other direct medical costs, as well as indirect costs (such as loss of productivity), were approximately $174 billion. Each year, more than 200,000 people die as a result of diabetes and its complications. For example, diabetes is the leading cause of new blindness in the United States and the leading cause of kidney failure requiring dialysis or organ transplant for survival (available from: http://www.cdc.gov/diabetes/pubs/pdf/ndfs_2011.pdf).

Medical nutrition therapy is integral to total diabetes care and management. The Diabetes Control and Complications Trial (DCCT) corroborated the significance of integrating nutrition and blood glucose self-management education in achieving and maintaining target blood glucose levels. Nutrition and meal planning are among the most challenging aspects of diabetes care for the person with diabetes and for the health care team. The major components of successful nutrition management are learning about nutrition therapy, altering eating habits, implementing new behaviors, participating in exercise, evaluating changes, and integrating this information into diabetes care. Observance of meal-planning principles requires people with diabetes to make demanding lifestyle changes. To be effective, the registered dietitian must be able to customize his or her approach to the personal lifestyle and diabetes management goals of the individual with diabetes. Cases 14–17 allow you to put this guideline into practice.

Case 14

Pediatric Type 1 Diabetes Mellitus

Objectives

After completing this case, the student will be able to:

1. Describe the pathophysiology of type 1 diabetes mellitus.
2. Develop a nutrition care plan—with appropriate measurable goals, interventions, and strategies for monitoring and evaluation—that addresses the nutrition diagnoses for this case.
3. Integrate an insulin regimen with nutrition therapy and provide appropriate recommendations for carbohydrate-to-insulin ratios and correction dosages.

4. Interpret laboratory parameters to assess fluid, electrolyte, and acid-base balance.
5. Prioritize survival skills and educational requirements for a newly diagnosed type 1 diabetic.

Rachel Roberts is a 12-year-old 7th grader previously in good health who is admitted through the ER with a new diagnosis of acute hyperglycemia—R/O diabetes mellitus.

Roberts, Rachel, Female, 12 y.o.
Allergies: NKA   **Code:** FULL   **Isolation:** None
Pt. Location: RM 744   **Physician:** M. Cho   **Admit Date:** 5/4

Patient Summary: Rachel Roberts is a 12-year-old female admitted with acute-onset hyperglycemia.

History:
Onset of disease: Patient presented to ER after fainting at soccer practice. During ER assessment, patient was noted to have serum glucose of 724 mg/dL.
Medical history: None—recently had strep throat
Surgical history: None
Medications at home: None
Tobacco use: Nonsmoker
Alcohol use: None
Family history: Father—HTN; Mother—hyperthyroidism; Sister—celiac disease

Demographics:
Marital status: Single 7th-grade female
Years education: 7 years
Language: English only
Occupation: Student
Hours of work: N/A
Household members: Mother, sister age 8 and brother age 4—parents divorced. Father lives in city and shares custody.
Ethnicity: Caucasian
Religious affiliation: Catholic

Admitting History/Physical:
Chief complaint: "I have just gotten over strep throat a few days ago. I felt like I was well enough to go to soccer practice today but after playing about 15 minutes, I just felt horrible. I sat down and they tell me I fainted. . . . I have been really thirsty—thirstier than I have ever been in my whole life and then I have had to use the bathroom a lot. . . . I even have to get up at night to go to the bathroom."
General appearance: Slim, healthy-appearing, 12-year-old female

Vital Signs: Temp: 98.6   Pulse: 101   Resp rate: 22
                              BP: 122/77   Height: 5'   Weight: 82 lbs

Heart: Regular rate and rhythm
HEENT: Head: WNL
                 Eyes: PERRLA
                 Ears: Clear
                 Nose: Clear
                 Throat: Dry mucous membranes without exudates or lesions
Genitalia: Deferred
Neurologic: Alert but slightly confused. Glasgow Coma Scale: 15

Roberts, Rachel, Female, 12 y.o.
Allergies: NKA   **Code:** FULL   **Isolation:** None
Pt. Location: RM 744   **Physician:** M. Cho   **Admit Date:** 5/4

Extremities: Noncontributory
Skin: Warm and dry
Chest/lungs: Respirations are rapid—clear to auscultation and percussion
Peripheral vascular: Pulse 4+ bilaterally, warm, no edema
Abdomen: Active bowel sounds ×4; tender, nondistended

Orders:

1. Regular insulin 1 unit/mL NS 40 mEq KCl/liter @ 135 mL/hr. Begin infusion at 0.1 unit/kg/hr = 3.7 units/hr and increase to 5 units/hr. Flush new IV tubing with 50 mL of insulin drip solution prior to connecting to patient and starting insulin infusion.

2. Labs: BMP Stat
   Phos Stat
   Calcium Stat
   UA with culture if indicated Stat Clean catch
   Bedside glucose Stat
   Bedside I-Stat: EG7 Stat
   Islet cell autoantibodies screen
   Thyroid peroxidase abs
   TSH
   Comp metabolic panel (CMP)
   Thyroglobulin antibodies
   C-peptide
   Immunoglobulin A level
   Hemoglobin A_{1c}
   Tissue transglutaminase

3. NPO except for ice chips and medications. After 12 hours, clear liquids if stable. Then, advance to consistent carbohydrate diet order—70–80 g breakfast and lunch; 85–95 g dinner; 3–15 gram snacks.

4. Consult diabetes education team for self-management training for patient and parents to begin education after stabilized.

Nursing Assessment	5/4
Abdominal appearance (concave, flat, rounded, obese, distended)	flat
Palpation of abdomen (soft, rigid, firm, masses, tense)	soft
Bowel function (continent, incontinent, flatulence, no stool)	continent
Bowel sounds (P=present, AB=absent, hypo, hyper)	
RUQ	P

(Continued)

Roberts, Rachel, Female, 12 y.o.
Allergies: NKA   **Code:** FULL   **Isolation:** None
Pt. Location: RM 744   **Physician:** M. Cho   **Admit Date:** 5/4

Nursing Assessment *(Continued)*

Nursing Assessment	5/4
LUQ	P
RLQ	P
LLQ	P
Stool color	light brown
Stool consistency	soft
Tubes/ostomies	NA
Genitourinary	
Urinary continence	yes
Urine source	clean specimen
Appearance (clear, cloudy, yellow, amber, fluorescent, hematuria, orange, blue, tea)	cloudy, amber
Integumentary	
Skin color	pale
Skin temperature (DI=diaphoretic, W=warm, dry, CL=cool, CLM=clammy, CD +=cold, M=moist, H=hot)	DI
Skin turgor (good, fair, poor, TENT=tenting)	good
Skin condition (intact, EC=ecchymosis, A=abrasions, P=petechiae, R=rash, W=weeping, S=sloughing, D=dryness, EX=excoriated, T=tears, SE=subcutaneous emphysema, B=blisters, V=vesicles, N=necrosis)	intact
Mucous membranes (intact, EC=ecchymosis, A=abrasions, P=petechiae, R=rash, W=weeping, S=sloughing, D=dryness, EX=excoriated, T=tears, SE=subcutaneous emphysema, B=blisters, V=vesicles, N=necrosis)	intact
Other components of Braden score: special bed, sensory pressure, moisture, activity, friction/shear (>18=no risk, 15–16=low risk, 13–14=moderate risk, ≤12=high risk)	21

Nutrition:

Meal type: NPO then progress to clear liquids and then consistent carbohydrate-controlled diet
Fluid requirement: 1840 mL
History: Parents present—patient states that she thinks she has lost weight recently: "My clothes are a little loose but I don't usually weigh myself." Mom states that the last weight she remembers was when they went to fast-care clinic for strep throat and that she weighed about 90 lbs. Patient confirms that that is what she usually weighs. Appetite has been normal—if anything patient states she has been more hungry than usual—but thought it was probably due to starting soccer season and exercising more. Relates history of increased thirst and increased urination.

Roberts, Rachel, Female, 12 y.o.
Allergies: NKA   **Code:** FULL   **Isolation:** None
Pt. Location: RM 744   **Physician:** M. Cho   **Admit Date:** 5/4

Usual intake (for past several months): Mom and Dad state that Rachel is kind of a picky eater. She eats only chicken and fish—eats salad, broccoli, carrots, tomatoes, and asparagus as her only vegetables. Breakfast—cereal and milk or Pop-Tart® with milk; packs lunch for school—peanut butter and jelly or turkey and cheese sandwich, chips, carrots, and usually drinks water. Has a cereal or granola bar before soccer practice—drinks water throughout practice. Dinner is usually prepared by Mom when she is at her house—always some salad, meat, and pasta, potato, or rice. Dad states that when the kids are with him he doesn't cook very often and they usually order in pizza or Chinese food. Snacks include cereal, ice cream, yogurt, some fruits (apples, bananas), popcorn, chips, or cookies.

MD Progress Note:
5/5 0820
Subjective: Rachel Roberts's previous 24 hours reviewed
Vitals: Temp: 99.5, Pulse: 82, Resp rate: 25, BP: 101/78
Urine Output: 2660 mL (71.8 mL/kg)

Physical Exam
General: Alert and oriented to person, place, and time
HEENT: WNL
Neck: WNL
Heart: WNL
Lungs: Clear to auscultation
Abdomen: Active bowel sounds
Assessment/Plan: Results: +ICA, GADA, IAA consistent with type 1 DM. Negative tTG.

Dx: New Diagnosis Type 1 Diabetes Mellitus

Plan: Change IVF to D_5.45NS with 40MEq K @ 135 mL/hr. Begin Apidra 0.5 u every 2 hours until glucose is 150–200 mg/dL. Tonight begin glargine 6 u at 9 pm. Progress Apidra using ICR 1:15. Continue bedside glucose checks hourly. Notify MD if blood glucose >200 or <80.
                                                                                    M Cho, MD

Roberts, Rachel, Female, 12 y.o.
Allergies: NKA   **Code:** FULL   **Isolation:** None
Pt. Location: RM 744   **Physician:** M. Cho   **Admit Date:** 5/4

Intake/Output

Date		5/4 0701–5/5 0700			
Time		0701–1500	1501–2300	2301–0700	Daily total
IN	P.O.	**NPO**	**NPO**	**320**	**320**
	I.V.	**1,080**	**1,080**	**1,080**	**3,240**
	(mL/kg/hr)	(3.6)	(3.6)	(3.6)	(3.6)
	I.V. piggyback				
	TPN				
	Total intake	**1,080**	**1,080**	**1,400**	**3,560**
	(mL/kg)	(30.0)	(30.0)	(37.6)	(95.5)
OUT	Urine	**600**	**480**	**1,580**	**2,660**
	(mL/kg/hr)	(2.01)	(1.61)	(5.30)	(2.97)
	Emesis output				
	Other				
	Stool				
	Total output	**600**	**480**	**1,580**	**2,660**
	(mL/kg)	(16.2)	(12.9)	(42.7)	(71.8)
Net I/O		**+480**	**+600**	**−180**	**+900**
Net since admission (5/4)		**+480**	**+1,080**	**+900**	**+900**

Laboratory Results

	Ref. Range	5/4 1780	5/5 1522
Chemistry			
Sodium (mEq/L)	136–145	126 !↓	131 !↓
Potassium (mEq/L)	3.5–5.5	4.3	4.0
Chloride (mEq/L)	95–105	101	100
Carbon dioxide (CO_2, mEq/L)	23–30	27	28
BUN (mg/dL)	8–18	15	12
Creatinine serum (mg/dL)	0.6–1.2	0.9	0.8
Glucose (mg/dL)	70–110	683 !↑	250 !↑
Phosphate, inorganic (mg/dL)	2.3–4.7	1.9 !↓	2.1 !↓
Magnesium (mg/dL)	1.8–3	1.9	2.1
Calcium (mg/dL)	9–11	10	9.8
Osmolality (mmol/kg/H_2O)	285–295	295.3 !↑	304 !↑

Roberts, Rachel, Female, 12 y.o.
Allergies: NKA   **Code:** FULL   **Isolation:** None
Pt. Location: RM 744   **Physician:** M. Cho   **Admit Date:** 5/4

Laboratory Results *(Continued)*

	Ref. Range	5/4 1780	5/5 1522
Bilirubin total (mg/dL)	≤1.5	0.8	
Bilirubin, direct (mg/dL)	<0.3	0.009	
Protein, total (g/dL)	6–8	6.1	
Albumin (g/dL)	3.5–5	3.7	
Prealbumin (mg/dL)	16–35	17	
Ammonia (NH$_3$, μmol/L)	9–33	11	
Alkaline phosphatase (U/L)	30–120	102	
ALT (U/L)	4–36	7.1	
AST (U/L)	0–35	18.2	
CPK (U/L)	30–135 F 55–170 M	31	
Lactate dehydrogenase (U/L)	208–378	208	
Cholesterol (mg/dL)	120–199	121	
Triglycerides (mg/dL)	35–135 F 40–160 M	55	
T$_4$ (μg/dL)	4–12	12	
T$_3$ (μg/dL)	75–98	77	
HbA$_{1C}$ (%)	3.9–5.2	14.6 !↑	
C-peptide (ng/mL)	0.51–2.72	0.10 !↓	
ICA	—	+ !↑	
GADA	—	+ !↑	
IA-2A	—	—	
IAA	—	+ !↑	
tTG	—	—	
Hematology			
WBC (× 10^3/mm^3)	4.8–11.8	9.6	
RBC (× 10^6/mm^3)	4.2–5.4 F 4.5–6.2 M	4.8	
Hemoglobin (Hgb, g/dL)	12–15 F 14–17 M	12.5	
Hematocrit (Hct, %)	37–47 F 40–54 M	37	
Hematology, Manual Diff			
Neutrophil (%)	50–70	55	
Lymphocyte (%)	15–45	18	

(Continued)

Roberts, Rachel, Female, 12 y.o.
Allergies: NKA   **Code:** FULL   **Isolation:** None
Pt. Location: RM 744   **Physician:** M. Cho   **Admit Date:** 5/4

Laboratory Results *(Continued)*

	Ref. Range	5/4 1780	5/5 1522
Monocyte (%)	3–10	4	
Eosinophil (%)	0–6	0	
Basophil (%)	0–2	1	
Blasts (%)	3–10	3	
Urinalysis			
Collection method	—	clean catch	
Color	—	yellow	
Appearance	—	clear	
Specific gravity	1.003–1.030	1.035 !↑	
pH	5–7	4.9 !↓	
Protein (mg/dL)	Neg	100 !↑	
Glucose (mg/dL)	Neg	+ !↑	
Ketones	Neg	+ !↑	
Blood	Neg	Neg	
Bilirubin	Neg	Neg	
Urobilinogen (EU/dL)	<1.1	Neg	
Leukocyte esterase	Neg	Neg	
Prot chk	Neg	+ !↑	
WBCs (/HPF)	0–5	3–4	

Case Questions

I. **Understanding the Diagnosis and Pathophysiology**

1. What are the current thoughts regarding the etiology of type 1 diabetes mellitus (T1DM)? No one else in Rachel's family has diabetes—is this unusual? Are there any other findings in her family medical history that would be important to note?

2. What are the standard diagnostic criteria for T1DM? Which are found in Rachel's medical record?

3. Using the information from Rachel's medical record, identify the factors that would allow the physician to distinguish between T1DM and T2DM.

4. Describe the metabolic events that led to Rachel's symptoms and subsequent admission to the ER (polyuria, polydipsia, polyphagia, fatigue, and weight loss), integrating the pathophysiology of T1DM into your discussion.

5. Describe the metabolic events that result in the signs and symptoms associated with DKA. Was Rachel in this state when she was admitted? What precipitating factors may lead to DKA?

6. Rachel will be started on a combination of Apidra prior to meals and snacks with glargine given in the a.m. and p.m. Describe the onset, peak, and duration for each of these types of insulin. Her discharge dosages are as follows: 7 u glargine with Apidra prior to each meal or snack—1:15 insulin:carbohydrate ratio. Rachel's parents want to know why she cannot take oral medications for her diabetes like some of their friends do. What would you tell them?

7. Rachel's physician explains to Rachel and her parents that Rachel's insulin dose may change due to something called a honeymoon phase. Explain what this is and how it might affect her insulin requirements.

8. How does physical activity affect blood glucose levels? Rachel is a soccer player and usually plays daily. What recommendations will you make to Rachel to assist with managing her glucose during exercise and athletic events?

9. Rachel's blood glucose records indicate that her levels have been consistently high when she wakes in the morning before breakfast. Describe the dawn phenomenon. Is Rachel experiencing this? How might it be prevented?

II.   Understanding the Nutrition Therapy

10. The MD ordered a consistent carbohydrate-controlled diet when Rachel begins to eat. Explain the rationale for monitoring carbohydrate in diabetes nutrition therapy.

11. Outline the basic principles for Rachel's nutrition therapy to assist in control of her T1DM.

III.   Nutrition Assessment

12. Assess Rachel's ht/age; wt/age; ht/wt; and BMI. What is her desirable weight?

13. Identify any abnormal laboratory values measured upon her admission. Explain how they may be related to her newly diagnosed T1DM.

14. Determine Rachel's energy and protein requirements. Be sure to explain what standards you used to make this estimation.

IV.   Nutrition Diagnosis

15. Prioritize two nutrition problems and complete the PES statement for each.

V.   Nutrition Intervention

16. Determine Rachel's initial nutrition prescription using her diet record from home as a guideline, as well as your assessment of her energy requirements.

17. What is an insulin:CHO ratio (ICR)? Rachel's physician ordered her ICR to start at 1:15. If her usual breakfast is 2 Pop-Tarts and 8 oz skim milk, how much Apidra should she take to cover the carbohydrate in this meal?

18. Dr. Cho set Rachel's fasting blood glucose goal at 90–180 mg/dL. If her total daily insulin dose is 33 u and her fasting a.m. blood glucose is 240 mg/dL, what would her correction dose be?

VI.   Nutrition Monitoring and Evaluation

19. Write an ADIME note for your initial nutrition assessment.

20. When Rachel comes back to the clinic, she brings the following food and blood glucose record with her.

    a.   Determine the amount of carbohydrates she is consuming at each meal.

    b.   Determine whether she is taking adequate amounts of Apidra for each meal according to her record.

    c.   Calculate a correction dose for her to use.

| | | | | | Insulin dosages | |
Time	Diet	Grams of CHO	Exercise	BG (mg/dL)	What patient took	What you would recommend
7:30 a.m.	2 Pop-Tarts 1 banana 16 oz skim milk with Ovaltine (2 tbsp)			(Pre) 150	5 u Apidra	
10:30 a.m.						
12 noon	2 slices of pepperoni pizza 2 chocolate chip cookies Water			(Pre) 180	6 u Apidra	
2 p.m.	Granola bar		PE class—30 minutes			

(Continued)

(Continued)

Time	Diet	Grams of CHO	Exercise	BG (mg/dL)	Insulin dosages	
					What patient took	What you would recommend
4:30 p.m.	Apple 6 saltines with 2 tbsp peanut butter			(Pre) 110		
5–6:30 p.m.	16 oz Gatorade		Soccer practice—1.5 hours	(Pre) 140		
6:30 p.m.	Chicken with broccoli stir-fry (1 c fried rice, 2 oz chicken, ½ c broccoli) Egg roll—1 2 c skim milk			(Pre) 80	5 u Apidra	
8:30 p.m.	2 c ice cream With 2 tbsp peanuts			(Pre) 150	4 u Apidra	
10:30 p.m.	Bed					

Bibliography

American Diabetes Association. Standards of Medical Care in Diabetes—2013. *Diabetes Care*. 2013;36:S11–S66.

Evert AB, Hess-Fischl A. *Pediatric Diabetes*. Chicago, IL: American Dietetic Association; 2006.

Hanas R, Donaghue KC, Klingensmith G, Swift PGF. ISPAD Clinical Practice Consensus Guidelines. *Pediatric Diabetes*. 2009;10:S1–210.

Haas L, Maryniuk M, Beck J, et al. National standards for diabetes self-management education and support. *Diabetes Educ*. 2012;38:619–29.

Roth S. Diseases of the endocrine system. In: Nelms M, Sucher K, Lacey K, Roth SL. *Nutrition Therapy and Pathophysiology*. 2nd ed. Belmont, CA: Wadsworth, Cengage Learning; 2011:471–519.

Internet Resources

ADA Evidence Analysis Library: http://www.adaevidencelibrary.com

American Diabetes Association: http://www.diabetes.org/

International Society for Pediatric and Adolescent Diabetes: http://www.ispad.org/

Pediatric Nutrition Care Manual: http://www.nutritioncaremanual.org

Case 15

Type 1 Diabetes Mellitus in the Adult

Objectives

After completing this case, the student will be able to:

1. Describe the pathophysiology of type 1, type 2, and latent autoimmune diabetes mellitus.
2. Discuss diagnostic criteria for type 1, type 2, and latent autoimmune diabetes mellitus.
3. Understand the acute complications of hyperglycemia and assess the appropriate medical and nutritional care for both HHS and DKA.
4. Develop a nutrition care plan—with appropriate measurable goals, interventions, and strategies for monitoring and evaluation—that addresses the nutrition diagnoses for this case.
5. Integrate an insulin regimen with nutrition therapy and provide appropriate recommendations for carbohydrate-to-insulin ratios and correction dosages.
6. Interpret laboratory parameters to assess fluid, electrolyte, and acid-base balance.
7. Prioritize survival skills and educational requirements for a newly diagnosed type 1 adult diabetic.

Armando Gutiérrez is a 32-year-old male admitted from the ER to the endocrinology service with acute uncontrolled hyperglycemia.

Gutiérrez, Armando, Male, 32 y.o.
Allergies: NKA
Pt. Location: 848

Code: FULL
Physician: C. Harris

Isolation: None
Admit Date: 7/1

Patient Summary: Armando Gutiérrez is a 32-year-old male admitted from the ER to the endocrinology service.

History:

Onset of disease: Patient transported to ER when found ill in his apartment by a friend. During ER assessment, patient was noted to have serum glucose of 610 mg/dL. Mr. Gutiérrez was diagnosed with T2DM one year ago and has been on metformin since that diagnosis. He does not take the medication regularly as he felt it really wasn't necessary.
Medical history: None
Surgical history: None
Medications at home: None
Tobacco use: Smoker 1 ppd × 10 years
Alcohol use: Daily
Family history: Father—MI; Mother—ovarian cancer, T2 DM

Demographics:

Marital status: Divorced
Years education: 16
Language: English/Spanish
Occupation: Computer software engineer
Hours of work: 8–7 M–F, some weekends
Ethnicity: Hispanic
Religious affiliation: Catholic

Admitting History/Physical:

Chief complaint: Friend states that Mr. Gutiérrez had not been feeling well at work the previous day. He thought he was fighting off a virus. When he didn't show up for work or answer his cell phone this morning, his friend went to check on him and found him groggy and almost unconscious at his apartment. He called 911 and patient was transported to University Hospital.
General appearance: Slim, Hispanic male, in obvious distress

Vital Signs: Temp: 99.6   Pulse: 100   Resp rate: 24
                   BP: 78/100   Height: 5'11"   Weight: 165 lbs

Heart: tachycardia
HEENT: Head: WNL
           Eyes: PERRLA
           Ears: Clear
           Nose: Clear
           Throat: Dry mucous membranes without exudates or lesions
Genitalia: Deferred
Neurologic: Lethargic but able to arouse. Follows commands appropriately. Glasgow Coma Scale: 13

Gutiérrez, Armando, Male, 32 y.o.

Allergies: NKA   **Code:** FULL   **Isolation:** None
Pt. Location: 848   **Physician:** C. Harris   **Admit Date:** 7/1

Extremities: + 4 ROM; DTR 2+
Skin: Warm and dry
Chest/lungs: Respirations are rapid—clear to auscultation and percussion
Peripheral vascular: Pulse 4+ bilaterally, warm, no edema
Abdomen: Active bowel sounds ×4; tender, nondistended

Orders:

1. Regular insulin 1 unit/mL NS 40 mEq KCl/liter @ 300 mL/hr. Begin infusion at 0.1 unit/kg/hr = 3.7 units/hr and increase to 5 units/hr. Flush new IV tubing with 50 mL of insulin drip solution prior to connecting to patient and starting insulin infusion.
2. Labs: BMP Stat
   Phos Stat
   Calcium Stat
   UA with culture if indicated Stat Clean catch
   Bedside glucose Stat
   Bedside I-Stat: EG7 Stat
   Islet cell autoantibodies screen (ICA)
   GADA
   TSH
   Comp metabolic panel (CMP)
   Thyroglobulin antibodies
   C-peptide
   Immunoglobulin A level
   Hemoglobin A_{1c}
   Tissue transglutaminase
3. NPO except for ice chips and medications. After 12 hours, clear liquids when stable. Then, advance to consistent carbohydrate diet order—70–80 g breakfast and lunch; 85–95 g dinner; 30-gram snack pm and HS.
4. Consult diabetes education team for self-management training for patient to begin education after stabilized.

Nursing Assessment	7/1
Abdominal appearance (concave, flat, rounded, obese, distended)	flat
Palpation of abdomen (soft, rigid, firm, masses, tense)	tense with guarding
Bowel function (continent, incontinent, flatulence, no stool)	continent
Bowel sounds (P=present, AB=absent, hypo, hyper)	
RUQ	P
LUQ	P
RLQ	P

(Continued)

Gutiérrez, Armando, Male, 32 y.o.
Allergies: NKA
Pt. Location: 848

Code: FULL
Physician: C. Harris

Isolation: None
Admit Date: 7/1

Nursing Assessment *(Continued)*

Nursing Assessment	7/1
LLQ	P
Stool color	light brown
Stool consistency	soft
Tubes/ostomies	NA
Genitourinary	
Urinary continence	catheter in place
Urine source	clean specimen
Appearance (clear, cloudy, yellow, amber, fluorescent, hematuria, orange, blue, tea)	cloudy, amber
Integumentary	
Skin color	pale
Skin temperature (DI=diaphoretic, W=warm, dry, CL=cool, CLM=clammy, CD+=cold, M=moist, H=hot)	DI; CLM
Skin turgor (good, fair, poor, TENT=tenting)	fair
Skin condition (intact, EC=ecchymosis, A=abrasions, P=petechiae, R=rash, W=weeping, S=sloughing, D=dryness, EX=excoriated, T=tears, SE=subcutaneous emphysema, B=blisters, V=vesicles, N=necrosis)	intact
Mucous membranes (intact, EC=ecchymosis, A=abrasions, P=petechiae, R=rash, W=weeping, S=sloughing, D=dryness, EX=excoriated, T=tears, SE=subcutaneous emphysema, B=blisters, V=vesicles, N=necrosis)	intact
Other components of Braden score: special bed, sensory pressure, moisture, activity, friction/shear (>18=no risk, 15–16=low risk, 13–14=moderate risk, ≤12=high risk)	20

Nutrition:

Meal type: NPO then progress to clear liquids and then consistent carbohydrate-controlled diet
Fluid requirement: 2200 mL
History: Usual intake (for past several months):
AM: Toast, jelly, coffee, and scrambled egg
Lunch: Subway sandwich, chips, diet soda
Dinner: Usually cooks pasta, rice, vegetables, and some type of meat; eats out 3–4 times per week at dinner.

MD Progress Note:

9/28 0700

Subjective: Armando Guitérrez previous 24 hours reviewed. Previously diagnosed with T2DM; treated with metformin but appears to not have taken it regularly.

Gutiérrez, Armando, Male, 32 y.o.
Allergies: NKA
Pt. Location: 848

Code: FULL
Physician: C. Harris

Isolation: None
Admit Date: 7/1

Vitals: Temp: 99.5, Pulse: 82, Resp rate: 25, BP: 101/78
Urine Output: 2660 mL (71.8 mL/kg)

Physical Exam
General: Alert and oriented to person, place, and time
HEENT: WNL
Neck: WNL
Heart: WNL
Lungs: Clear to auscultation
Abdomen: Active bowel sounds

Assessment/Plan: Results: + ICA, GADA, IAA consistent with type 1 DM vs LADA. Negative c-peptide.

Dx: Suspect Type 1 Diabetes Mellitus vs latent autoimmune diabetes of adult (LADA)

Plan: Change IVF to D_5.45NS with 40MEq K @ 135 mL/hr. Begin Novolog 0.5 u every 2 hours until glucose is 150–200 mg/dL. Tonight begin glargine 15 u at 9 pm. Progress Novolog using ICR 1:15. Continue bedside glucose checks hourly. Notify MD if blood glucose >200 or <80.

C. Harris, MD

Intake/Output

Date		7/1 0701 – 7/2 0700			
Time		0701–1500	1501–2300	2301–0700	Daily total
IN	P.O.	**NPO**	**NPO**	**720**	**720**
	I.V.	**2,400**	**2,400**	**2,400**	**7,200**
	(mL/kg/hr)	(4)	(4)	(4)	(4)
	I.V. piggyback	0	0	0	**0**
	TPN	0	0	0	**0**
	Total intake	**2,400**	**2,400**	**3,120**	**7,920**
	(mL/kg)	(32)	(32)	(41.6)	(105.6)
OUT	Urine	**2,150**	**2,671**	**3,000**	**7,821**
	(mL/kg/hr)	(3.58)	(4.45)	(5)	(4.34)
	Emesis output	150	0	0	**150**
	Other	0	0	0	**0**
	Stool	0	×1	0	×1
	Total output	**2,300**	**2,671**	**3,000**	**7,971**
	(mL/kg)	(30.7)	(35.6)	(40)	(106.3)
Net I/O		**+100**	**−271**	**+120**	**−51**
Net since admission (7/1)		**+100**	**−171**	**−51**	**−51**

Gutiérrez, Armando, Male, 32 y.o.

Allergies: NKA	**Code:** FULL	**Isolation:** None
Pt. Location: 848	**Physician:** C. Harris	**Admit Date:** 7/1

Laboratory Results

	Ref. Range	7/1 1780
Chemistry		
Sodium (mEq/L)	136–145	130 !↓
Potassium (mEq/L)	3.5–5.5	3.6
Chloride (mEq/L)	95–105	101
Carbon dioxide (CO_2, mEq/L)	23–30	31 !↑
BUN (mg/dL)	8–18	18
Creatinine serum (mg/dL)	0.6–1.2	1.1
Glucose (mg/dL)	70–110	683 !↑
Phosphate, inorganic (mg/dL)	2.3–4.7	2.1 !↓
Magnesium (mg/dL)	1.8–3	1.9
Calcium (mg/dL)	9–11	10
Osmolality (mmol/kg/H_2O)	285–295	306 !↑
Bilirubin total (mg/dL)	≤1.5	0.2
Bilirubin, direct (mg/dL)	<0.3	0.01
Protein, total (g/dL)	6–8	6.9
Albumin (g/dL)	3.5–5	4.4
Prealbumin (mg/dL)	16–35	32
Ammonia (NH_3, µmol/L)	9–33	9
Alkaline phosphatase (U/L)	30–120	110
ALT (U/L)	4–36	6.2
AST (U/L)	0–35	21
CPK (U/L)	30–135 F 55–170 M	61
Lactate dehydrogenase (U/L)	208–378	229
Cholesterol (mg/dL)	120–199	210 !↑
Triglycerides (mg/dL)	35–135 F 40–160 M	175 !↑
T_4 (µg/dL)	4–12	8
T_3 (µg/dL)	75–98	81
HbA_{1C} (%)	3.9–5.2	12.5 !↑
C-peptide (ng/mL)	0.51–2.72	0.09 !↓
ICA	—	+ !↑
GADA	—	+ !↑
IA-2A	—	—

Gutiérrez, Armando, Male, 32 y.o.
Allergies: NKA
Pt. Location: 848

Code: FULL
Physician: C. Harris

Isolation: None
Admit Date: 7/1

Laboratory Results *(Continued)*

	Ref. Range	7/1 1780
IAA	—	+ !↑
tTG	—	—
Hematology		
WBC ($\times 10^3/mm^3$)	4.8–11.8	10.6
RBC ($\times 10^6/mm^3$)	4.2–5.4 F 4.5–6.2 M	5.8
Urinalysis		
Collection method	—	catheter
Color	—	yellow
Appearance	—	clear
Specific gravity	1.003–1.030	1.008
pH	5–7	4.9 !↓
Protein (mg/dL)	Neg	+1 !↑
Glucose (mg/dL)	Neg	+3 !↑
Ketones	Neg	+4 !↑
Blood	Neg	Neg
Bilirubin	Neg	Ncg
Nitrites	Neg	Neg
Urobilinogen (EU/dL)	<1.1	Neg
Leukocyte esterase	Neg	Neg
Prot chk	Neg	tr !↑
WBCs (/HPF)	0–5	0
RBCs (/HPF)	0–5	0
Bact	0	0
Mucus	0	0
Crys	0	0
Casts (/LPF)	0	0
Yeast	0	0
Arte		
pH	7.35–7.45	7.31 !↓
pCO_2	35–45	35
SO_2 (%)	95	97
CO_2 content (m	25–30	28
O_2 content (%)	15–22	21

(Continued)

Gutiérrez, Armando, Male, 32 y.o.
Allergies: NKA   **Code:** FULL   **Isolation:** None
Pt. Location: 848   **Physician:** C. Harris   **Admit Date:** 7/1

Laboratory Results *(Continued)*

	Ref. Range	7/1 1780
pO$_2$ (mm Hg)	≥80	89
Base excess (mEq/L)	>3	–
Base deficit (mEq/L)	<3	–
HCO$_3$$^-$ (mEq/L)	24–28	22 !↓
COHb (%)	<2	1.1

Case Questions

I.   Understanding the Diagnosis and Pathophysiology

1.   What are the differences among T1DM, T2DM, and LADA?

2.   What are the standard diagnostic criteria for each of these diagnoses?

3.   Why do you think he was originally diagnosed with T2DM? Why does the MD now suspect he may actually have T1DM or LADA?

4.   Describe the metabolic events that led to Armando's symptoms and subsequent admission to the ER (polyuria, polydipsia, polyphagia, fatigue, and weight loss), integrating the pathophysiology of T1DM into your discussion.

5.   Describe the metabolic events that result in the signs and symptoms associated with DKA. Was Armando in this state when he was admitted? What precipitating factors may lead to DKA?

6.   Armando will be started on a combination of Novolog prior to meals and snacks with glargine given in the a.m. and p.m. Describe the onset, peak, and duration for each of these types of insulin.

7.   Using his current weight of 165 lbs, determine the discharge dose of glargine as well as an appropriate ICR for Armando to start with.

8.   Intensive insulin therapy requires frequent blood glucose self-monitoring. What are some of the barriers to success for patients who begin this type of therapy? Give suggestions on how you might work with Armando to support his compliance.

9.   Armando tells you that he is very frightened of having his blood sugar drop too low. What is hypoglycemia? What are the symptoms? What information would you give to Armando to make sure he is well prepared to prevent or treat hypoglycemia?

10. Armando's mother has T2DM. She is currently having problems with vision and burning in her feet. What is she most likely experiencing? Describe the pathophysiology of these complications. You can tell that he is worried not only about his mother but also about his own health. Explain, using the foundation research of the Diabetes Control and Complications Trial (DCCT) as well as any other pertinent research data, how he can prevent these complications.

II.   Understanding the Nutrition Therapy

11. Outline the basic principles for Armando's nutrition therapy to assist in control of his DM.

III. Nutrition Assessment

12. Assess Armando's height and weight. Calculate his BMI.

13. Identify any abnormal laboratory values measured upon his admission. Explain how they may be related to his newly diagnosed DM.

14. Determine Armando's energy and protein requirements. Be sure to explain what standards you used to make this estimation. Would you recommend that he either gain or lose weight in the future?

IV.   Nutrition Diagnosis

15. Prioritize two nutrition problems and complete the PES statement for each.

V.   Nutrition Intervention

16. Determine Armando's initial CHO prescription using his diet record from home as a guideline, as well as your assessment of his energy requirements. What nutrition education material would you use to teach Armando CHO counting?

17. Amando's usual breakfast consists of 2 slices of toast, butter, 2 tbsp jelly, 2 scrambled eggs, and orange juice (~1 c). Using the ICR that you calculated in question #7, how much Novolog should he take to cover the carbohydrate in this meal?

18. Using the ADA guidelines, what would be appropriate fasting and postprandial target glucose levels for Armando?

VI. Nutrition Monitoring and Evaluation

19. Write an ADIME note for your initial nutrition assessment.

20. Armando comes back to the clinic 2 weeks after his diagnosis. List the important questions you will ask him in order to plan the next steps for providing the additional education that he might need.

21. Armando states that he would like to start exercising again as he is feeling better. He is used to playing tennis several times per week as well as cycling at least 2 days per week for over 20 miles each time. Again, he expresses his concern regarding low blood sugar. How would you counsel Armando regarding physical activity, his diet, and his blood glucose monitoring?

22. Armando states that one of his friends has talked about using the glycemic index as a way to manage his diabetes. He says that he has also seen some nutrition programs advertise their food products as being "low glycemic index" on TV. Explain glycemic index, glycemic load, and how he might use this information within his nutrition therapy plans.

Bibliography

American Diabetes Association. Standards of Medical Care in Diabetes—2013. *Diabetes Care.* 2013;36:S11–S66.

Craig J. Carbohydrate counting, glycemic index, and glycemic load. Putting them all together. *Diabetes Self Manag.* 2012;29:41–50.

Diabetes Complication and Control Trial Study Research Group. Intensive diabetes treatment and cardiovascular disease in patients with type 1 diabetes. *NEJM.* 2005;353:2643–2653.

Esfahani A, Wong JM, Mirrahimi A, et al. The application of the glycemic index and glycemic load in weight loss: A review of the clinical evidence. *IUBMB Life.* 2011;63:7–13.

Haas L, Maryniuk M, Beck J, et al. National standards for diabetes self-management education and support. *Diabetes Educ.* 2012;38:619–29.

Hayes C, Kriska A. Role of physical activity in diabetes management and prevention. *J Am Diet Assoc.* 2008;108:S19–S23.

Hortensius K, Kars MC, Wierenga WS, et al. Perspectives of patients with type 1 or insulin-treated type 2 diabetes on self-monitoring of blood glucose: A qualitative study. *BMC Public Health.* 2012;12:167–79.

Malanda UL, Welschen LM, Riphagen II, et al. Self-monitoring of blood glucose in patients with type 2 diabetes mellitus who are not using insulin. *Cochrane Database Syst Rev.* 2012;1:AN:CD004060; PMID:22258959.

Roth S. Diseases of the Endocrine System. In: Nelms M, Sucher K, Lacey K, Roth SL. *Nutrition Therapy and Pathophysiology.* 2nd ed. Belmont, CA: Wadsworth, Cengage Learning; 2011:471–519.

Internet Resources

ADA Evidence Analysis Library: http://www.adaevidencelibrary.com

American Diabetes Association: http://www.diabetes.org/

Council for the Advancement of Diabetes Research and Education: http://www.cadre-diabetes.org/

Nutrition Care Manual: http://www.nutritioncaremanual.org

Adult Type 1 Blogs and Social Networks

A Sweet Life: asweetlife.org

Close Concerns: www.closeconcerns.com

Diabetes Daily: www.diabetesdaily.com

Case 16

Type 2 Diabetes Mellitus—Pediatric Obesity

Objectives

After completing this case, the student will be able to:

1. Describe the pathophysiology of type 2 diabetes mellitus in children.
2. Identify the risk factors for development of type 2 DM in children.
3. Outline the treatment options for type 2 diabetes in children.
4. Identify the potential complications for children with type 2 diabetes.
5. Develop a nutrition care plan—with appropriate measurable goals, interventions, and strategies for monitoring and evaluation—that addresses the nutrition diagnoses for this case.

Adane Ross is a 9-year-old 3rd grader who is seen in her pediatrician's office after diagnosis of type 2 DM during her routine school physical.

Ross, Adane, Female, 9 y.o.
Allergies: NKA
Pt. Location: Pediatric Clinic

Code: FULL
Physician: R. Fahey

Isolation: None
Admit Date: 8/3

Patient Summary: Adane Ross was recently diagnosed with T2 DM during her school physical.

History:
Onset of disease: Patient was a full-term infant with birthweight of 10 lbs 4 oz and 20" in length. Her mother had gestational diabetes during the pregnancy.
Medical history: Frequent ear infections as infant and toddler
Surgical history: None
Medications at home: None
Tobacco use: Nonsmoker
Alcohol use: None
Family history: Mother and grandmother—type 2 DM; Grandfather—high cholesterol and hypertension

Demographics:
Marital status: NA—3rd-grade female
Years education: 3 years
Language: English only
Occupation: Student
Hours of work: N/A
Household members: Mother, grandparents, sisters ages 14, 12, and brother age 10
Ethnicity: African-American
Religious affiliation: African Methodist Episcopal

Admitting History/Physical:
Chief complaint: Mother and grandmother present—they state they are here because they received a phone call from the office after Adane's school physical.
General appearance: Overweight

Vital Signs:	Temp: 98.6	Pulse: 72	Resp rate: 19	
	BP: 100/59	Height: 52"	Weight: 140 lbs	BMI 36.4

Heart: Regular rate and rhythm
HEENT: Head: WNL
            Eyes: PERRLA
            Ears: Clear
            Nose: Clear
            Throat: Dry mucous membranes without exudates or lesions
Genitalia: Deferred
Neurologic: Alert and oriented
Extremities: Noncontributory
Skin: Warm and dry
Chest/lungs: Respirations WNL—clear to auscultation and percussion

Ross, Adane, Female, 9 y.o.
Allergies: NKA
Pt. Location: Pediatric Clinic

Code: FULL
Physician: R. Fahey

Isolation: None
Admit Date: 8/3

Peripheral vascular: Pulse 4+ bilaterally, warm, no edema
Abdomen: Active bowel sounds ×4

Assessment/Plan: Results: HbA_{1c} 6.9% EAG 151
C-peptide 2.75 ng/mL A_{1c}

Dx: New Diagnosis Type 2 Diabetes Mellitus; obesity

Plan: Refer to clinic RD/CDE for family education
.... R Fahey, MD

Nutrition:

24-hour recall:

Breakfast:	Fruit punch or Kool-Aid—1 c; frosted flakes—2 c with whole milk
Midmorning:	2 slices toast with butter and jam
Snacks:	Cookies, crackers, chips, soda, fruit punch, iced tea or Kool-Aid, popsicles. Yesterday she had several chocolate chip cookies, 2 small bags of Cheetos, multiple glasses of fruit punch, and 2 popsicles.
Lunch:	Peanut butter sandwich with banana and mayonnaise—usually 2 sandwiches, 2 tbsp peanut butter, 1 tbsp mayo, and ½ banana on each sandwich, 2 c fruit punch, chips
Dinner:	Fried pork chop, greens, potatoes, cornbread with butter, iced tea made with sugar
Bedtime:	Pizza rolls and Coke

Laboratory Results

	Ref. Range	8/3 0800	8/4 0940
Chemistry			
Sodium (mEq/L)	136–145	137	
Potassium (mEq/L)	3.5–5.5	4.1	
Chloride (mEq/L)	95–105	101	
Carbon dioxide (CO_2, mEq/L)	23–30	25	
BUN (mg/dL)	8–18	9	
Creatinine serum (mg/dL)	0.6–1.2	0.7	
Glucose (mg/dL)	70–110	171 !↑	155 !↑
Phosphate, inorganic (mg/dL)	2.3–4.7	2.5	

(Continued)

Ross, Adane, Female, 9 y.o.
Allergies: NKA   **Code:** FULL   **Isolation:** None
Pt. Location: Pediatric Clinic   **Physician:** R. Fahey   **Admit Date:** 8/3

Laboratory Results *(Continued)*

	Ref. Range	8/3 0800	8/4 0940
Magnesium (mg/dL)	1.8–3	1.9	
Calcium (mg/dL)	9–11	9.2	
Osmolality (mmol/kg/H$_2$O)	285–295	286.7	
Bilirubin total (mg/dL)	≤1.5	0.8	
Bilirubin, direct (mg/dL)	<0.3	0.004	
Protein, total (g/dL)	6–8	6.9	
Albumin (g/dL)	3.5–5	4.2	
Prealbumin (mg/dL)	16–35	22	
Ammonia (NH$_3$, µmol/L)	9–33	9	
Alkaline phosphatase (U/L)	30–120	110	
ALT (U/L)	4–36	21	
AST (U/L)	0–35	18	
CPK (U/L)	30–135 F 55–170 M	33	
Lactate dehydrogenase (U/L)	208–378	210	
Cholesterol (mg/dL)	<170	210 !↑	
Triglycerides (mg/dL)	<150	175 !↑	
T$_4$ (µg/dL)	4–12	4.2	
T$_3$ (µg/dL)	75–98	81	
HbA$_{1C}$ (%)	3.9–5.2	6.9 !↑	
EAG	—	151 !↑	
C-peptide (ng/mL)	0.51–2.72	2.75 !↑	
ICA	—	Neg	
GADA	—	Neg	
IA-2A	—	Neg	
IAA	—	Neg	
tTG	—	Neg	
Hematology			
WBC (×10^3/mm^3)	4.8–11.8	5.6	
RBC (×10^6/mm^3)	4.2–5.4 F 4.5–6.2 M	4.8	
Hemoglobin (Hgb, g/dL)	12–15 F 14–17 M	13.5	
Hematocrit (Hct, %)	37–47 F 40–54 M	37	

Ross, Adane, Female, 9 y.o.
Allergies: NKA
Pt. Location: Pediatric Clinic

Code: FULL
Physician: R. Fahey

Isolation: None
Admit Date: 8/3

Laboratory Results *(Continued)*

	Ref. Range	8/3 0800	8/4 0940
Hematology, Manual Diff			
Neutrophil (%)	50–70	55	
Lymphocyte (%)	15–45	18	
Monocyte (%)	3–10	4	
Eosinophil (%)	0–6	0	
Basophil (%)	0–2	1	
Blasts (%)	3–10	3	
Urinalysis			
Collection method	—	clean catch	
Color	—	yellow	
Appearance	—	clear	
Specific gravity	1.003–1.030	1.016	
pH	5–7	5.6	
Protein (mg/dL)	Neg	tr !↑	
Glucose (mg/dL)	Neg	+ !↑	
Ketones	Neg	Neg	
Blood	Neg	Neg	
Bilirubin	Neg	Neg	
Urobilinogen (EU/dL)	<1.1	Neg	
Leukocyte esterase	Neg	Neg	
Prot chk	Neg	+ !↑	
WBCs (/HPF)	0–5	0	

Case Questions

I. Understanding the Diagnosis and Pathophysiology

1. What are the risk factors for developing type 2 DM as a child? What do the current ADA standards of medical care recommend concerning screening at-risk children?

2. Evaluate Adane's medical record. Identify which risk factors most likely led to the routine screening for DM during her school physical.

3. What are the ADA standard diagnostic criteria for T2DM? Which are included in Adane's medical record?

4. Adane's physician requested additional testing that included autoantibody levels and C-peptide. Explain why these tests were done and what the results indicate for Adane.

5. Insulin resistance is a major component of T2DM. Explain this pathophysiology. How could you determine whether Adane is exhibiting insulin resistance?

6. Children with T2DM are at high risk for early cardiovascular disease. Why does this complication occur with diabetes? Evaluate Adane's lipid profile. How does this compare to the lipid goals for children with diabetes?

7. Adane's grandmother asks about medication for treating high cholesterol as her husband is on this medicine. What are the recommendations for the use of statin drugs in children?

8. Adane's urinalysis is positive for protein. What does this mean and how may this be related to her diabetes?

9. Should Adane and her family be taught about self-monitoring of blood glucose (SMBG)? If so, what are the standard recommendations for daily frequency of testing? What would be the appropriate fasting and postprandial target glucose levels for Adane?

II.   Understanding the Nutrition Therapy

10. Outline the basic principles for Adane's nutrition therapy to assist in control of her T2DM.

III.   Nutrition Assessment

11. Using the charts on pp. 188–189, assess Adane's ht/age; wt/age; ht/wt; and BMI. What is her desirable weight?

12. Identify any abnormal laboratory values measured upon her admission. Explain how they may be related to her newly diagnosed T2DM.

13. Determine Adane's energy and protein requirements. Be sure to explain what standards you used to make these estimations. Should weight loss be a component of your estimation of energy requirements?

14. Using Adane's diet history, assess the approximate number of kilocalories her intake provided, as well as the energy distribution of calories for protein, carbohydrate, and fat, using the exchange system. Compare this to the recommendations that you made in question #10.

IV.   Nutrition Diagnosis

15. Prioritize two nutrition problems and complete the PES statement for each.

V.   Nutrition Intervention

16. Determine Adane's initial nutrition therapy prescription using her diet record from home as a guideline, as well as your assessment of her energy requirements.

17. Outline the initial steps you would use to teach Adane and her family about nutrition and diabetes. What education materials could you use?

18. Considering that Adane will not be started on medication, is it necessary to teach her and her family about hypoglycemia, sick-day rules, and exercise?

Stature-for-Age and Weight-for-Age Percentiles: Girls, 2 to 20 Years

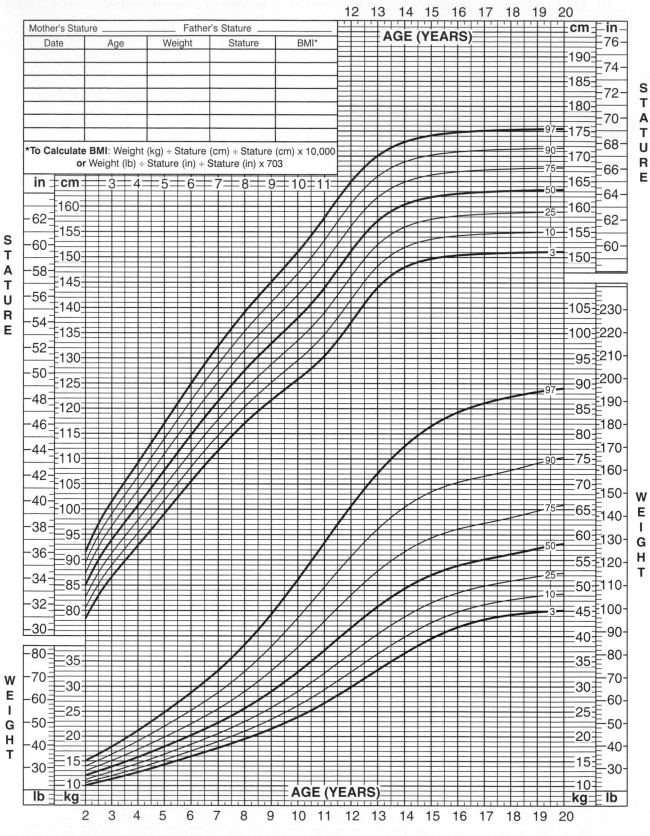

Body Mass Index-for-Age Percentiles: Girls, 2 to 20 Years

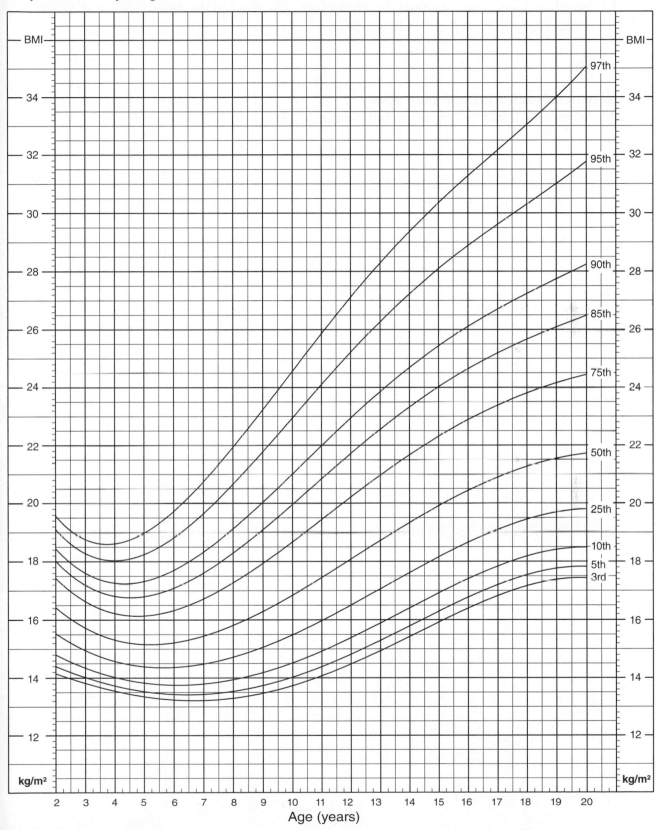

19. Adane's mom is worried that none of the children will ever be able to have birthday cake or other sweet treats. She feels that she cannot offer these to the other children if Adane cannot have them. What would you tell her?

VI. Nutrition Monitoring and Evaluation

20. Write an ADIME note for your initial nutrition assessment.

21. Adane's grandmother suggests that perhaps Adane should have "stomach surgery" so that she will lose weight more quickly. What are the recommendations for pediatric bariatric surgery?

Bibliography

American Diabetes Association. Standards of Medical Care in Diabetes—2013. *Diabetes Care.* 2013;36:S11–S66.

Badaru A, Pihoker C. Type 2 diabetes in childhood: Clinical characteristics and role of β-cell autoimmunity. *Curr Diab Rep.* 2012;12:75–81.

Evert AB, Hess-Fischl A. *Pediatr. Diabetes.* Chicago IL: American Dietetic Association; 2006.

Gidding S, Lichtenstein A, Faith M, et al. Implementing American Heart Association Pediatric and Adult Nutrition Guidelines: A scientific statement from the American Heart Association Nutrition Committee of the Council on Nutrition, Physical Activity and Metabolism, Council on Cardiovascular Disease in the Young, Council on Arteriosclerosis, Thrombosis and Vascular Biology, Council on Cardiovascular Nursing, Council on Epidemiology and Prevention, and Council for High Blood Pressure Research. *Circulation.* 2009;119:1161–1175.

Gungor N, Bacha F, Saad R, Janosky J, Arslanian S. Youth type 2 diabetes: Insulin resistance, β-cell failure, or both? *Diabetes Care.* 2005;28(3):638–644.

Hanas R, Donaghue KC, Klingensmith G, Swift PGF. ISPAD Clinical Practice Consensus Guidelines. *Pediatr. Diabetes.* 2009;10:S1–210.

Haas L, Maryniuk M, Beck J, et al. National standards for diabetes self-management education and support. *Diabetes Educ.* 2012;38:619–29.

Kong AP, Chan RS, Nelson EA, et al. Role of low-glycemic index diet in management of childhood obesity. *Obes Rev.* 2011;12:492–8.

Rosenbloom AL, Silverstein JH, Amemiya S, Zeitler P, Kingensmith GJ. Type 2 diabetes in children and adolescents. *Pediatr. Diabetes.* 2009;10(Suppl 12):17–32.

Roth S. Diseases of the endocrine system. In: Nelms M, Sucher K, Lacey K, Roth SL. *Nutrition Therapy and Pathophysiology.* 2nd ed. Belmont, CA: Wadsworth, Cengage Learning; 2011:471–519.

Smart C, Salander-van Vliet E, Waldron S. Nutritional management in children and adolescents with diabetes. *Pediatr. Diabetes.* 2009;10(Suppl 12):100–117.

Sunni M, Mays R, Kaup T, et al. Recognizing and managing type 2 diabetes mellitus in children: An algorithm for primary care providers. *Minn Med.* 2011;94:34–9.

Internet Resources

ADA Evidence Analysis Library: http://www.adaevidencelibrary.com

American Diabetes Association: http://www.diabetes.org/

International Society for Pediatric and Adolescent Diabetes: http://www.ispad.org/

Pediatric Nutrition Care Manual: http://www.nutritioncaremanual.org

Adult Type 2 Diabetes Mellitus: Transition to Insulin

Objectives

After completing this case, the student will be able to:

1. Describe the pathophysiology of type 2 diabetes mellitus and hyperglycemic hyperosmolar state.
2. Explain the pharmacology of oral medications for type 2 DM.
3. Integrate an insulin regimen with nutrition therapy and provide appropriate recommendations for carbohydrate-to-insulin ratios and correction dosages.
4. Develop a nutrition care plan—with appropriate measurable goals, interventions, and strategies for monitoring and evaluation—that addresses the nutrition diagnoses for this case.
5. Explain the appropriate use and interpretation of self-monitoring of blood glucose to adjust rapid-acting insulin.

Mr. Fagan is a 53-year-old with type 2 diabetes mellitus who has a long history of noncompliance. He is admitted through the ER with severe hyperglycemia and dehydration.

Fagan, Mitchell, Male, 53 y.o.
Allergies: NKA
Pt. Location: MICU Bed #5

Code: FULL
Physician: R. Petersen

Isolation: None
Admit Date: 4/12

Patient Summary: Mitchell Fagan is a 53-year-old male admitted with acute hyperglycemia.

History:
Onset of disease: Patient's coworker became concerned when patient did not report to work or answer his phone when called. Coworker went to patient's home and found him drowsy and confused. Took patient to ER, where patient was noted to have serum glucose of 1524 mg/dL.
Medical history: Type 2 DM × 1 year—prescribed glyburide and metformin but admits that he has not taken the medications regularly; HTN; hyperlipidemia; gout
Surgical history: ORIF R ulna; hernia repair
Medications at home: Glyburide 20 mg daily; 500 mg metformin twice daily; Dyazide once daily (25 mg hydrochlorothiazide and 37.5 mg triamterene); Lipitor 20 mg daily
Tobacco use: 1 ppd × 20 years—now quit
Alcohol use: 3–4 drinks per week
Family history: Father—HTN, CAD; mother—type 2 DM

Demographics:
Marital status: Single
Years education: 16
Language: English only
Occupation: Retired military—now works as consultant to military equipment company
Hours of work: 8–5 daily
Household members: NA—lives alone
Ethnicity: Caucasian
Religious affiliation: NA

Admitting History/Physical:
Chief complaint: "I had a lot of vomiting that I thought at first was food poisoning but I just kept getting worse." When questioned about medications, patient admits that he has not taken medications for the diabetes regularly—"I hate how they make me feel but I almost always take my other medications for blood pressure and cholesterol."
General appearance: Mildly obese, 53-year-old male.

Vital Signs: Temp: 100.5   Pulse: 105   Resp rate: 26
                    BP: 90/70   Height: 5'9"   Weight: 214 lbs

Heart: Regular rate and rhythm
HEENT: Head: WNL
          Eyes: PERRLA
          Ears: Clear
          Nose: Clear
          Throat: Dry mucous membranes without exudates or lesions
Genitalia: Deferred

Fagan, Mitchell, Male, 53 y.o.
Allergies: NKA   **Code:** FULL   **Isolation:** None
Pt. Location: MICU Bed #5   **Physician:** R. Petersen   **Admit Date:** 4/12

Neurologic: Alert but previously drowsy with mild confusion
Extremities: Noncontributory
Skin: Warm and dry; poor turgor
Chest/lungs: Respirations are rapid—clear to auscultation and percussion
Peripheral vascular: Pulse 4 + bilaterally, warm, no edema
Abdomen: Active bowel sounds × 4; tender, nondistended

Orders:

1. Replace 1 L NS stat. Then begin regular insulin. 1 unit/kg/h in NS 40 mEq KCl/liter @ 500 mL/
    hr × 3 hours. Then regular insulin 1 unit/mL NS 10 mEq KCl/liter @135 mL/hr. Begin infusion at
    0.1 unit/kg/hr = 3.7 units/hr and increase to 5 units/hr. Flush new IV tubing with 50 mL of insulin
    drip solution prior to connecting to patient and starting insulin infusion.

2. Labs: BMP Stat
    Phos Stat
    Calcium Stat
    UA with culture if indicated Stat Clean catch
    Bedside glucose Stat
    Bedside I-Stat: EG7 Stat
    Islet cell autoantibodies screen
    Thyroid peroxidase abs
    TSH
    Comp metabolic panel (CMP); Hematology panel
    Thyroglobulin antibodies
    C-peptide
    Hemoglobin A_{1c}
    Tissue transglutaminase

3. NPO except for ice chips and medications. After 12 hours, clear liquids if stable. Then, advance
    to consistent-carbohydrate diet. Consult dietitian for advancement, total carbohydrate Rx, and
    distribution.

4. Consult diabetes education team for self-management training for patient after stabilized and
    when transferred to floor.

Nursing Assessment	4/12
Abdominal appearance (concave, flat, rounded, obese, distended)	obese
Palpation of abdomen (soft, rigid, firm, masses, tense)	tense
Bowel function (continent, incontinent, flatulence, no stool)	continent
Bowel sounds (P=present, AB=absent, hypo, hyper)	

(Continued)

Fagan, Mitchell, Male, 53 y.o.
Allergies: NKA   **Code:** FULL   **Isolation:** None
Pt. Location: MICU Bed #5   **Physician:** R. Petersen   **Admit Date:** 4/12

Nursing Assessment *(Continued)*

Nursing Assessment	4/12
RUQ	P
LUQ	P
RLQ	P
LLQ	P
Stool color	light brown
Stool consistency	soft
Tubes/ostomies	NA
Genitourinary	
Urinary continence	yes
Urine source	clean specimen
Appearance (clear, cloudy, yellow, amber, fluorescent, hematuria, orange, blue, tea)	cloudy, amber
Integumentary	
Skin color	pale
Skin temperature (DI=diaphoretic, W=warm, dry, CL=cool, CLM=clammy, CD+=cold, M=moist, H=hot)	DI
Skin turgor (good, fair, poor, TENT=tenting)	good
Skin condition (intact, EC=ecchymosis, A=abrasions, P=petechiae, R=rash, W=weeping, S=sloughing, D=dryness, EX=excoriated, T=tears, SE=subcutaneous emphysema, B=blisters, V=vesicles, N=necrosis)	intact
Mucous membranes (intact, EC=ecchymosis, A=abrasions, P=petechiae, R=rash, W=weeping, S=sloughing, D=dryness, EX=excoriated, T=tears, SE=subcutaneous emphysema, B=blisters, V=vesicles, N=necrosis)	intact
Other components of Braden score: special bed, sensory pressure, moisture, activity, friction/shear (>18=no risk, 15–16=low risk, 13–14=moderate risk, ≤12=high risk)	20

Nutrition:

Meal type: NPO then progress to clear liquids and then consistent carbohydrate-controlled diet
Fluid requirement: 2000–2500 mL after rehydration
History: Patient states that he really doesn't follow any strict diet except for not adding salt—tries to avoid high-cholesterol foods and stays away from high-sugar desserts. Most recently he had experienced vomiting for approximately 12–24 hours and so had not eaten anything and only had sips of water. He has never seen anyone for diabetes-teaching, beyond what his physician has told him.

Fagan, Mitchell, Male, 53 y.o.
Allergies: NKA
Pt. Location: MICU Bed #5

Code: FULL
Physician: R. Petersen

Isolation: None
Admit Date: 4/12

Usual intake (for past several months):
AM: Coffee with half and half
Midmorning: Bagel with cream cheese, 2–3 c of coffee
Lunch: Out at restaurant—usually Jimmy John's or fast-food sandwich, chips, and diet soda
Dinner: Cooks sometimes at home—this would be grilled chicken or beef, salad, and pota-
                toes or rice. Often will meet friends for dinner—likes all foods and especially likes
                to try different ethnic foods such as Chinese, Mexican, Indian, or Thai.

MD Progress Note:

4/13 0750
Subjective: Mitchell Fagan's previous 24 hours reviewed
Vitals: Temp: 99.6, Pulse: 83, Resp rate: 25, BP: 129/92

Physical Exam
General: Alert and oriented to person, place, and time
HEENT: WNL
Neck: WNL
Heart: WNL
Lungs: Clear to auscultation
Abdomen: Active bowel sounds

Assessment/Plan:

Results: –ICA, GADA, IAA, Negative tTG, +C-peptide with insulin level indicating T2DM but now
requiring insulin at home

Dx: Type 2 DM uncontrolled with HHS

Plan: Change IVF to D$_5$.45NS with 20MEq K @ 135 mL/hr. Begin Lispro 0.5 u every 2 hours until
glucose is 150–200 mg/dL. Tonight begin glargine 19 u at 9 pm. Progress Lispro using ICR 1:15.
Continue bedside glucose checks hourly. Notify MD if blood glucose >200 or <80.

.... R. Peterson, MD

Fagan, Mitchell, Male, 53 y.o.
Allergies: NKA
Pt. Location: MICU Bed #5

Code: FULL
Physician: R. Petersen

Isolation: None
Admit Date: 4/12

Intake/Output

Date			4/12 0701–4/13 0700			
Time			0701–1500	1501–2300	2301–0700	Daily total
IN	P.O.		**NPO**	**NPO**	**NPO**	
	I.V.		**2,175**	**1,080**	**1,080**	**4,335**
	(mL/kg/hr)		(2.80)	(1.39)	(1.39)	(1.86)
	I.V. piggyback		**0**	**0**	**0**	**0**
	TPN		**0**	**0**	**0**	**0**
	Total intake		**2,175**	**1,080**	**1,080**	**4,335**
	(mL/kg)		(22.4)	(11.1)	(11.1)	(44.6)
OUT	Urine		**1,100**	**450**	**525**	**2,075**
	(mL/kg/hr)		(1.41)	(0.58)	(0.67)	(0.89)
	Emesis output		**120**	**0**	**0**	**120**
	Other		**0**	**0**	**0**	**0**
	Stool		**0**	**X1**	**0**	
	Total output		**1,220**	**450**	**525**	**2,195**
	(mL/kg)		(12.5)	(4.6)	(5.4)	(22.6)
Net I/O			**+955**	**+630**	**+555**	**+2,140**
Net since admission (4/12)			**+955**	**+1,585**	**+2,140**	**+2,140**

Laboratory Results

	Ref. Range	4/12 1780	4/13 1522
Chemistry			
Sodium (mEq/L)	136–145	132 !↓	134 !↓
Potassium (mEq/L)	3.5–5.5	3.9	4.0
Chloride (mEq/L)	95–105	101	100
Carbon dioxide (CO_2, mEq/L)	23–30	27	28
BUN (mg/dL)	8–18	31 !↑	20 !↑
Creatinine serum (mg/dL)	0.6–1.2	1.9 !↑	1.3 !↑
Glucose (mg/dL)	70–110	1524 !↑	475 !↑
Phosphate, inorganic (mg/dL)	2.3–4.7	1.8 !↓	2.1 !↓
Magnesium (mg/dL)	1.8–3	1.9	2.1
Calcium (mg/dL)	9–11	10	9.8
Osmolality (mmol/kg/H_2O)	285–295	360 !↑	304 !↑
Bilirubin total (mg/dL)	≤1.5	0.9	

Fagan, Mitchell, Male, 53 y.o.
Allergies: NKA
Pt. Location: MICU Bed #5

Code: FULL
Physician: R. Petersen

Isolation: None
Admit Date: 4/12

Laboratory Results *(Continued)*

	Ref. Range	4/12 1780	4/13 1522
Bilirubin, direct (mg/dL)	<0.3	0.019	
Protein, total (g/dL)	6–8	7.1	
Albumin (g/dL)	3.5–5	4.9	
Prealbumin (mg/dL)	16–35	33	
Ammonia (NH_3, µmol/L)	9–33	15	
Alkaline phosphatase (U/L)	30–120	112	
ALT (U/L)	4–36	21	
AST (U/L)	0–35	17	
CPK (U/L)	30–135 F 55–170 M	145	
Lactate dehydrogenase (U/L)	208–378	275	
Cholesterol (mg/dL)	120–199	205 !↑	
HDL-C (mg/dL)	>55 F, >45 M	55	
LDL (mg/dL)	<130	123	
LDL/HDL ratio	<3.22 F <3.55 M	2.26	
Triglycerides (mg/dL)	35–135 F 40–160 M	185 !↑	
T_4 (µg/dL)	4–12	12	
T_3 (µg/dL)	75–98	77	
HbA$_{1C}$ (%)	3.9–5.2	15.2 !↑	
C-peptide (ng/mL)	0.51–2.72	1.10	
ICA	—	–	
GADA	—	–	
IA-2A	—	–	
IAA	—	–	
TTG	—	–	
Hematology			
WBC (×10^3/mm^3)	4.8–11.8	13.5 !↑	
RBC (×10^6/mm^3)	4.2–5.4 F 4.5–6.2 M	6.1	
Hemoglobin (Hgb, g/dL)	12–15 F 14–17 M	14.5	
Hematocrit (Hct, %)	37–47 F 40–54 M	57 !↑	

(Continued)

Fagan, Mitchell, Male, 53 y.o.
Allergies: NKA
Pt. Location: MICU Bed #5

Code: FULL
Physician: R. Petersen

Isolation: None
Admit Date: 4/12

Laboratory Results *(Continued)*

	Ref. Range	4/12 1780	4/13 1522
Urinalysis			
Collection method	—	clean catch	
Color	—	yellow	
Appearance	—	clear	
Specific gravity	1.003–1.030	1.045 !↑	
pH	5–7	5.0 !↓	
Protein (mg/dL)	Neg	10 !↑	
Glucose (mg/dL)	Neg	+ !↑	
Ketones	Neg	+ !↑	
Blood	Neg	Neg	
Bilirubin	Neg	Neg	
Urobilinogen (EU/dL)	<1.1	Neg	
Leukocyte esterase	Neg	Neg	
Prot chk	Neg	+ !↑	
WBCs (/HPF)	0–5	3–4	

Case Questions

I.   Understanding the Diagnosis and Pathophysiology

1. What are the standard diagnostic criteria for T2DM? Which are found in Mitch's medical record?

2. Mitch was previously diagnosed with T2DM. He admits that he often does not take his medications. What types of medications are metformin and glyburide? Describe their mechanisms as well as their potential side effects/drug–nutrient interactions.

3. What other medications does Mitch take? List their mechanisms and potential side effects/ drug–nutrient interactions.

4. Describe the metabolic events that led to Mitch's symptoms and subsequent admission to the ER with the diagnosis of uncontrolled T2DM with HHS.

5. HHS and DKA are the common metabolic complications associated with diabetes. Discuss each of these clinical emergencies. Describe the information in Mitch's chart that supports the diagnosis of HHS.

6. HHS is often associated with dehydration. After reading Mitch's chart, list the data that are consistent with dehydration. What factors in Mitch's history may have contributed to his dehydration?

7. Assess Mitch's intake/output record for the first 24 hours of his admission. What does this tell you? Assuming that Mitch tells you that his usual weight is 228 lbs, can you estimate the volume of his dehydration?

8. Mitch was started on normal saline with potassium as well as an insulin drip. Why are these fluids a component of his rehydration and correction of the HHS?

9. Describe the insulin therapy that was started for Mitch. What is Lispro? What is glargine? How likely is it that Mitch will need to continue insulin therapy?

II. Understanding the Nutrition Therapy

10. Mitch was NPO when admitted to the hospital. What does this mean? What are the signs that will alert the RD and physician that Mitch may be ready to eat?

11. Outline the basic principles for Mitch's nutrition therapy to assist in control of his DM.

III. Nutrition Assessment

12. Assess Mitch's weight and BMI. What would be a healthy weight range for Mitch?

13. Identify and discuss any abnormal laboratory values measured upon his admission. How did they change after hydration and initial treatment of his HHS?

14. Determine Mitch's energy and protein requirements for weight maintenance. What energy and protein intakes would you recommend to assist with weight loss?

IV. Nutrition Diagnosis

15. Prioritize two nutrition problems and complete the PES statement for each.

V. Nutrition Intervention

16. Determine Mitch's initial CHO prescription using his diet history as well as your assessment of his energy requirements.

17. Identify two initial nutrition goals to assist with weight loss.

18. Mitch also has hypertension and high cholesterol levels. Describe how your nutrition interventions for diabetes can include nutrition therapy for his other conditions.

VI. Nutrition Monitoring and Evaluation

19. Write an ADIME note for your initial nutrition assessment.

Bibliography

American Diabetes Association. Standards of Medical Care in Diabetes—2013. *Diabetes Care.* 2013;36:S11–S66.

Chaithongdi N, Subauste JS, Koch CA, Geraci SA. Diagnosis and management of hyperglycemic emergencies. *Hormones.* 2011;10:250–260.

Hemphill RR. Hyperosmolar Hyperglycemic State. Available from: http://emedicine.medscape.com/article/1914705-overview. Accessed 8/24/12.

Kitabchi AE, Umpierrex GE, Fisher JN, Murph Stentz FB. Thirty years of personal experience in hyperglycemic crises: Diabetic ketoacidosis and hyperglycemic hyperosmolar state. *J Clin Endocrinol Metab.* 2008;93:1541–52.

McNaughten CD, Self WH, Slovis C. Diabetes in the emergency department: Acute care of diabetes patients. *Clinical Diabetes.* 2011;29:51–59.

Nyenwe EA, Kitabchi AE. Evidence-based management of hyperglycemic emergencies in diabetes mellitus. *Diabetes Res Clin Prac.* 2011;94:340–351.

Roth S. Diseases of the endocrine system. In: Nelms M, Sucher K, Lacey K, Roth SL. *Nutrition Therapy and Pathophysiology.* 2nd ed. Belmont, CA: Wadsworth, Cengage Learning; 2011:471–519.

Internet Resources

ADA Evidence Analysis Library: http://www.adaevidencelibrary.com

American Diabetes Association: http://www.diabetes.org/

Unit Seven

NUTRITION THERAPY FOR RENAL DISORDERS

There has been a noticeable growth in the field of medical nutrition therapy for patients with chronic kidney disease (CKD). The importance of nutrition in the care of patients with CKD is illustrated by the fact that indicators of nutritional status effectively predict morbidity and mortality in these patients. In 2011, over 20 million Americans had clinical evidence of kidney disease. Of this number, more than 85,000 Americans die each year because of kidney disease, and more than 485,000 suffer from advanced CKD and need renal replacement therapy to remain alive. Kidney disease is one of the costliest illnesses. In 2011, more than $42.5 billion was spent on renal replacement therapy (available at: http://kidney.niddk.nih.gov/KUDiseases/pubs/kustats/index.aspx#1, accessed December 12, 2012).

The primary cause of CKD is diabetes mellitus, which accounts for about 44 percent of all new cases each year. Uncontrolled hypertension is the second leading cause of CKD in the United States.

Cases 18 and 19 represent the progression of renal disease and the most common forms of renal replacement therapy—hemodialysis and peritoneal dialysis. Integrated into each case are factors that predispose an individual to CKD, such as diabetes mellitus and ethnicity. Incidence in African Americans is three times the rate seen in Caucasians. Researchers from the National Institutes of Health have established that CKD caused by diabetes mellitus is anywhere from 10 to 75 times more prevalent in Native Americans than in Caucasians, and the prevalence differs among tribes; 50 percent of Pima Indians age 35 years and over have type 2 diabetes mellitus—the highest rate in the world. Fundamental principles such as modification of nutrient composition in impaired renal function, CKD, and renal replacement therapy are included within these cases. Case 20 focuses on acute kidney injury (AKI). There are a large number of critically ill patients who experience AKI as a complication of their illness. You will use nutrition support and evidenced-based guidelines to determine the appropriate nutritional care for this patient.

Chronic Kidney Disease (CKD) Treated with Dialysis

Objectives

After completing this case, the student will be able to:

1. Describe the pathophysiology of chronic kidney disease (CKD).
2. Describe the stages of CKD.
3. Differentiate the mechanisms of peritoneal dialysis and hemodialysis.
4. Identify and explain common nutritional problems associated with CKD.
5. Interpret laboratory parameters for nutritional implications and significance.
6. Analyze nutrition assessment data to evaluate nutritional status and identify specific nutrition problems.
7. Determine nutrition diagnoses and write appropriate PES statements.
8. Develop a nutrition care plan—with appropriate measurable goals, interventions, and strategies for monitoring and evaluation—that addresses the nutrition diagnoses of this case.
9. Integrate sociocultural and ethnic food consumption issues within a nutrition care plan.
10. Make appropriate documentation in the medical record.

Enez Joaquin is a 24-year-old Pima Indian who has had type 2 diabetes mellitus since age 13. Mrs. Joaquin has experienced a declining glomerular filtration rate for the past two years. She is being admitted in preparation for kidney replacement therapy.

Joaquin, Enez, Female, 24 y.o.
Allergies: NKA   **Code:** FULL   **Isolation:** None
Pt. Location: RM 207   **Physician:** L. Nila   **Admit Date:** 3/5

Patient Summary: Mrs. Joaquin is a 24-yo Native American woman who was diagnosed with type 2 DM when she was 13 years old and has been poorly compliant with prescribed treatment.

History:

Onset of disease: Diagnosed with Stage 3 chronic kidney disease 2 years ago. Her acute symptoms have developed over the last 2 weeks.

Medical history: Gravida 1/ para 1. Infant weighed 10 lbs at birth 7 years ago. Pt admits she recently stopped taking a prescribed hypoglycemic agent, and she has never filled her prescription for antihypertensive medication. Progressive decompensation of kidney function has been documented by declining GFR, increasing creatinine and urea concentrations, elevated serum phosphate, and normochromic, normocytic anemia. She is being admitted for preparation for kidney-replacement therapy.

Surgical history: No surgeries

Medications at home: Glucophage (metformin), 850 mg twice daily

Tobacco use: No

Alcohol use: Yes, 12 oz beer daily

Family history: What? T2DM. Who? Parents

Demographics:

Marital status: Married—lives with husband and daughter; *Spouse name:* Eddie
Number of children: 1
Years education: High school
Language: English and Akimel O'odham (Pima)
Occupation: Secretary
Hours of work: 9–5
Ethnicity: Pima Indian
Religious affiliation: Catholic

Admitting History/Physical:

Chief complaint: Pt complains of anorexia; N/V; 4-kg weight gain in the past 2 weeks; edema in extremities, face, and eyes; malaise; progressive SOB with 3-pillow orthopnea; pruritus; muscle cramps; and inability to urinate

General appearance: Overweight Native American female who appears her age; lethargic, complaining of N/V.

Vital Signs: Temp: 98.6   Pulse: 86   Resp rate: 25
                          BP: 220/80   Height: 5'0"   Weight: 170 lbs

Heart: S4, S1, and S2, regular rate and rhythm. I/VI systolic ejection murmur, upper-left sternal border.
HEENT: Head: Normocephalic, equal carotid pulses, neck supple, no bruits
             Eyes: PERRLA
             Ears: Noncontributory

Joaquin, Enez, Female, 24 y.o.
Allergies: NKA
Pt. Location: RM 207

Code: FULL
Physician: L. Nila

Isolation: None
Admit Date: 3/5

Nose: Noncontributory
Throat: Noncontributory
Genitalia: Normal female
Neurologic: Oriented to person, place, and time; intact, mild asterixis
Extremities: Muscle weakness; 3+ pitting edema to the knees, no cyanosis
Skin: Dry and yellowish-brown
Chest/lungs: Generalized rhonchi with rales that are mild at the bases (Pt breathes with poor effort)
Peripheral vascular: Normal pulse (3+) bilaterally L
Abdomen: Bowel sounds positive, soft; generalized mild tenderness; no rebound

Nursing Assessment	3/5
Abdominal appearance (concave, flat, rounded, obese, distended)	rounded, obese
Palpation of abdomen (soft, rigid, firm, masses, tense)	soft
Bowel function (continent, incontinent, flatulence, no stool)	continent
Bowel sounds (P=present, AB=absent, hypo, hyper)	
RUQ	P
LUQ	P
RLQ	P
LLQ	P
Stool color	brown
Stool consistency	formed
Tubes/ostomies	N/A
Genitourinary	
Urinary continence	N/A
Urine source	N/A
Appearance (clear, cloudy, yellow, amber, fluorescent, hematuria, orange, blue, tea)	N/A
Integumentary	
Skin color	light brown
Skin temperature (DI=diaphoretic, W=warm, dry, CL=cool, CLM=clammy, CD += cold, M=moist, H=hot)	W
Skin turgor (good, fair, poor, TENT=tenting)	good
Skin condition (intact, EC=ecchymosis, A=abrasions, P=petechiae, R=rash, W=weeping, S=sloughing, D=dryness, EX=excoriated, T=tears, SE=subcutaneous emphysema, B=blisters, V=vesicles, N=necrosis)	intact, A

(Continued)

Joaquin, Enez, Female, 24 y.o.
Allergies: NKA   **Code:** FULL   **Isolation:** None
Pt. Location: RM 207   **Physician:** L. Nila   **Admit Date:** 3/5

Nursing Assessment *(Continued)*

Nursing Assessment	3/5
Mucous membranes (intact, EC=ecchymosis, A=abrasions, P=petechiae, R=rash, W=weeping, S=sloughing, D=dryness, EX=excoriated, T=tears, SE=subcutaneous emphysema, B=blisters, V=vesicles, N=necrosis)	intact
Other components of Braden score: special bed, sensory pressure, moisture, activity, friction/shear (>18 = no risk, 15–16 = low risk, 13–14 = moderate risk, ≤12 = high risk)	activity, 16

Orders:
Evaluate for kidney replacement therapy
Capoten/captopril 25 mg twice daily
Erythropoietin (r-HuEPO) 30 units/kg
Sodium bicarbonate 2 g daily
Renal caps—1 daily
Renvela—three times daily with each meal
Hectorol 2.5 μg four times daily 3 times/week
Glucophage (metformin) 850 mg twice daily
35 kcal/kg, 1.2 g protein/kg, 2 g K, 1 g phosphorus, 2 g Na, 1000 mL fluid + urine output per day
CBC, chemistry
Stool softener
Occult fecal blood
Nutrition consult

Nutrition:
History: Intake has been poor due to anorexia, N&V. Patient states that she tried to follow the diet that she was taught two years ago. "It went pretty well for a while, but it was hard to keep up with."
Usual dietary intake:
Breakfast: Cold cereal
             Bread or fried potatoes
             Fried egg (occasionally)
Lunch: Bologna sandwich
             Potato chips
             Coke
Dinner: Chopped meat
             Fried potatoes
Snacks: Crackers and peanut butter
Food allergies/intolerances/aversions: None

Joaquin, Enez, Female, 24 y.o.

Allergies: NKA	**Code:** FULL	**Isolation:** None
Pt. Location: RM 207	**Physician:** L. Nila	**Admit Date:** 3/5

Previous nutrition therapy? Yes. If yes, when: 2 years ago when Pt Dx with Stage 3 chronic kidney disease. Where? Reservation Health Service.
Current diet: Low simple sugar, 0.8 g protein/kg, 2–3 g Na
Food purchase/preparation: Self
Vit/min intake: None

Intake/Output

Date		3/5 0701–3/6 0700				3/6 0701–3/7 0700			
Time		0701–1500	1501–2300	2301–0700	Daily total	0701–1500	1501–2300	2301–0700	Daily total
IN	P.O.	0	50	0	50	NPO	NPO	NPO	NPO
	I.V. (mL/kg/hr)								
	I.V. piggyback								
	TPN								
	Total intake (mL/kg)	0 (0)	50 (0.7)	0 (0)	50 (0.7)	0 (0)	0 (0)	0 (0)	0 (0)
OUT	Urine (mL/kg/hr)	0 (0)	100 (0.171)	0 0	100 (0.057)	200 (0.341)	800 (1.365)	0 (0)	1000 (0.569)
	Emesis output	0	50	0	50	100	50	0	150
	Other								
	Stool						×1		
	Total output (mL/kg)	0 (0)	150 (2.0)	0 (0)	150 (2.0)	300 (4.1)	850 (11.6)	0 (0)	1150 (15.7)
Net I/O		0	−100	0	−100	−300	−850	0	−1150
Net since admission (3/5)		0	−100	−100	−100	−400	−1250	−1250	−1250

Laboratory Results

	Ref. Range	3/5
Chemistry		
Sodium (mEq/L)	136–145	130 !↓
Potassium (mEq/L)	3.5–5.5	5.8 !↑

(Continued)

Joaquin, Enez, Female, 24 y.o.
Allergies: NKA   **Code:** FULL   **Isolation:** None
Pt. Location: RM 207   **Physician:** L. Nila   **Admit Date:** 3/5

Laboratory Results *(Continued)*

	Ref. Range	3/5
Chloride (mEq/L)	95–105	91 !↓
Carbon dioxide (CO_2, mEq/L)	23–30	20 !↓
BUN (mg/dL)	8–18	69 !↑
Creatinine serum (mg/dL)	0.6–1.2	12.0 !↑
Glucose (mg/dL)	70–110	282 !↑
Phosphate, inorganic (mg/dL)	2.3–4.7	9.5 !↑
Magnesium (mg/dL)	1.8–3	2.9
Calcium (mg/dL)	9–11	8.2 !↓
Albumin (g/dL)	3.5–5	3.7
Prealbumin (mg/dL)	16–35	20
Alkaline phosphatase (U/L)	30–120	90
ALT (U/L)	4–36	26
AST (U/L)	0–35	28
Lactate dehydrogenase (U/L)	208–378	315
Cholesterol (mg/dL)	120–199	220 !↑
HDL-C (mg/dL)	>55 F, >45 M	50 !↓
LDL (mg/dL)	<130	135 !↑
LDL/HDL ratio	<3.22 F <3.55 M	2.7
Triglycerides (mg/dL)	35–135 F 40–160 M	200 !↑
HbA_{1c} (%)	3.9–5.2	8.9 !↑
Urinalysis		
Collection method	—	Random specimen
Color	—	Straw
Appearance	—	Slightly hazy
Specific gravity	1.003–1.030	1.025
pH	5–7	7.9 !↑
Protein (mg/dL)	Neg	2+ !↑
WBCs (/HPF)	0–5	20 !↑

Case Questions

I.   Understanding the Disease and Pathophysiology

1. Describe the physiological functions of the kidneys.

2. What diseases/conditions can lead to chronic kidney disease (CKD)? Explain the relationship between diabetes and CKD.

3. Outline the stages of CKD, including the distinguishing signs and symptoms.

4. From your reading of Mrs. Joaquin's history and physical, what signs and symptoms did she have that correlate with her chronic kidney disease?

5. What are the treatment options for Stage 5 CKD? Explain the differences between hemodialysis and peritoneal dialysis.

II.   Understanding the Nutrition Therapy

6. Explain the reasons for the following components of Mrs. Joaquin's medical nutrition therapy:

Nutrition Therapy	Rationale
35 kcal/kg	
1.2 g protein/kg	
2 g K	
1 g phosphorus	
2 g Na	
1000 mL fluid + urine output	

III.   Nutrition Assessment

7. Calculate and interpret Mrs. Joaquin's BMI. How does edema affect your interpretation?

8. What is edema-free weight? Calculate Mrs. Joaquin's edema-free weight.

9. What are the energy requirements for CKD?

10. Calculate what Mrs. Joaquin's energy needs will be once she begins hemodialysis.

11. What will Mrs. Joaquin's protein requirements be when she begins hemodialysis? What standard guidelines have you used to make these recommendations?

12. What are the considerations for differences in protein requirements among predialysis, hemodialysis, and peritoneal dialysis patients?

13. Mrs. Joaquin has a PO_4 restriction. Why? What foods have the highest levels of phosphorus?

14. Mrs. Joaquin tells you that one of her friends can drink only certain amounts of liquids and wants to know if that is the case for her. What foods are considered to be fluids? What recommendations can you make for Mrs. Joaquin? If a patient must follow a fluid restriction, what can be done to help reduce his or her thirst?

15. Several biochemical indices are used to diagnose chronic kidney disease. One is glomerular filtration rate (GFR). What does GFR measure? What is a normal GFR? Mrs. Joaquin's GFR is 28 mL/min. Interpret her value.

16. Evaluate Mrs. Joaquin's chemistry report. What labs support the diagnosis of Stage 5 CKD?

17. Which of Mrs. Joaquin's other symptoms would you expect to begin to improve?

18. Explain why the following medications were prescribed by completing the following table.

Medication	Indications/Mechanism	Nutritional Concerns
Capoten/ captopril		
Erythropoietin		
Sodium bicarbonate		
Renal caps		
Renvela		
Hectorol		
Glucophage		

19. What health problems have been identified in the Pima Indians through epidemiological data? Explain what is meant by the "thrifty gene" theory. Are the Pima at higher risk for complications of diabetes? Explain.

IV. Nutrition Diagnosis

20. Choose two high-priority nutrition problems and complete a PES statement for each.

V. Nutrition Intervention

21. For each PES statement, establish an ideal goal (based on the signs and symptoms) and appropriate intervention (based on the etiology).

22. Why is it recommended for patients to have at least 50% of their protein from sources that have high biological value?

23. What resources would you use to teach Mrs. Joaquin about her diet?

24. Using Mrs. Joaquin's typical intake and the prescribed diet, write a sample menu. Make sure you can justify your changes and that it is consistent with her nutrition prescription.

Diet PTA		Sample Menu
Breakfast:	Cold cereal (¾ c unsweetened)	
	Bread (2 slices) or fried potatoes (1 medium potato)	
	1 fried egg (occasionally)	
Lunch:	Bologna sandwich (2 slices white bread, 2 slices bologna, mustard)	
	Potato chips (1 oz)	
	1 can Coke	
Dinner:	Chopped meat (3 oz beef)	
	Fried potatoes (1½ medium)	
HS Snack:	Crackers (6 saltines) and peanut butter (2 tbsp)	

25. After evaluating Mrs. Joaquin's typical diet, what other recommendations can you make?

26. Write an initial ADIME note for your consultation with Mrs. Joaquin.

Bibliography

Academy of Nutrition and Dietetics Evidence Analysis Library. Chronic Kidney Disease. Executive Summary of Recommendations. http://andevidencelibrary.com/topic.cfm?cat=3929. Accessed 10/20/12.

American Nutrition and Dietetics *Nutrition Care Manual*. http://www.nutritioncaremanual.org/topic.cfm?ncm_heading=Diseases%2FConditions&ncm_toc_id=23081. Accessed 10/17/12.

Brown TL. Ethnic populations. In: Ross TA, Boucher JL, O'Connell BS. *American Dietetic Association Guide to Diabetes Medical Nutrition Therapy and Education*. Chicago, IL: American Dietetic Association; 2005:227–238.

Lacey K, Nahikian-Nelms M. Diseases of the renal system. In: Nelms M, Sucher K, Lacey K, Roth SL. *Nutrition Therapy and Pathophysiology*. 2nd ed. Belmont, CA: Wadsworth, Cengage Learning; 2011:520–561.

National Institute of Diabetes and Digestive and Kidney Diseases (NIDDK). Diabetes in American Indians and Alaska Natives. http://diabetes.niddk.nih.gov/dm/pubs/americanindian/. Accessed 10/15/12.

National Kidney Foundation. K/DOQI Clinical Practice Guidelines for Chronic Kidney Disease: Evaluation, Classification and Stratification. *Am J Kidney Dis*. 2002(Suppl 1):39:S1–S266.

Nelms MN. Nutrition assessment: Foundations of the nutrition care process. In: Nelms M, Sucher K, Lacey K, Roth SL. *Nutrition Therapy and Pathophysiology*. 2nd ed. Belmont, CA: Wadsworth, Cengage Learning; 2011:34–65.

Nelson RG, Bennett PH, Beck GJ, et al. Development and progression of renal disease in Pima Indians with non-insulin dependent diabetes mellitus. *N Engl J Med*. 1996;335:1636–1642.

Pronsky ZM. *Food-Medication Interactions*. 18th ed. Birchrunville, PA: Food-Medication Interactions; 2012.

The *Pima Indians: Pathfinders for Health*. http://diabetes.niddk.nih.gov/dm/pubs/pima/. Accessed 10/20/12.

Internet Resources

American Association of Kidney Patients: http://www.aakp.org/

Cook's Thesaurus: http://www.foodsubs.com/

Culinary Kidney Cooks: http://www.culinarykidneycooks.com/

eMedicineHealth: http://www.emedicinehealth.com/chronic_kidney_disease/article_em.htm

Kidney School: http://kidneyschool.org/

National Institute of Diabetes and Digestive and Kidney Diseases (NIDDK): http://www2.niddk.nih.gov/

National Kidney Foundation Council on Renal Nutrition: http://www.kidney.org/professionals/CRN/index.cfm

National Kidney Foundation KDOQI Guidelines: http://www.kidney.org/professionals/kdoqi/guidelines_ckd/toc.htm

The Nephron Information Center: http://www.nephron.com/

Renal Dialysis . . . A Team Effort: http://www.nufs.sjsu.edu/renaldial/index.html

Renal Web: http://www.renalweb.com/

United States Renal Data System (USRDS): http://www.usrds.org/

Case 19

Chronic Kidney Disease: Peritoneal Dialysis

Objectives

After completing this case, the student will be able to:

1. Describe the pathophysiology of chronic kidney disease (CKD).
2. Describe the basic concepts of kidney transplant.
3. Differentiate the physiology of peritoneal dialysis and hemodialysis.
4. Identify and explain common nutritional problems associated with CKD.
5. Interpret laboratory parameters for nutritional implications and significance.
6. Analyze nutrition assessment data to evaluate nutritional status, identify specific nutrition problems, and document corresponding PES statements.
7. Develop a nutrition care plan—with appropriate measurable goals, interventions, and strategies for monitoring and evaluation—that addresses the nutrition diagnoses of this case.

Mrs. Caldwell is a 49-year-old female who has a history of membranoproliferative glomerulonephritis. She has a history of kidney transplant ×2 and is now experiencing acute rejection with chronic allograft nephropathy. Her creatinine level has been rising with worsening of fluid overload. She is admitted to the hospital for further evaluation.

Caldwell, Mona, Female, 49 y.o.

Allergies: fish	**Code:** FULL	**Isolation:** None
Pt. Location: 789	**Physician:** K. Wolf	**Admit Date:** 1/21

Patient Summary: Patient is a 49-year-old female who has history of membranoproliferative glomerulonephritis resulting in kidney transplant ×2 and is now experiencing acute rejection with chronic allograft nephropathy. She has previously been treated with both hemodialysis and peritoneal dialysis before and between transplants. She is requesting to restart peritoneal dialysis as she would like to continue to work at the U.S. Postal Service. She is admitted for insertion of PD catheter and plans to use cycler at night so that she can continue to work.

History:

Onset of disease: Diagnosed with glomerulonephritis 15 years ago with resulting end-stage renal disease. Initiated hemodialysis at that time and received first transplant 1 year later. Acute rejection resulted in initiation of peritoneal dialysis. She received second transplant 2 years ago.

Medical history: Membranoproliferative glomerulonephritis; allograft transplant ×2; hypertension; dyslipidemia; anemia of chronic kidney disease

Surgical history: s/p unilateral oophorectomy; allograft kidney transplant ×2; umbilical hernia repair

Medications at home: Procardia, carvedilol, Catapres (clonidine), CellCept, fish oil, Lasix, prednisone, Gengraf, Prinivil, sodium bicarbonate, calcitriol, renal caps, Renvela

Tobacco use: 1 pack per day, 20 years of use; stopped tobacco use 5–6 years ago

Alcohol use: None

Family history: Mother—cervical cancer; father—lung cancer

Demographics:

Marital status: Married

Years education: 14+

Language: English only

Occupation: Postal clerk

Hours of work: 8–5

Household members: Lives with husband

Ethnicity: Caucasian

Religious affiliation: None

Admitting History/Physical:

Chief complaint: "They tell me that my transplant is failing—I am very short of breath and I have a lot of fluid—I am here to have a catheter placed so I can start dialysis again."

General appearance: Well-developed female in no acute distress; alert and oriented ×3.

Vital Signs:

Temp: 98.4	Pulse: 62	Resp rate: 12	
BP: 161/92	Height: 157.4 cm	Weight: 77.1 kg	UBW: ~74 kg

Heart: RRR. No clicks, rubs, murmurs, or gallops noted

HEENT: Head: WNL

Caldwell, Mona, Female, 49 y.o.

Allergies: fish	**Code:** FULL	**Isolation:** None
Pt. Location: 789	**Physician:** K. Wolf	**Admit Date:** 1/21

Eyes: PERRLA
Ears: Clear
Nose: Dry mucous membranes
Throat: Dry mucous membranes

Genitalia: Deferred
Neurologic: Alert and oriented
Extremities: WNL
Skin: Warm and dry
Chest/lungs: Respirations are rapid but clear to auscultation and percussion
Peripheral vascular: No peripheral edema noted
Abdomen: Soft—no incisional hernias; signs of prior kidney transplants and umbilical hernia repair noted

Nursing Assessment	1/22
Abdominal appearance (concave, flat, rounded, obese, distended)	rounded
Palpation of abdomen (soft, rigid, firm, masses, tense)	soft
Bowel function (continent, incontinent, flatulence, no stool)	continent
Bowel sounds (P=present, AB=absent, hypo, hyper)	
RUQ	P
LUQ	P
RLQ	P
LLQ	P
Stool color	none
Stool consistency	
Tubes/ostomies	NA
Genitourinary	
Urinary continence	yes
Urine source	clean catch
Appearance (clear, cloudy, yellow, amber, fluorescent, hematuria, orange, blue, tea)	cloudy, amber
Integumentary	
Skin color	pale
Skin temperature (DI=diaphoretic, W=warm, dry, CL=cool, CLM=clammy, CD +=cold, M=moist, H=hot)	W
Skin turgor (good, fair, poor, TENT=tenting)	good
Skin condition (intact, EC=ecchymosis, A=abrasions, P=petechiae, R=rash, W=weeping, S=sloughing, D=dryness, EX=excoriated, T=tears, SE=subcutaneous emphysema, B=blisters, V=vesicles, N=necrosis)	intact

(Continued)

Caldwell, Mona, Female, 49 y.o.
Allergies: fish
Pt. Location: 789

Code: FULL
Physician: K. Wolf

Isolation: None
Admit Date: 1/21

Nursing Assessment *(Continued)*

Nursing Assessment	1/22
Mucous membranes (intact, EC=ecchymosis, A=abrasions, P=petechiae, R=rash, W=weeping, S=sloughing, D=dryness, EX=excoriated, T=tears, SE=subcutaneous emphysema, B=blisters, V=vesicles, N=necrosis)	intact, D
Other components of Braden score: special bed, sensory pressure, moisture, activity, friction/shear (>18=no risk, 15–16=low risk, 13–14=moderate risk, ≤12=high risk)	20

Admission Orders:

Laboratory: CBC, Chemistry
Vital Signs: Every 4 hrs
I & O: Recorded every 8 hrs
Diet: 1500 kcal, 75 g pro, 3000 mg Na, 3500 mg K, 1000 mg P, 2000 cc fluid
Activity: Bed rest

Nutrition:

Meal type: 1500 kcal, 75 g pro, 3000 mg Na, 3500 mg K, 1000 mg P, 2000 cc fluid
History: Patient states she has noticed her appetite has not been as good lately. Describes mild nausea but no vomiting. Relates that food has a bad taste. Since her transplant she has only monitored her salt intake—no other restrictions.

24-hour recall:
AM: Egg McMuffin™, 6 oz orange juice, coffee
Lunch: Cheeseburger, fries, apple pie, and 12 oz Coke
Dinner: Roast beef au jus; 3 small oven-browned potatoes, ½ c broccoli, iced tea, roll with
          butter

Laboratory Results

	Ref. Range	1/21 0830
Chemistry		
Sodium (mEq/L)	136–145	130 !↓
Potassium (mEq/L)	3.5–5.5	3.8
Chloride (mEq/L)	95–105	96
Carbon dioxide (CO₂, mEq/L)	23–30	26
Bicarbonate (mEq/L)	21–32	16 !↓
BUN (mg/dL)	8–18	124 !↑
Creatinine serum (mg/dL)	0.6–1.2	6.8 !↑

(Continued)

Laboratory Results *(Continued)*

	Ref. Range	1/21 0830
BUN/Crea ratio	10.0–20.0	18.2
Est GFR, non-Afr Amer (mL/min/1.73 m^2)	61–589	6 !↓
Est GFR, Afr-Amer (mL/min/1.73 m^2)	61–714	8 !↓
Glucose (mg/dL)	70–110	80
Phosphate, inorganic (mg/dL)	2.3–4.7	11.9 !↑
Magnesium (mg/dL)	1.8–3	1.9
Calcium (mg/dL)	9–11	8.3 !↓
Anion gap (mmol/L)	10–20	22 !↑
Protein, total (g/dL)	6–8	5.9 !↓
Albumin (g/dL)	3.5–5	3.4 !↓
Prealbumin (mg/dL)	16–35	17
Ammonia (NH$_3$, µmol/L)	9–33	11
Alkaline phosphatase (U/L)	30–120	106
ALT (U/L)	4–36	5
AST (U/L)	0–35	9
CPK (U/L)	30–135 F 55–170 M	35
Lactate dehydrogenase (U/L)	208–378	209
Cholesterol (mg/dL)	120–199	154
HDL-C (mg/dL)	>55 F, >45 M	56
VLDL (mg/dL)	7–32	15
LDL (mg/dL)	<130	83
LDL/HDL ratio	<3.22 F <3.55 M	1.48
Triglycerides (mg/dL)	35–135 F 40–160 M	77
Coagulation (Coag)		
PT (sec)	12.4–14.4	16.9 !↑
INR	0.9–1.1	1.4 !↑
Hematology		
WBC (×10^3/mm^3)	4.8–11.8	8.27
RBC (×10^6/mm^3)	4.2–5.4 F 4.5–6.2 M	2.33 !↓

(Continued)

Caldwell, Mona, Female, 49 y.o.
Allergies: fish
Pt. Location: 789

Code: FULL
Physician: K. Wolf

Isolation: None
Admit Date: 1/21

Laboratory Results *(Continued)*

	Ref. Range	1/21 0830
Hemoglobin (Hgb, g/dL)	12–15 F 14–17 M	6.6 !↓
Hematocrit (Hct, %)	37–47 F 40–54 M	19.0 !↓
Mean cell volume (μm³)	80–96	65.3 !↓
Mean cell Hgb (pg)	26–32	21.5 !↓
Mean cell Hgb content (g/dL)	31.5–36	19.5 !↓
RBC distribution (%)	11.6–16.5	16.8 !↑
Platelet count (×10³/mm³)	140–440	274
Transferrin (mg/dL)	250–380 F 215–365 M	219 !↓
Ferritin (mg/mL)	20–120 F 20–300 M	5 !↓

Intake/Output

Date		1/22 0701–1/23 0700			
Time		0701–1500	1501–2300	2301–0700	Daily total
IN	P.O.	650	400	750	1800
	I.V. (mL/kg/hr)	0	0	0	0
	I.V. piggyback				
	TPN				
	Total intake (mL/kg)	650 (8.4)	400 (5.19)	750 (9.74)	1800 (23.37)
OUT	Urine (mL/kg/hr)	812 (1.32)	410 (0.66)	310 (0.50)	1532 (0.83)
	Emesis output				
	Other				
	Stool	×1	0	0	0
	Total output (mL/kg)	812 (10.53)	410 (5.32)	310 (4.02)	1532 (19.9)
Net I/O		−162	−10	+440	+268
Net since admission (1/21)		−162	−172	+268	+268

Case Questions

I. Understanding the Diagnosis and Pathophysiology

1. Describe the major exocrine and endocrine functions of the kidney.

2. What is glomerulonephritis and how can it lead to kidney failure?

3. What laboratory values or other tests support Mrs. Caldwell's diagnosis of chronic kidney disease? List all abnormal values and explain the likely cause for each abnormal value.

4. This patient has had two previous kidney transplants. What are the potential sources for a donor kidney? How is rejection prevented after a kidney transplant? What does it mean when the physician states she is experiencing acute rejection?

5. Based on the admitting history and physical, what signs and symptoms does this patient have that are consistent with acute rejection of the transplant?

6. Mrs. Caldwell has requested that she restart peritoneal dialysis. Describe the basic concepts of this medical treatment and how it differs from hemodialysis.

II. Understanding the Nutrition Therapy

7. This patient was prescribed the following diet in the hospital:

   1500 kcal, 75 g pro, 3000 mg Na, 3500 mg K, 1000 mg P, 2000 cc fluid

   Explain the rationale for each component of her nutrition therapy Rx. How might this change once she has started peritoneal dialysis?

III. Nutrition Assessment

8. Assess Mrs. Caldwell's height and weight. Calculate her BMI and her % usual body weight. How would edema affect your interpretation of this information? Using the KDOQI guidelines, what is Mrs. Caldwell's adjusted body weight?

9. Determine Mrs. Caldwell's energy and protein requirements. Explain the rationale for the method you used to calculate these requirements.

10. List all medications that Mrs. Caldwell is receiving. Determine the action of each medication and identify any drug–nutrient interactions that you should monitor for.

11. Mrs. Caldwell's laboratory values that you discussed previously in this case indicate she has anemia. Why do renal patients suffer from anemia? How is this typically treated in dialysis patients?

12. What factors in Mrs. Caldwell's history may affect her ability to eat? What are the most likely causes of these symptoms? Can you expect that they will change?

13. Evaluate Mrs. Caldwell's diet history and 24-hour recall. Is her usual diet consistent with her inpatient diet order?

IV.   Nutrition Diagnosis

14. Identify the pertinent nutrition problems and the corresponding nutrition diagnoses.

15. Write a PES statement for each high-priority nutrition problem.

V.   Nutrition Intervention

Mrs. Caldwell was discharged from the hospital and was prescribed the following regimen of peritoneal dialysis to begin at home:

CCPD daily. Ca 2.50; Mg 0.5, Dextrose 2.5%. Total fills (or exchanges) = 3 (3 fills/cycle @2500 mL). Total fill volume/24 hours: 10000 mL.

16. Determine the amount of energy that Mrs. Caldwell's PD prescription will provide each day. How will this affect your nutrition recommendations?

17. Using the KDOQI adult guidelines for peritoneal dialysis patients, determine Mrs. Caldwell's nutrition prescription for outpatient use. (Include energy, protein, phosphorus, calcium, potassium, sodium, and fluid.)

18. Using the identified nutrition problems (and with the understanding that Mrs. Caldwell has received a significant amount of nutrition education in the past), what would you determine to be the most important topics for nutrition education when she returns to the PD clinic?

VI.   Nutrition Monitoring and Evaluation

19. List factors that you would monitor to assess Mrs. Caldwell's nutritional status when she returns to the PD clinic.

Bibliography

Kopple JD, Massry SG, eds. *Nutrition Management of Renal Disease.* 2nd ed. Philadelphia, PA: Lippincott Williams & Wilkins; 2004.

Lacey K, Nahikian-Nelms ML. Diseases of the renal system. In: Nelms M, Sucher K, Lacey K, Roth SL. *Nutrition Therapy and Pathophysiology.* 2nd ed. Belmont, CA: Wadsworth, Cengage Learning; 2011:520–565.

McCann L, ed. Pocket Guide to Nutrition Assessment of the Patient with Chronic Kidney Disease. 3rd ed. New York, NY: National Kidney Foundation Council on Renal Nutrition; 2002.

Moore L. Implications for Nutrition Practice in the Mineral-Bone Disorder of Chronic Kidney Disease. *Nutr Clin Pract.* 2011;26:391–400.

National Kidney Foundation. KDOQI Clinical Practice Guidelines. Available at: http://www.kidney.org /professionals/kdoqi/guidelines_commentaries.cfm

Internet Resources

ADA Evidence Analysis Library: http://www.adaevidencelibrary.com

American Association of Kidney Patients: http://www.aakp.org

International Society for Peritoneal Dialysis: http://www.ispd.org

National Institute of Diabetes, Digestive and Kidney Diseases of the National Institute of Health: http://www.niddk.nih.gov

National Kidney Foundation: http://www.kidney.org

Nutrition Care Manual: http://www.nutritioncaremanual.org

Case 20

Acute Kidney Injury (AKI)

Objectives

After completing this case, the student will be able to:

1. Define and describe the classifications of acute kidney injury (AKI) and its potential etiologies.
2. Describe the pathophysiology of AKI.
3. Identify and explain common nutritional problems associated with AKI.
4. Interpret laboratory parameters for nutritional implications and significance.
5. Analyze nutrition assessment data to evaluate nutritional status, identify specific nutrition problems, and document corresponding PES statements.

6. Develop a nutrition care plan—with appropriate measurable goals, interventions, and strategies for monitoring and evaluation—that addresses the nutrition diagnoses of this case.

Mr. Randall Maddox is a 67-year-old male admitted for a planned coronary bypass surgery 7 days ago. Postoperatively he experienced respiratory distress and infection, and now his urine output has suddenly decreased. It is now day 7 and a nutrition consult has been ordered for Mr. Maddox.

Maddox, Randall, Male, 67 y.o.
Allergies: penicillin, sulfa
Pt. Location: CICU

Code: FULL
Physician: C. Taylor

Isolation: Contact
Admit Date: 10/8

Patient Summary: Patient is a 67-year-old male who is s/p CABG on 10/9. Surgery was successful with 3-vessel bypass. He experienced a hypotensive event in recovery but responded to IV fluids. Further postoperative recovery has been complicated by respiratory distress, subsequent intubation, and infection.

History:
Onset of disease: Diagnosed with acute kidney injury postoperative day 7
Medical history: CAD, s/p MI 15 years ago, hyperlipidemia, and type 2 DM
Surgical history: s/p 3-vessel CABG 6 days ago
Medications at home: Lovostatin, Lasix, Lopressor
Tobacco use: 2 packs per day, 20 years of use; stopped tobacco use 1 year ago
Alcohol use: Socially
Family history: Mother—diabetes, breast cancer; father—heart disease, lung cancer

Demographics:
Marital status: Divorced
Years education: 14+
Language: English only
Occupation: Retired computer sales
Hours of work: N/A
Household members: Lives by self
Ethnicity: Caucasian
Religious affiliation: Methodist

MD Progress Note 10/16:
Patient is now POD#7. Experienced continued drop in urine output—+5295 mL since admission. Urine output last 24 hours—125 mL.

Vital Signs: Temp: 98.4   Pulse: 77   Resp rate: 18
                    BP: 116/88   Height: 6'2"   Weight: 225 lbs (admission wt: 208 lbs)

Cardiovascular: s/p CABG × 3
Pulmonary: Postoperative atelectasis: Encourage aggressive respiratory therapy
Neurologic: Alert and oriented ×3; chronic pain control—scheduled oxycodone
Extremities: Warm with normal pulses
Skin: Cool, pale
Chest/lungs: Respirations are rapid but clear to auscultation and percussion. CXR viewed.
Peripheral vascular: Peripheral edema noted +3
Abdomen: Soft—nt, nd. Tolerated sips of oral intake.

Maddox, Randall, Male, 67 y.o.
Allergies: penicillin, sulfa
Pt. Location: CICU

Code: FULL
Physician: C. Taylor

Isolation: Contact
Admit Date: 10/8

Consult: Nephrology for acute kidney injury—will begin on CRRT today
Consult: Nutrition for enteral feeding—7 days post-op with negligible nutritional intake
.... C. Taylor, MD

Nursing Assessment	10/16
Abdominal appearance (concave, flat, rounded, obese, distended)	obese
Palpation of abdomen (soft, rigid, firm, masses, tense)	soft
Bowel function (continent, incontinent, flatulence, no stool)	continent
Bowel sounds (P=present, AB=absent, hypo, hyper)	
RUQ	P
LUQ	P
RLQ	P
LLQ	P
Stool color	dark brown
Stool consistency	
Tubes/ostomies	NA
Genitourinary	
Urinary continence	catheter
Urine source	catheter
Appearance (clear, cloudy, yellow, amber, fluorescent, hematuria, orange, blue, tea)	cloudy, amber
Integumentary	
Skin color	pale
Skin temperature (DI=diaphoretic, W=warm, dry, CL=cool, CLM=clammy, CD+=cold, M=moist, H=hot)	CL
Skin turgor (good, fair, poor, TENT=tenting)	good
Skin condition (intact, EC=ecchymosis, A=abrasions, P=petechiae, R=rash, W=weeping, S=sloughing, D=dryness, EX=excoriated, T=tears, SE=subcutaneous emphysema, B=blisters, V=vesicles, N=necrosis)	EC, P
Mucous membranes (intact, EC=ecchymosis, A=abrasions, P=petechiae, R=rash, W=weeping, S=sloughing, D=dryness, EX=excoriated, T=tears, SE=subcutaneous emphysema, B=blisters, V=vesicles, N=necrosis)	intact
Other components of Braden score: special bed, sensory pressure, moisture, activity, friction/shear (>18=no risk, 15–16=low risk, 13–14=moderate risk, ≤12=high risk)	13

Maddox, Randall, Male, 67 y.o.
Allergies: penicillin, sulfa
Pt. Location: CICU

Code: FULL
Physician: C. Taylor

Isolation: Contact
Admit Date: 10/8

Intake/Output

Date		10/15 0701–10/16 0700			
Time		0701–1500	1501–2300	2301–0700	Daily total
IN	P.O.	**490**	**240**	**0**	**730**
	I.V. (mL/kg/hr)				
	I.V. piggyback				
	TPN				
	Total intake (mL/kg)	**490** (5.2)	**240** (2.5)	**0** (0)	**730** (7.7)
OUT	Urine (mL/kg/hr)	**0** (0)	**75** (0.10)	**50** (0.07)	**125** (0.05)
	Emesis output				
	Other				
	Stool	**170**	**0**	**0**	**170**
	Total output (mL/kg)	**170** (1.8)	**75** (0.8)	**50** (0.5)	**295** (3.1)
Net I/O		**+320**	**+165**	**−50**	**+435**
Net since admission (1/21)		**+5,180.4**	**+5,345.4**	**+5,295.4**	**+5,295.4**

Laboratory Results

	Ref. Range	10/21 0600
Chemistry		
Sodium (mEq/L)	136–145	138
Potassium (mEq/L)	3.5–5.5	5.7 !↑
Chloride (mEq/L)	95–105	102
Carbon dioxide (CO_2, mEq/L)	23–30	24
Bicarbonate (mEq/L)	21–32	16 !↓
BUN (mg/dL)	8–18	38 !↑
Creatinine serum (mg/dL)	0.6–1.2	6.62 !↑
BUN/Crea ratio	10.0–20.0	18.2
Est GFR, non-Afr Amer (mL/min/1.73 m^2)	61–589	8 !↓

Maddox, Randall, Male, 67 y.o.
Allergies: penicillin, sulfa
Pt. Location: CICU

Code: FULL
Physician: C. Taylor

Isolation: Contact
Admit Date: 10/8

Laboratory Results *(Continued)*

	Ref. Range	10/21 0600
Est GFR, Afr-Amer(mL/min/1.73 m^2)	61–714	10 !↓
Glucose (mg/dL)	70–110	123 !↑
Phosphate, inorganic (mg/dL)	2.3–4.7	5.3 !↑
Magnesium (mg/dL)	1.8–3	2.9
Calcium (mg/dL)	9–11	8.6 !↓
Anion gap (mmol/L)	10–20	17
Protein, total (g/dL)	6–8	5.0 !↓
Albumin (g/dL)	3.5–5	2.5 !↓
Prealbumin (mg/dL)	16–35	13 !↓
Ammonia (NH$_3$, µmol/L)	9–33	11
Alkaline phosphatase (U/L)	30–120	141 !↑
ALT (U/L)	4–36	37 !↑
AST (U/L)	0–35	41 !↑
CPK (U/L)	30–135 F 55–170 M	190 !↑
Lactate dehydrogenase (U/L)	208–378	382 !↑
Cholesterol (mg/dL)	120–199	245 !↑
HDL-C (mg/dL)	>55 F, >45 M	30 !↓
VLDL (mg/dL)	7–32	44 !↑
LDL (mg/dL)	<130	171 !↑
LDL/HDL ratio	<3.22 F <3.55 M	5.7 !↑
Triglycerides (mg/dL)	35–135 F 40–160 M	220 !↑
Coagulation (Coag)		
PT (sec)	12.4–14.4	13.7
INR	0.9–1.1	1.0
Hematology		
WBC (×10^3/mm^3)	4.8–11.8	12.8 !↑
RBC (×10^6/mm^3)	4.2–5.4 F 4.5–6.2 M	2.95 !↓

(Continued)

Maddox, Randall, Male, 67 y.o.
Allergies: penicillin, sulfa
Pt. Location: CICU

Code: FULL
Physician: C. Taylor

Isolation: Contact
Admit Date: 10/8

Laboratory Results *(Continued)*

	Ref. Range	10/21 0600
Hemoglobin (Hgb, g/dL)	12–15 F 14–17 M	9.1 !↓
Hematocrit (Hct, %)	37–47 F 40–54 M	28.6 !↓
Mean cell volume (µm³)	80–96	94.5
Mean cell Hgb (pg)	26–32	25.9 !↓
Mean cell Hgb content (g/dL)	31.5–36	19.5 !↓
RBC distribution (%)	11.6–16.5	16.8 !↑
Platelet count (×10³/mm³)	140–440	390

Case Questions

I.   Understanding the Diagnosis and Pathophysiology

1.   Summarize the major physiological functions of the kidney.

2.   Define and describe Mr. Maddox's diagnosis of acute kidney injury.

3.   Explain the major causes of prerenal, postrenal, and intrarenal AKI. What do you suspect is the etiology of Mr. Maddox's AKI?

4.   Explain the major nutrient/metabolic changes (glucose, lipid, protein, and energy expenditure) that may occur during AKI and that would potentially affect your nutrition recommendations.

5.   What laboratory values or other tests support Mr. Maddox's diagnosis of AKI? List all abnormal values and explain the likely cause for each abnormal value.

6.   Mr. Maddox will be started on continuous renal replacement therapy (CRRT). Describe the basic mechanisms of this therapy.

7.   Explain how nutrition therapy recommendations for an AKI patient may differ if he is receiving CRRT versus not receiving any dialysis treatment.

II.   Understanding the Nutrition Therapy

8.   Mr. Maddox has not eaten since his surgery. He was started on clear liquids several days ago but has taken in very little. Using the ASPEN guidelines, justify his requirement for nutrition support.

III.   Nutrition Assessment

9.   Assess Mr. Maddox's height, weight, and BMI. What factors from his medical record will affect your interpretation of his weight?

10. Determine Mr. Maddox's energy and protein requirements. Explain the rationale for the method you used to calculate these requirements.

IV. Nutrition Diagnosis

11. Identify the pertinent nutrition problems and the corresponding nutrition diagnoses.

12. Write a PES statement for each high-priority nutrition problem.

V. Nutrition Intervention

13. Outline the appropriate nutrition support plan for Mr. Maddox.

VI. Nutrition Monitoring and Evaluation

14. Write your ADIME note for the nutrition support recommendations.

Bibliography

Bellomo R, Kellum JA, Ronco C. Acute kidney injury. *Lancet.* 2012;380:756–766.

Chiolero R, Berger MM. Nutritional support during renal replacement therapy. *Contrib Nephrol.* 2007;156:267–274.

Dirkes S. Acute kidney injury: Not just acute renal failure anymore. *Crit Care Nurs.* 2011;31:37–50.

Fiaccadori E, Regolisti G, Cabassi A. Specific nutritional problems in acute kidney injury, treated with non-dialysis and dialytic modalities. *NDT Plus.* 2010;3:1–7.

Gervasio JM, Garmon WP, Holowatyj M. Nutrition support in acute kidney injury. *Nutrition in Clinical Practice.* 2011;26:374–381.

Kariyawasam D. Nutritional requirements in acute kidney injury. *J Ren Nurs.* 2012;4:231–5.

Kopple JD, Massry SG, eds. *Nutrition Management of Renal Disease.* 2nd ed. Philadelphia, PA: Lippincott Williams & Wilkins; 2004.

Lacey K, Nahikian-Nelms ML. Diseases of the renal system. In: Nelms M, Sucher K, Lacey K, Roth SL. *Nutrition Therapy and Pathophysiology.* 2nd ed. Belmont, CA: Wadsworth, Cengage Learning; 2011:525–561.

Li Y, Tang X, Zhang J, Wu T. Nutritional support for acute kidney injury. *Cochrane Database Syst Rev.* 2010;(1):CD005426.

Maursetter L, Kight CE, Mennig J, Hofmann RM. Review of the mechanism and nutrition recommendations for patients undergoing continuous renal replacement therapy. *Nutr Clin Prac.* 2011;26:382–390.

Patel N, Rogers CA, Angelini GD, Murphy GJ. Pharmacological therapies for the prevention of acute kidney injury following cardiac surgery: A systematic review. Heart Failure Review. 2011;16:553–567.

Tranter S. Cochrane Nursing Care Corner: Nutritional support for acute kidney injury. *Ren Soc Aust J.* 2011;7:36–37.

Valencia E, Marin A, Hardy G. Nutrition therapy for acute renal failure: A new approach based on "risk, injury, failure, loss and end-stage kidney" classification (RIFLE). *Curr Opin Clin Nutr Metabolic Care.* 2009;12:241–244.

Ukleja A., et al. Standards for nutrition support: Adult hospitalized patients. *Nutr Clin Pract.* 2010;25:403–414.

Wiesen P, Van Overmeire L, Delanaye P, Dubois B, Preiser JC. Nutrition disorders during acute renal failure and renal replacement therapy. *JPEN J Parenter Enteral Nutr.* 2010;35:217–222.

Internet Resources

ADA Evidence Analysis Library: http://www.adaevidencelibrary.com

American Association of Kidney Patients: http://www.aakp.org

Cleveland Clinic Center for Continuing Education: http://www.clevelandclinicmeded.com/medicalpubs /diseasemanagement/nephrology/acute-kidney-injury/

National Institute of Diabetes and Digestive and Kidney Diseases of the National Institutes of Health: http://www2.niddk.nih.gov/

National Kidney Foundation: http://www.kidney.org

Nutrition Care Manual: http://www.nutritioncaremanual.org

NUTRITION THERAPY FOR HEMATOLOGICAL DISORDERS

This unit features two cases new to this edition. These patient scenarios provide opportunities to apply your nutrition knowledge of vitamins and minerals within the context of a deficiency state. These cases are examples of nutritional anemias—both microcytic and macrocytic. In the first case, the context of pregnancy allows for review of nutritional needs during gestation as well as nutrition therapy for microcytic anemia. The second case discusses the physiological roles of B_{12} and folate, as well as the interrelationships between the two that make it difficult to diagnose deficiencies of these vitamins. Each case reviews all components of the nutrition care process with a particular emphasis on supplementation of these nutrients via food sources and through medications. Important drug–nutrient interactions are highlighted within each case.

Anemia in Pregnancy

Objectives

After completing this case, the student will be able to:

1. Describe the physiological changes in pregnancy and apply those principles for the nutrition care process.
2. Define the pathophysiology of iron-deficiency anemia.
3. Evaluate the signs and symptoms consistent with iron-deficiency anemia.
4. Evaluate a pregnant woman's dietary intake for adequacy, and make recommendations for enhancing dietary iron intake and absorption.
5. Identify risks of anemia for maternal and fetal health outcomes.

6. Develop a nutrition care plan—with appropriate measurable goals, interventions, and strategies for monitoring and evaluation—that addresses the nutrition diagnoses for this case.

Amber Morris, a 31-year-old pregnant woman, is admitted through the ER after falling on the ice. She is admitted to rule out premature labor, but because of her low hemoglobin levels, a complete hematological workup is completed. She is diagnosed with hypochromic, microcytic anemia.

Morris, Amber, Female, 31 y.o.

Allergies: NKA	**Code:** FULL	**Isolation:** None
Pt. Location: 732	**Physician:** F. Bowman	**Admit Date:** 1/17

Patient Summary: Amber Morris is a 31-year-old pregnant woman, gravida 3, para 2, who presented to the ER in her 23rd week of gestation. She has experienced vaginal spotting and some abdominal pain. She is now admitted for observation and to rule out premature labor secondary to her fall.

History:

Onset of disease: Mrs. Morris is a 31-year-old pregnant woman, gravida 3, para 2, who presented to the ER in her 23rd week of gestation. She has experienced vaginal spotting and some abdominal pain. She is now admitted for observation and to rule out premature labor secondary to her fall. Patient states that she is much more tired with this pregnancy but has attributed it to having two small children. She also describes being short of breath, which she experienced with other pregnancies, but it has started earlier with this pregnancy.

Medical history: Two previous pregnancies delivered at 38 and 37 weeks, respectively. No other contributory history

Surgical history: s/p appendectomy age 12; cesarean section—18 months previously

Medications at home: Prenatal vitamins

Tobacco use: Yes—but trying to cut back; 1–2 cigarettes/day

Alcohol use: None

Family history: Mother—type 2 diabetes; father—hypertension, CAD

Demographics:

Marital status: Married

Years education: 13

Language: English only

Occupation: Stay-at-home mom

Hours of work: N/A

Household members: Husband—age 31; sons aged 12 months and 3 years

Ethnicity: Caucasian

Religious affiliation: Methodist

Admitting History/Physical:

Chief complaint: "I went out to get the mail and slipped on the ice. After I got back in I noticed a small amount of bleeding when I went to the bathroom. Over the next hour, I had some abdominal pain. I called my doctor and the office said I should come here to be checked out."

General appearance: 31-year-old pregnant female, pale, in no acute distress

Vital Signs:	Temp: 98.6	Pulse: 88	Resp rate: 19	Prepregnancy weight:
	BP: 118/72	Height: 5'5"	Weight: 142 lbs	135 lbs

Heart: RRR, heart sounds normal

Morris, Amber, Female, 31 y.o.

Allergies: NKA	**Code:** FULL	**Isolation:** None
Pt. Location: 732	**Physician:** F. Bowman	**Admit Date:** 1/17

HEENT: Head: WNL

   Eyes: Sclera pale, PEERLA, fundi without lesions

   Ears: Clear

   Nose: Clear

   Throat: Pharynx clear without postnasal drainage

Genitalia: Normal

Neurologic: Alert and oriented

Extremities: No edema, DTR 2+ and symmetrical throughout

Skin: Pale, warm and dry

Chest/lungs: Clear to auscultation and percussion

Peripheral vascular: Diminished pulses bilaterally

Abdomen: Bowel sounds ×4

Nursing Assessment	1/17
Abdominal appearance (concave, flat, rounded, obese, distended)	rounded with pregnancy
Palpation of abdomen (soft, rigid, firm, masses, tense)	soft
Bowel function (continent, incontinent, flatulence, no stool)	continent
Bowel sounds (P=present, AB=absent, hypo, hyper)	
RUQ	P
LUQ	P
RLQ	P
LLQ	P
Stool color	none
Stool consistency	
Tubes/ostomies	NA
Genitourinary	
Urinary continence	yes
Urine source	clean catch
Appearance (clear, cloudy, yellow, amber, fluorescent, hematuria, orange, blue, tea)	clear, yellow
Integumentary	
Skin color	pale
Skin temperature (DI=diaphoretic, W=warm, dry, CL=cool, CLM=clammy, CD+=cold, M=moist, H=hot)	W
Skin turgor (good, fair, poor, TENT=tenting)	good

(Continued)

Morris, Amber, Female, 31 y.o.
Allergies: NKA
Pt. Location: 732

Code: FULL
Physician: F. Bowman

Isolation: None
Admit Date: 1/17

Nursing Assessment *(Continued)*

Nursing Assessment	1/17
Skin condition (intact, EC=ecchymosis, A=abrasions, P=petechiae, R=rash, W=weeping, S=sloughing, D=dryness, EX=excoriated, T=tears, SE=subcutaneous emphysema, B=blisters, V=vesicles, N=necrosis)	intact
Mucous membranes (intact, EC=ecchymosis, A=abrasions, P=petechiae, R=rash, W=weeping, S=sloughing, D=dryness, EX=excoriated, T=tears, SE=subcutaneous emphysema, B=blisters, V=vesicles, N=necrosis)	intact
Other components of Braden score: special bed, sensory pressure, moisture, activity, friction/shear (>18=no risk, 15–16=low risk, 13–14=moderate risk, ≤12=high risk)	21

Admission Orders:
Laboratory: CBC, RPR, Chem 16
Repeat CBC, Amylase, Lipase in 12 hrs
Repeat Chem 7 every 6 hrs

Radiology:
Abdominal U/S: Pregnancy > 1st

Vital Signs: Every 4 hrs
Monitor fetal heart tones and contractions
I & O recorded every 8 hrs

Diet: NPO

Activity: Bed rest

IVF: LR @ 100 mL/hr

Nutrition:
Meal type: NPO
Fluid requirement: 2000–2400 mL/day
History: Patient states appetite is good right now. She suffered a lot of morning sickness during her first trimester but is better now. States that there are a lot of foods she doesn't like. Describes herself as a picky eater. Had been on WIC during last pregnancy. Does not always take prenatal vitamins because they make her stomach hurt. States that she gained 15 lbs with her first pregnancy and almost 20 with her second.

Morris, Amber, Female, 31 y.o.

Allergies: NKA	**Code:** FULL	**Isolation:** None
Pt. Location: 732	**Physician:** F. Bowman	**Admit Date:** 1/17

Usual dietary intake:
AM: Coffee, cold cereal, occasionally toast
Lunch: Sandwich or soup
Dinner: Casserole such as Hamburger Helper, hot dogs, soup; sometimes she cooks a full meal
   with meat and vegetables. Her husband works nights so she doesn't always cook except
   on his days off.

24-hour-recall (PTA):
AM: 2 c Frosted Flakes, ½ c whole milk, black coffee
Lunch: Hot dog on bun, ½ c macaroni and cheese, iced tea
Dinner: 3 oz Salisbury steak, 1 c green beans, 1 c mashed potatoes with gravy, 1 roll with butter,
            iced tea

MD Progress Note:
1/18 0640
Subjective: Amber Morris previous 24 hours reviewed
Vitals: Temp: 98.6 Pulse: 82 Resp rate: 20 BP: 120/82
Urine Output: 4344 mL (67.3 mL/kg)

Physical Exam
General: 23 week gestation—no contractions; no further vaginal spotting
HEENT: WNL
Neck: WNL
Heart: WNL
Lungs: Clear to auscultation
Abdomen: WNL

Dx: Hypochromic Microcytic Anemia; 23-week gestation with normal ultrasound. Fetal heart
sounds WNL.

Plan: D/C IVF. Begin 40 mg ferrous sulfate; nutrition consult; discharge to home.

.... F. Bowman, MD

Morris, Amber, Female, 31 y.o.
Allergies: NKA
Pt. Location: 732

Code: FULL
Physician: F. Bowman

Isolation: None
Admit Date: 1/17

Laboratory Results

	Ref. Range	1/17 1540
Chemistry		
Sodium (mEq/L)	136–145	142
Potassium (mEq/L)	3.5–5.5	3.8
Chloride (mEq/L)	95–105	104
Carbon dioxide (CO_2, mEq/L)	23–30	26
BUN (mg/dL)	8–18	8
Creatinine serum (mg/dL)	0.6–1.2	0.7
BUN/Crea ratio	10.0–20.0	11.4
Uric acid (mg/dL)	2.8–8.8 F 4.0–9.0 M	3.2
Glucose (mg/dL)	70–110	105
Phosphate, inorganic (mg/dL)	2.3–4.7	3.1
Magnesium (mg/dL)	1.8–3	2.2
Calcium (mg/dL)	9–11	10.2
Osmolality (mmol/kg/H_2O)	285–295	292
Bilirubin total (mg/dL)	≤1.5	0.4
Bilirubin, direct (mg/dL)	<0.3	0.1
Protein, total (g/dL)	6–8	6.2
Albumin (g/dL)	3.5–5	3.9
Prealbumin (mg/dL)	16–35	33
Ammonia (NH_3, µmol/L)	9–33	9
Alkaline phosphatase (U/L)	30–120	45
ALT (U/L)	4–36	8
AST (U/L)	0–35	2
CPK (U/L)	30–135 F 55–170 M	31
Lactate dehydrogenase (U/L)	208–378	210
Lipase (U/L)	0–110	5
Amylase (U/L)	25–125	26
CRP (mg/dL)	<1	0.004
Cholesterol (mg/dL)	120–199	145
HDL-C (mg/dL)	>55 F, >45 M	62
VLDL (mg/dL)	7–32	13

Morris, Amber, Female, 31 y.o.
Allergies: NKA
Pt. Location: 732

Code: FULL
Physician: F. Bowman

Isolation: None
Admit Date: 1/17

Laboratory Results *(Continued)*

	Ref. Range	1/17 1540
LDL (mg/dL)	<130	70
LDL/HDL ratio	<3.22 F <3.55 M	1.12
Triglycerides (mg/dL)	35–135 F 40–160 M	75
Coagulation (Coag)		
PT (sec)	12.4–14.4	13.2
PTT (sec)	24–34	27
Hematology		
WBC ($\times 10^3$/mm^3)	4.8–11.8	7.2
RBC ($\times 10^6$/mm^3)	4.2–5.4 F 4.5–6.2 M	3.8 ↓!
Hemoglobin (Hgb, g/dL)	12–15 F 14–17 M	9.1 ↓!
Hematocrit (Hct, %)	37–47 F 40–54 M	33 ↓!
Mean cell volume (µm^3)	80–96	72 ↓!
Retic (%)	0.8–2.8	0.2 ↓!
Mean cell Hgb (pg)	26–32	23 ↓!
Mean cell Hgb content (g/dL)	31.5–36	28 ↓!
RBC distribution (%)	11.6–16.5	22 ↑!
Platelet count ($\times 10^3$/mm^3)	140–440	282
Total iron binding capacity (µg/dL)	240–450	465 ↑!
Ferritin (µg/dL)	20–120 F 20–300 M	10 ↓!
ZPP (µmol/mol)	30–80	84 ↑!
Vitamin B$_{12}$ (ng/mL)	24.4–100	95
Folate (ng/dL)	5–25	2 ↓!

Morris, Amber, Female, 31 y.o.
Allergies: NKA
Pt. Location: 732

Code: FULL
Physician: F. Bowman

Isolation: None
Admit Date: 1/17

Intake/Output

Date		1/17 0701 – 1/18 0700			
Time		0701–1500	1501–2300	2301–0700	Daily total
IN	P.O.	0	550	600	1150
	I.V.	800	800	800	2400
	(mL/kg/hr)	(1.55)	(1.55)	(1.55)	(1.55)
	I.V. piggyback				
	TPN				
	Total intake	800	1350	1400	3550
	(mL/kg)	(12.4)	(20.9)	(21.7)	(55)
OUT	Urine	1312	1410	1622	4344
	(mL/kg/hr)	(2.54)	(2.73)	(3.14)	(2.80)
	Emesis output	0	0	0	0
	Other	0	0	0	0
	Stool	0	0	0	0
	Total output	1312	1410	1622	4344
	(mL/kg)	(20.3)	(21.8)	(25.1)	(67.3)
Net I/O		−512	−60	−222	−794
Net since admission (1/17)		−512	−572	−794	−794

Case Questions

I.   Understanding the Diagnosis and Pathophysiology

1.   Evaluate the patient's admitting history and physical. Are there any signs or symptoms that support the diagnosis of anemia?

2.   What laboratory values or other tests support this diagnosis? List all abnormal values and explain the likely cause for each abnormal value.

3.   Mrs. Morris's physician ordered additional lab work when her admitting CBC revealed a low hemoglobin. Why is this a concern? Are there normal changes in hemoglobin associated with pregnancy? If so, what are they? What other hematological values, if any, normally change in pregnancy?

4.   There are several classifications of anemia. Define each of the following: megaloblastic anemia, pernicious anemia, normocytic anemia, microcytic anemia, sickle cell anemia, and hemolytic anemia.

5.   What is the role of iron in the body? Are there additional functions of iron during fetal development?

6.   Several stages of iron deficiency actually precede iron-deficiency anemia. Discuss these stages—including the symptoms—and identify the laboratory values that would be affected during each stage.

7.   What potential risk factor(s) for the development of iron-deficiency anemia can you identify from Mrs. Morris's history?

8.   What is the relationship between the health of the fetus and maternal iron status? Is there a risk for the infant if anemia continues?

II.   Understanding the Nutrition Therapy

9.   Discuss the specific nutritional requirements during pregnancy. Be sure to address all macro- and micronutrients that are altered during pregnancy.

10. What are best dietary sources of iron? Describe the differences between heme and nonheme iron.

11. Explain the digestion and absorption of dietary iron.

III.   Nutrition Assessment

12. Assess Mrs. Morris's height and weight. Calculate her BMI and % usual body weight.

13. Check Mrs. Morris's prepregnancy weight. Plot her weight gain on the maternal weight gain curve. Is her weight gain adequate? How does her weight gain compare to the current recommendations? Was the weight gain from her previous pregnancies WNL?

14. Determine Mrs. Morris's energy and protein requirements. Explain the rationale for the method you used to calculate these requirements.

15. Using her 24-hour recall, compare her dietary intake to the energy and protein requirements that you calculated in Question 14.

16. Again using her 24-hour recall, assess the patient's daily iron intake. How does it compare to the recommendations for this patient (which you provided in question #9)?

IV.   Nutrition Diagnosis

17. Identify the pertinent nutrition problems and the corresponding nutrition diagnoses.

18. Write a PES statement for each nutrition problem.

V.   Nutrition Intervention

19. Mrs. Morris was discharged on 40 mg of ferrous sulfate three times daily. Are there potential side effects from this medication? Are there any drug–nutrient interactions? What instructions might you give her to maximize the benefit of her iron supplementation?

20. Mrs. Morris says she does not take her prenatal vitamin regularly. What nutrients does this vitamin provide? What recommendations would you make to her regarding her difficulty taking the vitamin supplement?

VI.   Nutrition Monitoring and Evaluation

21. List factors that you would monitor to assess her pregnancy, nutritional, and iron status.

22. You note in Mrs. Morris's history that she received nutrition counseling from the WIC program. What is WIC? Should you refer her back to that program? What are the qualifications for enrollment? Are there any you can confirm for her referral?

Bibliography

American Dietetic Association. Position of the American Dietetic Association: Nutrition and lifestyle for a healthy pregnancy outcome. *J Am Diet Assoc.* 2008;108:553–561.

Arnold DL, Williams MA, Miller RS, Qiu C, Sorrensen TK. Iron deficiency anemia, cigarette smoking and risk of abruptio placentae. *J Obstet Gynaecol Res.* 2009;35:446–52.

Heuberger RA. Diseases of the hematological system. In: Nelms M, Sucher K, Lacey K, Roth SL. *Nutrition Therapy and Pathophysiology.* 2nd ed. Belmont, CA: Wadsworth, Cengage Learning; 2011:562–608.

Krafft A, Murray-Kolb L, Milman N. Anemia and iron deficiency in pregnancy. *J Pregnancy.* 2012;2012:241869. Published online August 28, 2012. doi: 10.1155/2012/241869.

Milman N. Oral iron prophylaxis during pregnancy: Not too little and not too much! *J Pregnancy.* 2012;2012:514345. Epub July 24, 2012.

Nahikian-Nelms ML, Habash D. Nutrition assessment: Foundation of the nutrition care process. In: Nelms M, Sucher K, Lacey K, Roth SL. *Nutrition Therapy and Pathophysiology.* 2nd ed. Belmont, CA: Wadsworth, Cengage Learning; 2011:14–65.

Pena-Rosas JP, De-Reqil LM, Dowswell T, Viteri FE. Intermittent oral iron supplementation during pregnancy. *Cochrane Database Syst Rev.* July 2012:7:CD009997.

Scholl TO. Iron status during pregnancy: Setting the stage for mother and infant. *Am J Clin Nutr.* 2005;81:1218S–1222S.

Internet Resources

ADA Evidence Analysis Library: http://www.adaevidencelibrary.com

Institute of Medicine: http://www.iom.edu/Reports/2009/Weight-Gain-During-Pregnancy-Reexamining-the-Guidelines.aspx

Nutrition Care Manual: http://www.nutritioncaremanual.org

Folate and Vitamin B$_{12}$ Deficiencies

Objectives

After completing this case, the student will be able to:

1. Demonstrate understanding of the physiological roles of folate and vitamin B$_{12}$.
2. Describe the digestion and absorption of folate and vitamin B$_{12}$ and discuss their interdependence.
3. Discuss the potential etiology of folate and vitamin B$_{12}$ deficiencies.
4. Evaluate the signs and symptoms consistent with macrocytic anemias.
5. Evaluate the current literature regarding nutrition support for acute pancreatitis.
6. Determine the appropriate nutrition interventions for both folate and vitamin B$_{12}$ deficiencies.
7. Develop a nutrition care plan—with appropriate measurable goals, interventions, and strategies for monitoring and evaluation—that addresses the nutrition diagnoses for this case.

Marie Hicks is a 72-year-old woman who has been suffering from severe fatigue as well as numbness and tingling in her hands and feet. She is admitted to the hospital after her family MD discovers a general pancytopenia that warrants further investigation.

Hicks, Marie, Female, 72 y.o.
Allergies: NKA
Pt. Location: 1014

Code: FULL
Physician: E. Yonas

Isolation: None
Admit Date: 3/15

Patient Summary: Patient is 72-year-old female who was admitted after her physician noted general pancytopenia from her recent outpatient laboratory work.

History:

Onset of disease: Patient is a 72-year-old retired guidance counselor who had enjoyed an active retirement until, over the last six months, she noted increasing fatigue and parathesias in hands and feet. Her physician noted pancytopenia in her laboratory work and ordered a hematology consult at University Hospital.

Medical history: Gravida 1 para 1—vaginal delivery age 32; vertebral compression fracture L1–L2 secondary to osteoporosis; osteoarthritis

Surgical history: s/p appendectomy age 12; s/p gastric bypass × 25 years

Medications at home: Fosamax 10 mg one time daily; Celebrex 200 mg one time daily; 800 mg calcium twice daily; 800 IU vitamin D

Tobacco use: None

Alcohol use: Occasional wine or cocktail socially

Family history: Mother—ovarian cancer; father—atherosclerosis, type 2 DM

Demographics:

Marital status: Widow
Years education: 16
Language: English; Spanish
Occupation: Retired
Household members: Lives with daughter, son-in-law, and granddaughter
Ethnicity: Caucasian
Religious affiliation: Catholic

Admitting History/Physical:

Chief complaint: "My doctor felt that my blood work needed a second opinion."
General appearance: Pale, obese female in no acute distress

Vital Signs: Temp: 98.5   Pulse: 94   Resp rate: 27
                    BP: 130/78   Height: 5'1"   Weight: 165 lbs

Heart: RRR, unremarkable
HEENT: Head: WNL
          Eyes: PERRLA
          Ears: Clear
          Nose: Moist mucous membranes
          Throat: Moist mucous membranes
Genitalia: Deferred
Neurologic: Alert and oriented

Hicks, Marie, Female, 72 y.o.

Allergies: NKA	**Code:** FULL	**Isolation:** None
Pt. Location: 1014	**Physician:** E. Yonas	**Admit Date:** 3/15

Extremities: WNL
Skin: Clammy, diaphoretic
Chest/lungs: Respirations are rapid but clear to auscultation and percussion
Peripheral vascular: Diminished pulses bilaterally
Abdomen: Hypoactive bowel sounds ×4

Nursing Assessment	3/15
Abdominal appearance (concave, flat, rounded, obese, distended)	obese
Palpation of abdomen (soft, rigid, firm, masses, tense)	soft
Bowel function (continent, incontinent, flatulence, no stool)	continent
Bowel sounds (P=present, AB=absent, hypo, hyper)	
RUQ	P, hypo
LUQ	P, hypo
RLQ	P, hypo
LLQ	P, hypo
Stool color	lt brown
Stool consistency	soft
Tubes/ostomies	NA
Genitourinary	
Urinary continence	continent
Urine source	clean catch
Appearance (clear, cloudy, yellow, amber, fluorescent, hematuria, orange, blue, tea)	clear, yellow
Integumentary	
Skin color	pale
Skin temperature (DI=diaphoretic, W=warm, dry, CL=cool, CLM=clammy, CD+=cold, M=moist, H=hot)	CLM, DI
Skin turgor (good, fair, poor, TENT=tenting)	fair
Skin condition (intact, EC=ecchymosis, A=abrasions, P=petechiae, R=rash, W=weeping, S=sloughing, D=dryness, EX=excoriated, T=tears, SE=subcutaneous emphysema, B=blisters, V=vesicles, N=necrosis)	intact
Mucous membranes (intact, EC=ecchymosis, A=abrasions, P=petechiae, R=rash, W=weeping, S=sloughing, D=dryness, EX=excoriated, T=tears, SE=subcutaneous emphysema, B=blisters, V=vesicles, N=necrosis)	intact
Other components of Braden score: special bed, sensory pressure, moisture, activity, friction/shear (>18=no risk, 15–16=low risk, 13–14=moderate risk, ≤12=high risk)	18

Hicks, Marie, Female, 72 y.o.
Allergies: NKA
Pt. Location: 1014

Code: FULL
Physician: E. Yonas

Isolation: None
Admit Date: 3/15

Admission Orders:

Laboratory: CBC, Chem 27, white count with differential, folate, B_{12}, MMA, Hcy, antiparietal cell antibodies; anti-intrinsic factor antibodies; Schilling test. Hematology consult for bone marrow aspirate.
Vital signs: Every 8 hrs
I & O recorded every 8 hrs
Diet: Regular
Activity: Ad lib
Scheduled medications:
Continue home medications
Colace (docusate) 100 mg po two times daily prn if no bowel movement
Milk of Magnesia (MOM) 30 mL po daily prn

Nutrition:

Meal type: Regular
Fluid requirement: 1800–2000 mL
History: Patient states that she lost over 150 lbs after her gastric bypass surgery (Roux-en-Y). Since that time her weight has fluctuated between 150 and 175. For the last five years, her weight has been stable at 165 lbs. She states that she eats mostly all foods but most frequently eats fruits and vegetables—some grains. Rarely eats meat but does like chicken, eggs, and dairy products.

24-hour recall:
AM: 2 slices cheddar cheese melted on 2 slices of English muffin; coffee with half and half; 1 c cantaloupe
Lunch: Broccoli and cheese soup—1 c; ½ c chicken salad with 12 whole-wheat crackers; iced tea with lemon and artificial sweetener
Dinner: Fettuccine Alfredo with chicken breast—2 c; romaine lettuce with tomatoes—1 c; iced tea
Snack: 1 c strawberry ice cream

MD Progress Note:

3/16 0640
Subjective: Marie Hicks's previous 24 hours reviewed
Vitals: Temp: 98.2 Pulse: 88 Resp rate: 18 BP: 110/82
Urine Output: 1562 (21 mL/kg)

Physical Exam
General: Complains of fatigue and continued numbness, burning and tingling in her feet
HEENT: WNL
Neck: WNL
Heart: WNL
Lungs: Clear to auscultation
Abdomen: WNL

Hicks, Marie, Female, 72 y.o.
Allergies: NKA   **Code:** FULL   **Isolation:** None
Pt. Location: 1014   **Physician:** E. Yonas   **Admit Date:** 3/15

Assessment/Plan:

Results from laboratory assessment indicate: low folate, low B$_{12}$, elevated MMA and normal Hcy consistent with mixed deficiency; BM aspirate consistent with megaloblastic anemia.

Dx: B$_{12}$ and folate deficiency with megaloblastic anemia secondary to gastric bypass and malabsorption combined with probable deficient dietary intake.

Plan: 1000 µg cyanocobalamin IM; 5000 µg folate. Discharge and follow weekly for additional B$_{12}$ injections, folate supplementation, and repeat laboratory values. Nutrition Consult.

.... E. Yonas, MD

Intake/Output

Date		3/15 0701–3/16 0700			
Time		0701–1500	1501–2300	2301–0700	Daily total
IN	P.O.	**450**	**600**	**520**	**1570**
	I.V.				
	(mL/kg/hr)				
	I.V. piggyback				
	TPN				
	Total intake	**450**	**600**	**520**	**1570**
	(mL/kg)	(6.0)	(8.0)	(6.9)	(20.9)
OUT	Urine	**412**	**550**	**600**	**1562**
	(mL/kg/hr)	(0.69)	(0.92)	(1.00)	(0.87)
	Emesis output				
	Other				
	Stool	**0**	**300**	**0**	**300**
	Total output	**412**	**850**	**600**	**1862**
	(mL/kg)	(5.5)	(11.3)	(8.0)	(24.8)
Net I/O		**+38**	**−250**	**−80**	**−292**
Net since admission (3/15)		**+38**	**−212**	**−292**	**−292**

Laboratory Results

	Ref. Range	3/15 0622
Chemistry		
Sodium (mEq/L)	136–145	142
Potassium (mEq/L)	3.5–5.5	3.9
Chloride (mEq/L)	95–105	97
Carbon dioxide (CO$_2$, mEq/L)	23–30	30

(Continued)

Hicks, Marie, Female, 72 y.o.
Allergies: NKA
Pt. Location: 1014

Code: FULL
Physician: E. Yonas

Isolation: None
Admit Date: 3/15

Laboratory Results *(Continued)*

	Ref. Range	3/15 0622
BUN (mg/dL)	8–18	9
Creatinine serum (mg/dL)	0.6–1.2	1.1
Uric acid (mg/dL)	2.8–8.8 F 4.0–9.0 M	5.2
Glucose (mg/dL)	70–110	101
Phosphate, inorganic (mg/dL)	2.3–4.7	3.1
Magnesium (mg/dL)	1.8–3	1.9
Calcium (mg/dL)	9–11	9.8
Osmolality (mmol/kg/H_2O)	285–295	292
Bilirubin total (mg/dL)	≤1.5	1.1
Bilirubin, direct (mg/dL)	<0.3	0.2
Protein, total (g/dL)	6–8	6.2
Albumin (g/dL)	3.5–5	3.7
Prealbumin (mg/dL)	16–35	25
Ammonia (NH_3, µmol/L)	9–33	10
Alkaline phosphatase (U/L)	30–120	44
ALT (U/L)	4–36	31
AST (U/L)	0–35	15
CPK (U/L)	30–135 F 55–170 M	131
Lactate dehydrogenase (U/L)	208–378	209
Cholesterol (mg/dL)	120–199	150
HDL-C (mg/dL)	>55 F, >45 M	58
VLDL (mg/dL)	7–32	29
LDL (mg/dL)	<130	70
LDL/HDL ratio	<3.22 F <3.55 M	1.2
Triglycerides (mg/dL)	35–135 F 40–160 M	110
T_4 (µg/dL)	4–12	4.1
T_3 (µg/dL)	75–98	76
HbA_{1C} (%)	3.9–5.2	4.9

Hicks, Marie, Female, 72 y.o.
Allergies: NKA
Pt. Location: 1014

Code: FULL
Physician: E. Yonas

Isolation: None
Admit Date: 3/15

Laboratory Results *(Continued)*

	Ref. Range	3/15 0622
Coagulation (Coag)		
PT (sec)	12.4–14.4	14.4
INR	0.9–1.1	0.97
PTT (sec)	24–34	25
Hematology		
WBC (×10^3/mm^3)	4.8–11.8	9.2
RBC (×10^6/mm^3)	4.2–5.4 F 4.5–6.2 M	4.2
Hemoglobin (Hgb, g/dL)	12–15 F 14–17 M	12
Hematocrit (Hct, %)	37–47 F 40–54 M	37
Mean cell volume (µm^3)	80–96	130 !↑
Mean cell Hgb (pg)	26–32	34 !↑
Mean cell Hgb content (g/dL)	31.5–36	38 !↑
RBC distribution (%)	11.6–16.5	17.8 !↑
Platelet count (×10^3/mm^3)	140–440	135 !↓
Transferrin (mg/dL)	250–380 F 215–365 M	379
Ferritin (mg/mL)	20–120 F 20–300 M	20
Iron (µg/dL)	65–165 F 75–175 M	66
Total iron binding capacity (µg/dL)	240–450	442
Iron saturation (%)	15–50% F 10–50% M	15
ZPP (µmol/mol)	30–80	45
Vitamin B$_{12}$ (ng/dL)	24.4–100	11 !↓
Folate (ng/dL)	5–25	3.2 !↓
MMA (mmol/L)	0.08–0.56	0.75 !↑
Hcy (µg/dL)	66–160 F 80–210 M	104

(Continued)

Hicks, Marie, Female, 72 y.o.
Allergies: NKA
Pt. Location: 1014

Code: FULL
Physician: E. Yonas

Isolation: None
Admit Date: 3/15

Laboratory Results *(Continued)*

	Ref. Range	3/15 0622
Anti-parietal cell antibodies	Neg	Neg
Anti-intrinsic factor antibodies	Neg	Neg
Hematology, Manual Diff		
Neutrophil (%)	50–70	55
Lymphocyte (%)	15–45	25
Monocyte(%)	3–10	6
Eosinophil(%)	0–6	4
Basophil(%)	0–2	1
Blasts (%)	3–10	4
Segs(%)	0–60	45
Bands (%)	0–10	4

Case Questions

I.   Understanding the Diagnosis and Pathophysiology

1. Describe the role of vitamin B$_{12}$ in normal metabolism.

2. Summarize the role of folate in normal metabolism.

3. Describe the digestion and absorption of vitamin B$_{12}$.

4. Describe the digestion and absorption of folate.

5. Vitamin B$_{12}$ and folate deficiencies are often difficult to distinguish from one another. Describe the interdependence of these two nutrients and how the deficiency of one may be related to the deficiency of the other.

6. List the most common causes of folate and B$_{12}$ deficiencies.

7. After reading the history and physical, determine the potential causes of Mrs. Hicks's vitamin deficiencies. Next, list the signs and symptoms that are consistent with her deficiencies.

8. List all abnormal laboratory values and explain the likely cause for each abnormal value. What laboratory values or other tests support Mrs. Hicks's diagnosis?

9. Explain why each of the following was assessed as a component of this patient's medical diagnostic workup:

    a. anti-intrinsic factor antibodies.

    b. antiparietal cell antibodies.

    c. methylmalonic acid.

    d. homocysteine.

    e. Schilling test.

10. Mrs. Hicks is diagnosed with megaloblastic anemia. What is pernicious anemia? How do these two diagnoses differ?

11. Mrs. Hicks has previously been diagnosed with osteoporosis. She has been prescribed Fosamax. What is this medication? Are there any important drug–nutrient interactions that Mrs. Hicks should be aware of?

II.   Understanding the Nutrition Therapy

12. Identify the best dietary sources of vitamin B_{12} and folate.

13. Mrs. Hicks takes both calcium and vitamin D. What are the current recommendations for supplementation of these nutrients for the treatment of osteoporosis?

III.   Nutrition Assessment

14. Assess Mrs. Hicks's height and weight. Calculate her BMI and % usual body weight.

15. Evaluate Mrs. Hicks's initial nutrition assessment. What factors revealed by her nutrition assessment will help to determine your nutrition recommendations?

16. Determine Mrs. Hicks's energy and protein requirements. Explain the rationale for the methods you have used to estimate her needs.

17. Assess Mrs. Hicks's usual dietary intake. Estimate the amount of B_{12} and folate that she may receive from her usual dietary intake. Compare her intake to the DRI for these nutrients.

IV.   Nutrition Diagnosis

18. Identify the pertinent nutrition problems and the corresponding nutrition diagnoses.

19. Write your PES statement for each nutrition problem.

V.   Nutrition Intervention

20. Determine your nutritional recommendations for Mrs. Hicks based on the identified nutrition problems.

21. Mrs. Hicks's physician ordered B$_{12}$ and folate supplementation. Why was she given a B$_{12}$ injection?

VI.   Nutrition Monitoring and Evaluation

22. List factors that you would monitor to assess tolerance and adequacy of supplementation and dietary interventions.

23. Write an ADIME note that includes your initial nutrition assessment and nutrition recommendations.

Bibliography

Anonymous. Vitamin B_{12} deficiency in older adults. *Clinical Advisor for Nurse Practitioners*. 2012;15:39.

Bailey LB and Gergory JF. Folate metabolism and requirements. *J Nutr*. 1999;129:779–782. Available From: http://jn.nutrition.org/content/129/4/779.full

Bryan RH. Are we missing vitamin B_{12} deficiency in the primary care setting? *Journal of Nurse Practitioners*. 2010;6:519–23.

Dewey M. Heuberger R. Vitamin D and calcium status and appropriate recommendations in bariatric surgery patients. *Gastroenterology Nursing*. 2011;34:367–74.

Donadelli SP, et al. Daily vitamin supplementation and hypovitaminosis after obesity surgery. *Nutrition*. 2012;28:391–6.

Hvas AM, Nexo E. Diagnosis and treatment of vitamin B_{12} deficiency. An update. *Haematologica*. 2006;91:1506–1512. Available at: http://haematologica.com/content/91/11/1506.full.pdf

Heuberger RA. Diseases of the hematological system. In: Nelms M, Sucher K, Lacey K, Roth SL. *Nutrition Therapy and Pathophysiology*. 2nd ed. Belmont, CA: Wadsworth, Cengage Learning, 2011:562–608.

Nahikian-Nelms ML, Habash D. Nutrition assessment: Foundation of the nutrition care process. In: Nelms M, Sucher K, Lacey K, Roth SL. *Nutrition Therapy and Pathophysiology*. 2nd ed. Belmont, CA: Wadsworth, Cengage Learning, 2011:14–65.

O'Donnell K. Severe micronutrient deficiencies in RYGB patients: Rare but potentially devastating. *Practical Gastroenterology*. November 2011. Available from: http://www.medicine.virginia.edu/clinical/departments/medicine/divisions/digestive-health/nutrition-support-team/nutrition-articles/ODonnell_November2011.pdf

Radigan AE. Post-gastrectomy: Managing the nutrition fall-out. *Practical Gastroenterology*. Available from: http://www.medicine.virginia.edu/clinical/departments/medicine/divisions/digestive-health/nutrition-support-team/nutrition-articles/index-to-articles.

Snow CF. Laboratory diagnosis of vitamin B_{12} and folate deficiency. *Arch Intern Med*. 1999;159:1289–1298.

Von Drygalski A, Andris DA. Anemia after bariatric surgery: More than just iron deficiency. *Nutrition in Clinical Practice*. 2009;24:217–26.

Internet Resources

ADA Evidence Analysis Library: http://www.adaevidencelibrary.com

Nutrition Care Manual: http://www.nutritioncaremanual.org

U.S. National Library of Medicine: http://www.ncbi.nlm.nih.gov/pubmedhealth/PMH0001586/

NUTRITION THERAPY FOR NEUROLOGICAL DISORDERS

Case 23 addresses neurological conditions through one of the most common diagnoses: stroke. According to the Centers for Disease Control and Prevention, stroke is the fourth leading cause of death in the United States (http://www.cdc.gov/nchs/fastats/stroke.htm, accessed December 12, 2012). The health consequences resulting from stroke are significant and, as in many neurological conditions, may involve impairment in the ability to maintain nutritional status. Symptoms, such as impaired vision or ambulation, may result in the inability to shop or prepare adequate meals. Depending on the severity of the stroke, symptoms may interfere with chewing, swallowing, or feeding oneself. These problems are often not easily identified or easily solved. Each situation is highly individualized and requires a comprehensive nutrition assessment. Throughout the course of the disease and rehabilitation, nutrition therapy plays a crucial role in the maintenance of nutritional status and quality of life.

Parkinson's disease and Alzheimer's disease are the additional diagnoses for which you will plan nutrition interventions for neurological disease. The prevalence of Alzheimer's disease (AD) continues to increase as the U.S. population ages, and it is the sixth leading cause of death in the United States (available from: http://www.cdc.gov/nchs/fastats/alzheimer.htm, accessed December 12, 2012). Consequences of these devastating diseases can interfere with all phases of obtaining, eating, and enjoying food. Cases 24 and 25 examine these consequences and allow for discussion about the role of the registered dietitian during terminal illness.

Ischemic Stroke

Objectives

After completing this case, the student will be able to:

1. Apply a working knowledge of stroke pathophysiology to the nutrition care process.
2. Analyze nutrition assessment data to establish baseline nutritional status.
3. Identify and explain the role of nutrition support in recovery and rehabilitation from stroke.
4. Assess and identify nutritional risks in dysphagia.
5. Establish the nutrition diagnosis and compose a PES statement.
6. Create strategies to maximize calorie and protein intake.
7. Apply current recommendations for nutritional supplementation and determine appropriate nutrition interventions.

Mrs. Ruth Noland, a 77-year-old woman, is transported to the emergency room of University Hospital with the following symptoms: slurred speech, numbness on the right side of her face, and weakness of her right arm and leg.

Noland, Ruth, Female, 77 y.o.
Allergies: NKA   **Code:** FULL   **Isolation:** None
Pt. Location: RM 926   **Physician:** S. Young   **Admit Date:** 8/12

Patient Summary: Patient is brought to the hospital with right-sided hemiparesis and slurred speech.

History:
Onset of disease: N/A
Medical history: HTN × 10 years, hyperlipidemia × 2 years
Surgical history: Hysterectomy 10 years ago
Medications at home: Captopril 25 mg twice daily; lovastatin 20 mg once daily
Tobacco use: No
Alcohol use: No
Family history: What? Noncontributory

Demographics:
Marital status: Married—lives with husband; *Spouse name:* Robert
Number of children: Children are grown and do not live at home.
Years education: High school diploma
Language: English only
Occupation: Retired hairdresser
Hours of work: N/A
Ethnicity: European American
Religious affiliation: Protestant

Admitting History/Physical:
Chief complaint: Mr. Noland states that his wife woke up this morning with everything pretty normal, but midmorning she became dizzy and then she couldn't talk or move one side of her body.
General appearance: Elderly female who is unable to speak; unable to move right side

Vital Signs: Temp: 98.8   Pulse: 91   Resp rate: 19
                    BP: 138/88   Height: 5'2"   Weight: 165 lbs

Heart: Regular rate and rhythm, no gallops or rubs, point of maximal impulse at the fifth intercostal space in the midclavicular line
HEENT: *Head:* Normocephalic
          Eyes: Wears glasses for myopia
          Ears: Tympanic membranes normal
          Nose: WNL
          Throat: Slightly dry mucous membranes w/out exudates or lesions
Genitalia: Normal w/out lesions
Neurologic: New onset weakness of the right side involving right arm and leg. Face and arm weakness is disproportionate to leg weakness and sensation is impaired on the contralateral side. Dysarthria with tongue deviation. Cranial nerves III, V, VII, XII impaired. Motor function tone and strength diminished. Plantar reflex decreased on right side. Blink reflex intact.

Noland, Ruth, Female, 77 y.o.

Allergies: NKA	**Code:** FULL	**Isolation:** None
Pt. Location: RM 926	**Physician:** S. Young	**Admit Date:** 8/12

Extremities: Reduced strength, bilaterally
Skin: Normal without lesions
Chest/lungs: Respirations normal; no crackles, rhonchi, wheezes, or rubs noted
Peripheral vascular: Bilateral, 3+ pedal pulses
Abdomen: Normal bowel sounds. No hepatomegaly, splenomegaly, masses, inguinal lymph nodes, or abdominal bruits.

Nursing Assessment	8/12
Abdominal appearance (concave, flat, rounded, obese, distended)	rounded, obese
Palpation of abdomen (soft, rigid, firm, masses, tense)	soft
Bowel function (continent, incontinent, flatulence, no stool)	incontinent
Bowel sounds (P=present, AB=absent, hypo, hyper)	
RUQ	P
LUQ	P
RLQ	P
LLQ	P
Stool color	brown
Stool consistency	formed
Tubes/ostomies	N/A
Genitourinary	
Urinary continence	catheter
Urine source	catheter
Appearance (clear, cloudy, yellow, amber, fluorescent, hematuria, orange, blue, tea)	yellow
Integumentary	
Skin color	pale
Skin temperature (DI=diaphoretic, W=warm, dry, CL=cool, CLM=clammy, CD+=cold, M=moist, H=hot)	CL
Skin turgor (good, fair, poor, TENT=tenting)	fair
Skin condition (intact, EC=ecchymosis, A=abrasions, P=petechiae, R=rash, W=weeping, S=sloughing, D=dryness, EX=excoriated, T=tears, SE=subcutaneous emphysema, B=blisters, V=vesicles, N=necrosis)	intact
Mucous membranes (intact, EC=ecchymosis, A=abrasions, P=petechiae, R=rash, W=weeping, S=sloughing, D=dryness, EX=excoriated, T=tears, SE=subcutaneous emphysema, B=blisters, V=vesicles, N=necrosis)	intact
Other components of Braden score: special bed, sensory pressure, moisture, activity, friction/shear (>18=no risk, 15–16=low risk, 13–14=moderate risk, ≤12=high risk)	sensory pressure, 12

Noland, Ruth, Female, 77 y.o.
Allergies: NKA   **Code:** FULL   **Isolation:** None
Pt. Location: RM 926   **Physician:** S. Young   **Admit Date:** 8/12

Orders:

Administer 0.6 mg/kg intravenous rtPA over one hour with 10% of total dose given as an initial intravenous bolus over one minute. Total dose of 67.5 mg.

Vital signs: q 15 minutes × 2 hours; then q 30 minutes × 6 hours; then q 1 hour × 16 hours.

Neuro checks: Level of consciousness and extremity weakness (use NIHSS scoring): q 30 minutes × 6 hours, then q 1 hour × 16 hours

IV: 0.9 NS at 75 cc/hr

O_2 at 2 liters/minute via nasal cannula (if needed to keep O_2 sats ≥ 95%)

Continuous cardiac monitoring

Strict Intake/Output records

Diet: NPO except medications for 24 hours

Noncontrast CT scan

Labs: Chem 16, coagulation times, CBC

Medications: Acetaminophen 650 mg po PRN for pain q 4 to 6 hours

No heparin, warfarin, or aspirin for 24 hours. After 24 hrs: CT to exclude intracranial hemorrhage before any anticoagulants.

Bedside swallowing assessment. Endoscopy with modified barium swallow.

Speech-language pathologist and dietitian to determine staged dysphagia diet

Nutrition:

History: Mr. Noland states that his wife has a good appetite. She has not followed any special diet except for trying to avoid fried foods, and she has stopped adding salt at the table. She made these changes several years ago.

24-hour recall: According to Mr. Noland, his wife ate the following:

Breakfast:	Orange juice–1 c
	Raisin bran–1 c with 6 oz 2% milk
	1 banana
	8 oz coffee with 2 tbsp 2% milk and sweetener
Lunch:	Chicken tortellini soup (cheese tortellini cooked in chicken broth)—2 c
	Saltine crackers—about 8
	Canned pears—2 halves
	6 oz iced tea with sweetener
Dinner:	Baked chicken (with skin)—4–6 oz. breast
	Baked potato—1 medium—with 2 tbsp margarine
	Seamed broccoli—approx. 1 c with 1 tsp.margarine
	Canned peaches in juice—6–8 slices
	6 oz iced tea with sweetener

Food allergies/intolerances/aversions: None

Previous nutrition therapy? No

Current diet: NPO

Noland, Ruth, Female, 77 y.o.
Allergies: NKA
Pt. Location: RM 926

Code: FULL
Physician: S. Young

Isolation: None
Admit Date: 8/12

Food purchase/preparation: Mrs. Noland and spouse
Vit/min intake: Multivitamin/mineral supplement daily, 500 mg calcium 3 × daily

Laboratory Results

	Ref. Range	8/12
Chemistry		
Sodium (mEq/L)	136–145	141
Potassium (mEq/L)	3.5–5.5	3.8
Chloride (mEq/L)	95–105	101
Carbon dioxide (CO_2, mEq/L)	23–30	27
BUN (mg/dL)	8–18	11
Creatinine serum (mg/dL)	0.6–1.2	0.9
Glucose (mg/dL)	70–110	82
Phosphate, inorganic (mg/dL)	2.3–4.7	4.2
Magnesium (mg/dL)	1.8–3	2.1
Calcium (mg/dL)	9–11	9.2
Albumin (g/dL)	3.5–5	4.2
Prealbumin (mg/dL)	16–35	22
Alkaline phosphatase (U/L)	30–120	119
ALT (U/L)	4–36	21
AST (U/L)	0–35	33
Lactate dehydrogenase (U/L)	208–378	241
Cholesterol (mg/dL)	120–199	210 !↑
HDL-C (mg/dL)	>55 F, >45 M	40 !↓
LDL (mg/dL)	<130	155 !↑
LDL/HDL ratio	<3.22 F <3.55 M	3.875 !↑
Triglycerides (mg/dL)	35–135 F 40–160 M	198 !↑
HbA_{1c} (%)	3.9–5.2	4.9
Hematology		
WBC (×10^3/mm³)	4.8–11.8	10
RBC (×10^6/mm³)	4.2–5.4 F 4.5–6.2 M	4.5
Hemoglobin (Hgb, g/dL)	12–15 F 14–17 M	12.7

(Continued)

Noland, Ruth, Female, 77 y.o.
Allergies: NKA   **Code:** FULL   **Isolation:** None
Pt. Location: RM 926   **Physician:** S. Young   **Admit Date:** 8/12

Laboratory Results *(Continued)*

	Ref. Range	8/12
Hematocrit (Hct, %)	37–47 F 40–54 M	38
Mean cell volume (µm³)	80–96	82
Mean cell Hgb (pg)	26–32	27
Mean cell Hgb content (g/dL)	31.5–36	33.2
Platelet count (×10³/mm³)	140–440	154

Case Questions

I.   Understanding the Disease and Pathophysiology

1.   Define *stroke*. Describe the differences between ischemic and hemorrhagic strokes.

2.   The noncontrast CT confirmed that Mrs. Noland had suffered a lacunar ischemic stroke—NIH Stroke Scale Score of 14. What does Mrs. Noland's score for the NIH stroke scale indicate?

3.   What factors place an individual at risk for stroke?

4.   What specific signs and symptoms noted with Mrs. Noland's exam and history are consistent with her diagnosis? Which symptoms place Mrs. Noland at nutritional risk? Explain your rationale.

5.   What is rtPA? Why was it administered?

II.   Understanding the Nutrition Therapy

6.   Define *dysphagia*. What is the primary nutrition implication of dysphagia?

7.   Describe the four phases of swallowing:

   a.   Oral preparation

   b.   Oral transit

   c.   Pharyngeal

   d.   Esophageal

8. The National Dysphagia Diet defines three levels of solid foods and four levels of fluid consistency to be used when planning a diet for someone with dysphagia. Describe each of these levels of diet modifications.

9. It is determined that Mrs. Noland's dysphagia is centered in the esophageal transit phase and she has reduced esophageal peristalsis. Which dysphagia diet level is appropriate to try with Mrs. Noland?

10. Describe a bedside swallowing assessment. What are the background and training requirements of a speech-language pathologist?

11. Describe a modified barium swallow or fiberoptic endoscopic evaluation of swallowing.

12. Thickening agents and specialty food products are often used to provide texture changes needed for the dysphagia diet. Describe one of these products and how it may be incorporated into the diet.

III. Nutrition Assessment

13. Mrs. Noland's usual body weight is approximately 165 lbs. Calculate and interpret her BMI.

14. Estimate Mrs. Noland's energy and protein requirements. Should weight loss or weight gain be included in this estimation? What is your rationale?

15. Using Mrs. Noland's usual dietary intake, calculate the total number of kilocalories she consumed as well as the energy distribution of kilocalories for protein, carbohydrate, and fat.

16. Compare this to the nutrient recommendations for an individual with hyperlipidemia and hypertension. Should these recommendations apply for Mrs. Noland during this acute period after her stroke?

17. Estimate Mrs. Noland's fluid needs using the following methods: weight; age and weight; and energy needs.

18. Which method of fluid estimation appears most reasonable for Mrs. Noland? Explain.

19. Review Mrs. Noland's labs upon admission. Identify any that are abnormal. For each abnormal value, explain the reason for the abnormality and describe the clinical significance and nutritional implications for Mrs. Noland.

IV.   Nutrition Diagnosis

20. Select two nutrition problems and complete the PES statement for each.

V.   Nutrition Intervention

21. For each of the PES statements that you have written, establish an ideal goal (based on the signs and symptoms) and an appropriate intervention (based on the etiology).

VI.   Nutrition Monitoring and Evaluation

22. To maintain or attain normal nutritional status while reducing danger of aspiration and choking, the texture (of foods) and/or viscosity (of fluids) are personalized for a patient with dysphagia. In the following table, define each term used to describe characteristics of foods and give an example.

Term	Definition	Example
Consistency		
Texture		
Viscosity		

23. Using Mrs. Noland's 24-hour recall, make suggestions for consistency changes or food substitutions (if needed) to Mrs. Noland and her family.

Orange juice _____

_____

Raisin bran _____

_____

2% milk _____

_____

Banana _____

_____

Coffee _____

_____

Sweetener _____

_____

Chicken tortellini soup _____

_____

Saltine crackers _____

_____

Canned pears _____

_____

Iced tea _____

_____

Baked chicken _____

_____

Baked potato _____

_____

Steamed broccoli _____

_____

Margarine _____

_____

Canned peaches _____

_____

24. Describe Mrs. Noland's potential nutritional problems upon discharge. What recommendations could you make to her husband to prevent each problem you identified? How would you monitor her progress?

25. Would Mrs. Noland be an appropriate candidate for a stroke rehabilitation program? Why or why not?

Bibliography

Academy of Nutrition and Dietetics. *Nutrition Care Manual.* Cerebrovascular Disease. Accessed 11/3/12 from http://nutritioncaremanual.org/topic.cfm?ncm_heading =Diseases%2FConditions&ncm_toc_id=8233.

Adams HP Jr, del Zoppo G, Alberts MJ, et al. Guidelines for the early management of adults with ischemic stroke: A guideline from the American Heart Association/ American Stroke Association Stroke Council, Clinical Cardiology Council, Cardiovascular Radiology and Intervention Council, and the Atherosclerotic Peripheral Vascular Disease and Quality of Care Outcomes in Research Interdisciplinary Working Groups: the American Academy of Neurology affirms the value of this guideline as an educational tool for neurologists. *Stroke.* 2007;38:1655–1711.

Brody RA, Tougher-Decker R, VonHagen S, et al. Role of registered dietitians in dysphagia screening. *J Am Diet Assoc.* 2000;100(9):1029–1037.

Bushnell CD, Johnston DCC, Goldstein LB. Retrospective assessment of initial stroke severity. Comparison of the NIH Stroke Scale and the Canadian Neurological Scale. *Stroke.* 2001;32:656.

Brynningsen PK, Damsgaard EM, Husted SE. Improved nutritional status in elderly patients 6 months after stroke. *J Nutr Health Aging.* 2007;11(1):75–9.

Dennis M, Lewis S, Cranswick G, Forbes J; FOOD Trial Collaboration. FOOD: a multicentre randomised trial evaluating feeding policies in patients admitted to hospital with a recent stroke. *Health Technol Assess.* 2006;10(2):iii–iv, ix–x, 1–120.

Grise EM, Adeoye O, Lindsell C, et al. Emergency department adherence to American Heart Association Guidelines for blood pressure management in acute ischemic stroke. *Stroke.* 2012;43:557–559.

Maasland L, Koudstaal PJ, Habbema JD, Dippel DW. Knowledge and understanding of disease process, risk factors and treatment modalities in patients with a recent TIA or minor ischemic stroke. *Cerebrovasc Dis.* 2007;23(5–6):435–40.

Morgenstern LB, Hemphill JC, Anderson C, et al. AHA/ ASA Guideline: Guidelines for the management of spontaneous intracerebral hemorrhage. *Stroke.* 2010;41:2108–2129.

Internet Resources

American Speech-Language-Hearing Association: http://www.asha.org

American Stroke Association: A Division of the American Heart Association: http://www.strokeassociation.org

National Institute of Neurological Disorders and Stroke: http://www.ninds.nih.gov/disorders/stroke/stroke.htm

NIH Stroke Scale and Scoring: http://www.strokecenter .org/trials/scales/nihss.html

Video Link for Evaluation of Stages of Swallowing: http://www.linkstudio.info/images/portfolio/medani /Swallow.swf

Progressive Neurological Disease: Parkinson's Disease

Objectives

After completing this case, the student will be able to:

1. Demonstrate a working knowledge of the pathophysiology of Parkinson's disease and current medical care.
2. Describe the potential drug–nutrient interactions for Parkinson's disease.
3. Discuss the current understanding of nutrient supplementation in treatment of Parkinson's disease.
4. Assess for and identify nutritional health risks in dysphagia.
5. Analyze nutrition assessment data to evaluate nutritional status, identify specific nutrition problems, and document corresponding PES statements.
6. Use current assessment data to assess for and diagnose malnutrition.
7. Interpret laboratory parameters for nutritional implications and significance.
8. Develop a nutrition care plan—with appropriate measurable goals, interventions, and strategies for monitoring and evaluation—that addresses the nutrition diagnoses of this case.

Rita McCormick is admitted to University Hospital with fever, cough, and respiratory symptoms for an assessment to R/O pneumonia. Her family states that she is experiencing coughing, choking, and difficulty eating. She was diagnosed with Parkinson's disease 10 years ago and has experienced continued progression of her disease.

McCormick, Rita, Female, 69 y.o.
Allergies: NKA   **Code:** FULL   **Isolation:** None
Pt. Location: 1101   **Physician:** S. Goldman   **Admit Date:** 2/13

Patient Summary: The patient is a 69-year-old female with Parkinson's disease. At her physician's office, she presented with fever and increased white blood cell count. Her family relates that she has had increasing difficulty eating. She often coughs and appears to choke during meals.

History:
Onset of disease: Diagnosed initially 10 years ago
Medical history: Parkinson's disease
Surgical history: Bilateral salpingo-oophorectomy
Medications at home: Sinemet: carbidopa/levodopa, 50/200 mg controlled-release tablet twice daily; citalopram 20 mg daily; esomeprazole 20 mg daily; omega-3-fatty acids 1000 mg daily
Tobacco use: Quit over 30 years ago
Alcohol use: Socially
Family history: Mother—Alzheimer's disease; father—CAD

Demographics:
Marital status: Widowed
Years education: 12 years
Language: English
Occupation: Retired hairdresser
Household members: Lives with son (45) and his wife (42)
Ethnicity: Caucasian
Religious affiliation: Methodist

Admitting History/Physical:
Chief complaint: "Every time I eat, something gets stuck in my throat. I cough and feel like I'm choking. It just scares me to eat."

Vital Signs: Temp: 101.5   Pulse: 80   Resp rate: 22
                    BP: 135/85   Height: 60"   Weight: 90 lbs
                    UBW: 110 lbs (six months previous)

Heart: RRR. No gallops or rubs, point of maximal impulse at the fifth intercostal space in the midclavicular line.
HEENT: Head: Normocephalic; dry, dull hair; sunken cheeks; evidence of temporal wasting
          Eyes: Glasses for myopia; bilateral redness, fissured eyelid corners; reduced subcutaneous fat within orbital area
          Ears: Tympanic membranes normal
          Nose: Dry mucous membranes without lesions
          Throat: Slightly dry mucous membranes without exudates or lesions
Genitalia: WNL
Neurologic: Alert and oriented × 3; decreased blink reflex; positive palmomental; diminished postural reflexes. Family reports three falls in past 6 months. UPDRS: Stage 3.

McCormick, Rita, Female, 69 y.o.
Allergies: NKA
Pt. Location: 1101

Code: FULL
Physician: S. Goldman

Isolation: None
Admit Date: 2/13

Extremities: Reduced strength, evidence of muscle loss in quadriceps and gastrocnemius; koilonychias, bilateral tremor
Skin: Warm, dry, poor turgor, angular stomatitis and cheilosis noted on lips
Chest/lungs: Respirations rapid; crackles, rhonchi noted
Peripheral vascular: Bilateral, 3 + pedal pulses
Abdomen: Normal bowel sounds—in all regions, without masses or splenomegaly

Nursing Assessment	2/13
Abdominal appearance (concave, flat, rounded, obese, distended)	flat
Palpation of abdomen (soft, rigid, firm, masses, tense)	soft
Bowel function (continent, incontinent, flatulence, no stool)	continent
Bowel sounds (P=present, AB=absent, hypo, hyper)	
RUQ	P
LUQ	P
RLQ	P
LLQ	P
Stool color	dark brown
Stool consistency	hard
Tubes/ostomies	none
Genitourinary	
Urinary continence	yes
Urine source	clean catch
Appearance (clear, cloudy, yellow, amber, fluorescent, hematuria, orange, blue, tea)	cloudy, amber
Integumentary	
Skin color	pale
Skin temperature (DI=diaphoretic, W=warm, dry, CL=cool, CLM=clammy, CD+=cold, M=moist, H=hot)	W
Skin turgor (good, fair, poor, TENT=tenting)	poor
Skin condition (intact, EC=ecchymosis, A=abrasions, P=petechiae, R=rash, W=weeping, S=sloughing, D=dryness, EX=excoriated, T=tears, SE=subcutaneous emphysema, B=blisters, V=vesicles, N=necrosis)	EC
Mucous membranes (intact, EC=ecchymosis, A=abrasions, P=petechiae, R=rash, W=weeping, S=sloughing, D=dryness, EX=excoriated, T=tears, SE=subcutaneous emphysema, B=blisters, V=vesicles, N=necrosis)	intact, D
Other components of Braden score: special bed, sensory pressure, moisture, activity, friction/shear (>18=no risk, 15–16=low risk, 13–14=moderate risk, ≤12=high risk)	15

McCormick, Rita, Female, 69 y.o.

Allergies: NKA	**Code:** FULL	**Isolation:** None
Pt. Location: 1101	**Physician:** S. Goldman	**Admit Date:** 2/13

Admission Orders:

Laboratory: CBC, Chem 27, CXR
Initiate azithromycin 500 mg IV once daily
Vital Signs: Per protocol
Diet: NPO—Nutrition consult
Gastroenterology consult for MBS
SLP Consult: Flexible endoscopic evaluation of swallowing

Nutrition:

Meal type: Previously regular at home—now NPO
History: Mostly liquids, because solids are difficult to swallow.

Usual dietary intake: Before developing current difficulty with swallowing, Mrs. McCormick usually ate the following:

Breakfast: ½ scrambled egg, ½ slice toast or English muffin, 1 tsp jelly, coffee with 2% milk and artificial sweetener
Lunch: ½ ham or turkey sandwich, 6–7 chips, iced tea with artificial sweetener
Dinner: ¾ c spaghetti with ½ c meat sauce, 2–3 tbsp green peas or other vegetable, ½ c canned fruit cocktail, ½ slice bread with 1 tsp butter, iced tea with artificial sweetener

Intake/Output

Date			2/13 0701–2/14 0700			
Time			0701–1500	1501–2300	2301–0700	Daily total
IN		P.O.	0	0	0	0
		I.V.	400	400	400	1200
		(mL/kg/hr)	(1.22)	(1.22)	(1.22)	(1.22)
		I.V. piggyback	250	0	0	250
		TPN				
		Total intake	650	400	400	1450
		(mL/kg)	(15.89)	(9.78)	(9.78)	(35.44)
OUT		Urine	320	310	200	830
		(mL/kg/hr)	(0.98)	(0.95)	(0.61)	(0.84)
		Emesis output				
		Other				
		Stool	×1			
		Total output	320	310	200	830
		(mL/kg)	(7.82)	(7.58)	(4.89)	(20.29)
Net I/O			+330	+90	+200	+620
Net since admission (2/13)			+330	+420	+620	+620

McCormick, Rita, Female, 69 y.o.
Allergies: NKA
Pt. Location: 1101

Code: FULL
Physician: S. Goldman

Isolation: None
Admit Date: 2/13

Laboratory Results

	Ref. Range	2/13 0830
Chemistry		
Sodium (mEq/L)	136–145	145
Potassium (mEq/L)	3.5–5.5	4.1
Chloride (mEq/L)	95–105	101
Carbon dioxide (CO_2, mEq/L)	23–30	24
Bicarbonate (mEq/L)	21–32	22
BUN (mg/dL)	8–18	14
Creatinine serum (mg/dL)	0.6–1.2	1.1
BUN/Crea ratio	10.0–20.0	12.7
Glucose (mg/dL)	70–110	78
Phosphate, inorganic (mg/dL)	2.3–4.7	2.9
Magnesium (mg/dL)	1.8–3	1.9
Calcium (mg/dL)	9–11	8.9 !↓
Protein, total (g/dL)	6–8	5.8 !↓
Albumin (g/dL)	3.5–5	3.2 !↓
Prealbumin (mg/dL)	16–35	15 !↓
Ammonia (NH_3, μmol/L)	9–33	9.1
Alkaline phosphatase (U/L)	30–120	110
ALT (U/L)	4–36	8
AST (U/L)	0–35	22
CPK (U/L)	30–135 F 55–170 M	45
Lactate dehydrogenase (U/L)	208–378	209
Cholesterol (mg/dL)	120–199	109 !↓
HDL-C (mg/dL)	>55 F, >45 M	42 !↓
VLDL (mg/dL)	7–32	11
LDL (mg/dL)	<130	56
LDL/HDL ratio	<3.22 F <3.55 M	3.09
Triglycerides (mg/dL)	35–135 F 40–160 M	55
Hematology		
WBC ($\times 10^3$/mm^3)	4.8–11.8	11.9 !↑
RBC ($\times 10^6$/mm^3)	4.2–5.4 F 4.5–6.2 M	3.9 !↓

(Continued)

McCormick, Rita, Female, 69 y.o.
Allergies: NKA
Pt. Location: 1101

Code: FULL
Physician: S. Goldman

Isolation: None
Admit Date: 2/13

Laboratory Results *(Continued)*

	Ref. Range	2/13 0830
Hemoglobin (Hgb, g/dL)	12–15 F 14–17 M	11.5 !↓
Hematocrit (Hct, %)	37–47 F 40–54 M	35 !↓
Mean cell volume (µm³)	80–96	74 !↓
Mean cell Hgb (pg)	26–32	23 !↓
Mean cell Hgb content (g/dL)	31.5–36	28 !↓
RBC distribution (%)	11.6–16.5	11.9
Platelet count (×10³/mm³)	140–440	210
Transferrin (mg/dL)	250–380 F 215–365 M	392 !↑
Ferritin (mg/mL)	20–120 F 20–300 M	11 !↓

Case Questions

I. Understanding the Diagnosis and Pathophysiology

1. Describe our current understanding of the pathophysiology of Parkinson's disease.

2. How does this pathophysiology translate into the cardinal signs and symptoms of Parkinson's? Which may contribute to nutritional risk? Which of these are noted in Mrs. McCormick's history and physical?

3. Currently the disease status is often described using the Unified Parkinson's Disease Rating Scale (UPDRS). What is this, and how would you interpret Mrs. McCormick's staging?

4. Identify and describe the primary medical interventions that are used for treatment of Parkinson's disease.

5. One of the major medications used to treat Parkinson's is levodopa. How does diet potentially play a role in this medication's efficacy? Identify all drug–nutrient interactions for Mrs. McCormick's prescribed medications.

II. Understanding the Nutrition Therapy

6. Define *dysphagia*. What medical and nutritional complications may be associated with dysphagia?

7. Mrs. McCormick is having an SLP consult and a FEES test completed. What training does a speech-language therapist have? What is a FEES? What type of information will this provide for the MD and the RD?

8. What is an MBS? What information will this test provide?

9. Using the Barichella (2009) review article, discuss the major nutrition concerns associated with Parkinson's. Provide evidence for any of these that Mrs. McCormick may be experiencing.

10. Mrs. McCormick's MD discusses the potential for the placement of a PEG tube for this patient. Provide any justification for nutrition support using the appropriate criteria.

11. There are a number of supplements that have been studied as components of Parkinson's treatment. Discuss the use of coenzyme Q, omega-3-fatty acids, and creatine and their potential efficacy for patients with Parkinson's.

III.   Nutrition Assessment

12. Evaluate Mrs. McCormick's weight by calculating her BMI and % UBW.

13. After examining Mrs. McCormick's history and physical, identify any clinical signs and symptoms that may alert you to a nutrient deficiency. What further assessments can you make to assess her risk for malnutrition?

14. Evaluate Mrs. McCormick's laboratory values. List all abnormal values and explain the likely cause for each abnormal value.

15. Determine Mrs. McCormick's energy and protein requirements. Explain the rationale for the method you used to calculate these requirements.

16. Assess Mrs. McCormick's diet prior to having difficulty swallowing. Compare her energy and protein intakes to her estimated nutrient needs.

IV.   Nutrition Diagnosis

17. Identify the pertinent nutrition problems and the corresponding nutrition diagnoses.

18. Write your PES statement for each nutrition problem.

V.   Nutrition Intervention

19. The National Dysphagia Diet defines three levels of solid foods and four levels of fluid consistency to be used when planning a diet for someone with dysphagia. Describe each of these levels of diet modifications.

20. The Dysphagia Outcome and Severity Scale (DOSS) is used to determine the nutrition prescription for a patient. Discuss this scale and how it corresponds to the level of dysphagia diet that is recommended.

21. The FEES and MBS indicate the following: "Patient demonstrates difficulty initiating the swallow and bolus was held in the mouth for an excessive amount of time. Spillage into the larynx is noted with some aspiration." Identify the diet you would recommend at this time.

22. Using the data collected during your nutrition assessment, what vitamin and mineral supplementation would you recommend?

VI. **Nutrition Monitoring and Evaluation**
23. Identify factors that you will need to monitor to ensure adequacy of her nutrition intervention.

24. What criteria would you use to determine whether Mrs. McCormick requires enteral feeding?

Bibliography

Bachmann CG, Trenkwalder C. Body weight in patients with Parkinson's disease. *Mov Disord.* 2006;21(11):1824–1830

Barichella M, Marczewska A, De Notaris R, et al. Special low-protein foods ameliorate postprandial off in patients with advanced Parkinson's disease. *Mov Disord.* 2006;21(10):1682–1687.

Bariachella M, Cerada E, Pezzoli G. Major nutritional issues in the management of Parkinson's disease. *Mov Disord.* 2009;24:1881–92.

Bender A, Koch W, Elstner M, et al. Creatine supplementation in Parkinson disease: A placebo-controlled randomized pilot trial. *Neurology.* 2006;67(7):1262–1264.

Chao J, Leung Y, Wang M, Chuen-Chung Chang R. Nutraceuticals and their preventive or potential value in Parkinson's disease. *Nutr Rev.* 2012;70:373–386.

Cushing ML, Traviss KA, Caine Sm. Parkinson's disease: Implications for nutritional care. *Canadian J Diet Prac Res.* 2002;63:81–7.

Delikanaki-Skaribas E, Trail M, Wong WW, Lai EC. Daily energy expenditure, physical activity, and weight loss in Parkinson's disease patients. *Mov Disord.* 2009;24(5):667–671.

Evatt ML. Nutritional therapies in Parkinson's disease. *Curr Treat Options Neurol.* 2007;9:198–204.

Marcason W. What are the primary nutritional issues for a patient with Parkinson's disease? *J Am Diet Assoc.* 2009;109:1316.

Nahikian-Nelms ML. Diseases of the upper gastrointestinal tract. In: Nelms M, Sucher K, Lacey K, Roth SL. *Nutrition Therapy and Pathophysiology.* 2nd ed. Belmont, CA: Wadsworth, Cengage Learning; 2011:340–375.

Nahikian-Nelms ML. Diseases and disorders of the neurological system. In: Nelms M, Sucher K, Lacey K, Roth SL. *Nutrition Therapy and Pathophysiology.* 2nd ed. Belmont, CA: Wadsworth, Cengage Learning; 2011:610–647.

NINDS NET-PD Investigators. A randomized clinical trial of coenzyme Q10 and GPI-1485 in early Parkinson disease. *Neurology.* 2007;68(1):20–28

O'Neil, Purdy M, Falk J, Gallo L. The dysphagia outcome and severity scale. *Dysphagia.* 1999;14:139–145.

Soo-Peary K, Hsu A. The pharmcokinetics and pharmocodynamics of levodopa in the treatment of Parkinson's disease. *Current Clinical Pharmacology.* 2007;2:234–243.

Storch A, Jost WH, Vieregge P, et al. Randomized, double-blind, placebo-controlled trial on symptomatic effects of coenzyme Q(10) in Parkinson disease. *Arch Neurol.* 2007;64(7):938–944.

Internet Resources

EMedicine–Overview of Aspiration Pneumonia: http://emedicine.medscape.com/article/296198-overview.

GI Motility Online: http://www.nature.com/gimo/contents/pt1/full/gimo28.html

Movement Disorders: http://www.movementdisorders.org/UserFiles/unified.pdf

National Institute of Neurological Disorders and Stroke: http://www.ninds.nih.gov/disorders/parkinsons_disease/parkinsons_disease.htm

Nutrition Care Manual: http://www.nutritioncaremanual.org

Parkinson's Disease Foundation: www.pdf.org

Alzheimer's Disease

Objectives

After completing this case, the student will be able to:

1. Identify and explain common nutritional problems associated with Alzheimer's disease.
2. Evaluate current recommendations for nutritional supplementation in wound healing.
3. Identify potential drug–nutrient interactions.
4. Analyze nutrition assessment data to evaluate nutritional status and identify specific nutrition problems.
5. Determine nutrition diagnoses and write appropriate PES statements.
6. Develop a nutrition care plan—with appropriate measurable goals, interventions, and strategies for monitoring and evaluation—that addresses the nutrition diagnoses of this case.

Ralph McCormick is an 89-year-old man admitted through the emergency room from the regional Veteran's Long-Term Care Facility/Alzheimer's Unit for multiple abrasions and a nonhealing wound on his right hip.

McCormick, Ralph, Male, 89 y.o.
Allergies: Penicillin   **Code:** DNR   **Isolation:** None
Pt. Location: RM 926   **Physician:** B. Byrd   **Admit Date:** 8/12

Patient Summary: Patient has a Stage III full thickness nonpressure wound (laceration) with purulent drainage and foul odor.

History:
Onset of disease: Patient history indicates that 4 years ago patient was having difficulty taking care of his life-long home and immediate medical needs. Reported some forgetfulness but could manage with assistance. His son moved him to an assisted living facility nearby and he was able to live there for approximately 1 year. His son reports that three years ago his "Alzheimer's" became acutely worse. His needs at the assisted care facility had significantly increased, and even with a nurse's aide coming in twice a day in addition to the assisted care, the patient's safety was questioned. Patient began wandering away from the facility and became combative. He was admitted to an Alzheimer's unit at a local nursing home, and then approximately 3 weeks later was transferred to the Veteran's Home when an opening was available.
Medical history: s/p MI × 2 at ages 45 and 62; HTN × 44 years
Surgical history: s/p 4 vessel CABG at age 62. R hip replacement 5 years ago.
Medications at home: Furosemide 80 mg daily, atenolol 25 mg daily, lisinopril 20 mg daily, Zocor 40 mg daily, haloperidol 0.5 mg AM and PM, warfarin 5 mg daily, donepezil 10 mg PM
Tobacco use: Yes, but quit over 20 years ago
Alcohol use: No
Family history: What? Cardiac disease, Alzheimer's; Who? Father, uncles, brother—All died before age 50 of MI. Mother had Alzheimer's or some type of dementia.

Demographics:
Marital status: Divorced—ex-spouse deceased. One adult child who lives in the area. Patient has been a resident of the Veteran's Long-Term Care Facility for past three years.
Years education: HS diploma
Language: English only
Occupation: Retired from phone company as telephone technician
Hours of work: N/A
Ethnicity: Caucasian
Religious affiliation: None

Admitting History/Physical:
Chief complaint: Transfer from Veteran's Long-Term Care Facility. Patient was in a combative episode with roommate. He fell and hit his hip on the corner of a bed. He is admitted for evaluation of this nonhealing wound.
General appearance: Frail, thin, elderly gentleman who is obviously confused and agitated.

Vital Signs: Temp: 100.3°F   Pulse: 85   Resp. rate: 32 (increases with minimal activity)
                          BP: 94/68   Height: 5'11"   Weight: 138 lbs

McCormick, Ralph, Male, 89 y.o.
Allergies: Penicillin   **Code:** DNR   **Isolation:** None
Pt. Location: RM 926   **Physician:** B. Byrd   **Admit Date:** 8/12

Heart: PMI sustained and displaced laterally; Normal S_1; S_2; $+S_3$ at apex
HEENT: Pupils are small and react to light sluggishly; ocular fundus is pale; negative thyromegaly and adenopathy. + JVD—increased 4 cm above sternal angle at 45°.
Genitalia: Normal
Rectal: Not performed
Neurologic: Disoriented to time, place, and person
Extremities: Cool to touch, pale with bruising. 2+ radial pulses, 1+ dorsalis pedis, and 1+ posterior tibial pulses bilaterally. DRT 2+ and symmetrical; strength 2/5 throughout.
Skin: Transparent with decreased turgor, pale, cool with multiple ecchymoses; open, draining purulent wound approximately 2 cm × 2 cm × 8 cm located on right posterior thigh
Chest/lungs: Clear to auscultation and percussion with no rubs
Abdomen: Nontender with bowel sounds present

Nursing Assessment	8/12
Abdominal appearance (concave, flat, rounded, obese, distended)	rounded
Palpation of abdomen (soft, rigid, firm, masses, tense)	soft
Bowel function (continent, incontinent, flatulence, no stool)	continent
Bowel sounds (P=present, AB=absent, hypo, hyper)	
RUQ	P
LUQ	P
RLQ	P
LLQ	P
Stool color	light brown
Stool consistency	soft to liquid
Tubes/ostomies	N/A
Genitourinary	
Urinary continence	catheter
Urine source	catheter
Appearance (clear, cloudy, yellow, amber, fluorescent, hematuria, orange, blue, tea)	clear, yellow
Integumentary	
Skin color	pale
Skin temperature (DI=diaphoretic, W=warm, dry, CL=cool, CLM=clammy, CD+=cold, M=moist, H=hot)	CL
Skin turgor (good, fair, poor, TENT=tenting)	fair

(Continued)

McCormick, Ralph, Male, 89 y.o.
Allergies: Penicillin   **Code:** DNR   **Isolation:** None
Pt. Location: RM 926   **Physician:** B. Byrd   **Admit Date:** 8/12

Nursing Assessment *(Continued)*

Nursing Assessment	8/12
Skin condition (intact, EC=ecchymosis, A=abrasions, P=petechiae, R=rash, W=weeping, S=sloughing, D=dryness, EX=excoriated, T=tears, SE=subcutaneous emphysema, B=blisters, V=vesicles, N=necrosis)	EC
Mucous membranes (intact, EC=ecchymosis, A=abrasions, P=petechiae, R=rash, W=weeping, S=sloughing, D=dryness, EX=excoriated, T=tears, SE=subcutaneous emphysema, B=blisters, V=vesicles, N=necrosis)	intact
Other components of Braden score: special bed, sensory pressure, moisture, activity, friction/shear (>18=no risk, 15–16=low risk, 13–14=moderate risk, ≤12=high risk)	activity, 13

Orders:
Culture wound exudate
Start pt on 1.5 grams of ampicillin-sulbactam IV every six hours
Scheduled for initial wound debridement with consult for wound management
Nutrition consult

Nutrition:
Usual dietary intake: (prior to current illness)
The inpatient RD at the hospital obtained the following information from a discussion with the dietary manager employed at the Veteran's Home. As a resident in a long-term care facility, the patient was prescribed a regular diet with specific modifications made for the Alzheimer's unit. These include finger foods and access to snacks for a minimum of three times daily.
Mr. McCormick did not have a good appetite and had difficulty attending to the task of eating. He required assistance with all meals. His best meal of the day was in the morning with cornflakes, banana, and high-calorie, high-protein shake. Intake after that was highly variable. Patient's son indicates his dad's weight had been about 170 lbs until four years ago. He lost quite a bit of weight during the transition to assisted living and then to a long-term care facility. Weight has been stable for the past six months, but overall he has lost over 30 lbs over the past four years, with most of that in the first year.

Previous nutrition therapy? None
Food purchase/preparation: Long-term care facility
Vitamin intake: None

McCormick, Ralph, Male, 89 y.o.
Allergies: Penicillin
Pt. Location: RM 926

Code: DNR
Physician: B. Byrd

Isolation: None
Admit Date: 8/12

Laboratory Results

	Ref. Range	8/12 0926	
Chemistry			
Sodium (mEq/L)	136–145	136	
Potassium (mEq/L)	3.5–5.5	3.5	
Chloride (mEq/L)	95–105	96	
Carbon dioxide (CO_2, mEq/L)	23–30	27	
BUN (mg/dL)	8–18	22	!↑
Creatinine serum (mg/dL)	0.6–1.2	1.3	!↑
Glucose (mg/dL)	70–110	82	
Phosphate, inorganic (mg/dL)	2.3–4.7	2.5	
Magnesium (mg/dL)	1.8 3	1.9	
Calcium (mg/dL)	9–11	9	
Bilirubin, direct (mg/dL)	<0.3	0.1	
Protein, total (g/dL)	6–8	5.5	!↓
Albumin (g/dL)	3.5–5	2.9	!↓
Prealbumin (mg/dL)	16–35	14	!↓
Ammonia (NH_3, μmol/L)	9–33	24	
Alkaline phosphatase (U/L)	30–120	80	
ALT (U/L)	4–36	25	
AST (U/L)	0–35	21	
C-reactive protein (mg/dL)	<1.0	5.1	!↑
CPK (U/L)	30–135 F 55–170 M	56	
Cholesterol (mg/dL)	120–199	155	
HDL-C (mg/dL)	>55 F, >45 M	33	!↓
LDL (mg/dL)	<130	121	
LDL/HDL ratio	<3.22 F <3.55 M	3.67	!↑
Triglycerides (mg/dL)	35–135 F 40–160 M	153	
HbA_{1c} (%)	3.9–5.2	4.6	

(Continued)

McCormick, Ralph, Male, 89 y.o.
Allergies: Penicillin
Pt. Location: RM 926

Code: DNR
Physician: B. Byrd

Isolation: None
Admit Date: 8/12

Laboratory Results *(Continued)*

	Ref. Range	8/12 0926	
Coagulation (Coag)			
PT (sec)	12.4–14.4	13.1	
Hematology			
WBC ($\times 10^3$/mm^3)	4.8–11.8	16.0	!↑
RBC ($\times 10^6$/mm^3)	4.2–5.4 F 4.5–6.2 M	5.1	
Hemoglobin (Hgb, g/dL)	12–15 F 14–17 M	13.5	!↓
Hematocrit (Hct, %)	37–47 F 40–54 M	39	!↓
Mean cell volume (µm^3)	80–96	77	!↓
Mean cell Hgb (pg)	26–32	24	!↓
Mean cell Hgb content (g/dL)	31.5–36	30	!↓
Platelet count ($\times 10^3$/mm^3)	140–440	145	
Transferrin (mg/dL)	250–380 F 215–365 M	165	!↓
Ferritin (mg/mL)	20–120 F 20–300 M	18	!↓
Hematology, Manual Diff			
Lymphocyte (%)	15–45	10	!↓
Monocyte (%)	3–10	5	
Eosinophil (%)	0–6	1	
Segs (%)	0–60	50	

Case Questions

I.   Understanding the Disease and Pathophysiology

1. Define *dementia*. Define *Alzheimer's disease* (AD). How do they differ?

2. What is the current theory regarding the etiology of AD? How is AD diagnosed?

3. Based on Mr. McCormick's medical record, what are his other concurrent diagnoses? Could any of these contribute to his symptoms?

4. What are the current medical interventions available for the management of AD? What are the goals of these interventions?

5. Mr. McCormick has a Stage III full thickness nonpressure wound. What does that mean?

6. Describe the normal stages of wound healing.

II.   Understanding the Nutrition Therapy

7. Name a minimum of three factors that support wound healing. Name a minimum of three factors that may impair wound healing. Identify the most probable factors that may have contributed to Mr. McCormick's poor wound healing.

8. Describe the potential roles of vitamin A, vitamin C, vitamin E, zinc, copper, glutamine, and arginine in wound healing.

III.   Nutrition Assessment

9. Assess this patient's available anthropometric data. Calculate % UBW and BMI. Which of these is the most pertinent in identifying the patient's nutrition risk? Why?

10. Discuss the progressive weight loss Mr. McCormick has experienced. Why is this of concern? What factors may have contributed to this weight loss?

11. Calculate energy and protein requirements for Mr. McCormick.

12. How would you determine the levels of micronutrients that Mr. McCormick needs?

13. Identify all medications that Mr. McCormick is prescribed. Describe the basic function of each.

14. a. Using his admission chemistry and hematology values, which biochemical measures are abnormal?

    b. Which values can be used to further assess his nutritional status? Explain.

    c. Which laboratory measures are related to his infection and wound?

    d. Which laboratory measures are related to any of Mr. McCormick's concurrent diagnoses? Explain.

15. Do you think this patient is malnourished? If so, why? Will this impact the success of interventions for wound healing?

16. Identify issues related to Mr. McCormick's primary diagnosis that could potentially interfere with his ability to consume an adequate diet.

17. Are residents in a long-term care facility at higher nutritional risk than elders living independently? Why or why not?

IV. Nutrition Diagnosis
18. Select two nutrition problems and complete the PES statement for each.

V.   Nutrition Intervention

19. For each of the PES statements that you have written, establish an ideal goal (based on the signs and symptoms) and an appropriate intervention (based on the etiology).

20. What specific dosage recommendations would you make about supplementing zinc, copper, vitamin A, vitamin C, and arginine to promote wound healing for Mr. McCormick?

21. Mr. McCormick drinks high-calorie, high-protein milkshakes at the Veteran's Home. What products might you recommend for him during his hospitalization?

22. What specific interventions might you recommend for a patient with Alzheimer's that could improve his oral intake during a hospitalization?

VI.   Nutrition Monitoring and Evaluation

23. What measures of the adequacy of oral intake would be appropriate to use during Mr. McCormick's hospitalization?

24. If Mr. McCormick's intake is inadequate, is he a candidate for enteral feeding? Outline the pros and cons for recommending nutrition support for this patient. What are the ethical considerations?

Bibliography

Academy of Nutrition and Dietetics. Nutrition Care Manual. Alzheimer's disease. Accessed 11/7/12 from http://nutritioncaremanual.org/topic.cfm?ncm_heading=Diseases%2FConditions&ncm_toc_id=32269.

Academy of Nutrition and Dietetics. Nutrition Care Manual. Dementia. Accessed 11/7/12 from http://nutritioncaremanual.org/topic.cfm?ncm_heading=Diseases%2FConditions&ncm_toc_id=32339.

Agency for Healthcare Research and Quality. National Guideline Clearinghouse. Pressure ulcer treatment recommendations. In: Prevention and treatment of pressure ulcers: Clinical practice guideline. Accessed 11/17/12 from http://www.guidelines.gov/content.aspx?id=25139&search=npuap.

Doley J. Nutrition management of pressure ulcers. *Nutr Clin Pract.* 2010;25:50–60.

Genius J, Klafki H, Benninghoff J, Esselmann H, Wiltfang. Current application of neurochemical biomarkers in the prediction and differential diagnosis of Alzheimer's disease and other neurodegenerative dementias. *Eur Arch Psychiatry Clin Neurosci.* 2012;262(Suppl 2):S71–S77.

Hyman BT, Phelps CH, Beach TG, et al. National Institute on Aging—Alzheimer's Association guidelines for the neuropathologic assessment of Alzheimer's disease. *Alzheimer's and Dementia.* 2012;8:1–13.

Kimchi E, Desai AK, Grossberg GT. New Alzheimer's disease guidelines: Implications for clinicians. *Current Psychiatry.* 2012;11:15–20.

Nelms MN, Habash D. Nutrition assessment: Foundation of the nutrition care process. In: Nelms M, Sucher K, Lacey K, Roth SL. *Nutrition Therapy and Pathophysiology.* 2nd ed. Belmont, CA: Wadsworth, Cengage Learning; 2011:34–65.

Nelms MN, Frazier C. Cellular and physiological response to injury: The role of the immune system. In: Nelms M, Sucher K, Lacey K, Roth SL. *Nutrition Therapyand Pathophysiology.* 2nd ed. Belmont, CA: Wadsworth, Cengage Learning; 2011:149–182.

Irwin KJ, Hansen-Petrik M. Diseases and disorders of the neurological system. In: Nelms M, Sucher K, Lacey K, Roth SL. *Nutrition Therapy and Pathophysiology.* 2nd ed. Belmont, CA: Wadsworth, Cengageg Learning; 2011:609–647.

Skipper A. Challenges in nutrition, pressure ulcers, and wound healing. *Nutr Clin Pract.* 2010;25:13–15.

Internet Resources

Abbott Nutrition: Pressure ulcer prevention and treatment: the relationship between lean body mass, nutrition, and healing: http://abbottnutrition.com/downloads/resourcecenter/pressure-ulcer-prevention-and-treatment.pdf

Alzheimer's Association: www.alz.org

Alzheimer Research Forum: http://www.alzforum.org

Mayo Clinic: http://www.mayoclinic.com/health/alzheimers-disease/DS00161

National Institute of Neurological Disorders and Stroke: http://www.ninds.nih.gov/disorders/alzheimersdisease/alzheimersdisease.htm

Unit Ten

NUTRITION THERAPY FOR PULMONARY DISORDERS

The two cases in this section portray the interrelationship between nutrition and the respiratory system. In a healthy individual, the respiratory system receives oxygen for cellular metabolism and expires waste products—primarily carbon dioxide. Fuels—carbohydrate, protein, and lipid—are metabolized, using oxygen and producing carbon dioxide. The type of fuel an individual receives can affect physiological conditions and interfere with normal respiratory function.

Nutritional status and pulmonary function are interdependent. Malnutrition can evolve from pulmonary disorders and can contribute to declining pulmonary status. The incidence of malnutrition is common for people with COPD, ranging anywhere from 25 percent to 50 percent. In respiratory disease, maintaining nutritional status improves muscle strength needed for breathing, decreases risk of infection, facilitates weaning from mechanical ventilation, and improves ability for physical activity.

The American Thoracic Society defines chronic obstructive pulmonary disease (COPD) as a disease process of chronic airway obstruction caused by chronic bronchitis, emphysema, or a combination of both. These conditions place a significant burden on the health care systems in the United States.

In Cases 26 and 27, nutritional assessment and evaluation demonstrate the effects of COPD on nutritional status. As the patient is started on nutrition support in Case 27, you will examine the impact of nutrition on declining respiratory status.

Chronic Obstructive Pulmonary Disease

Objectives

After completing this case, the student will be able to:

1. Identify and explain common nutritional problems associated with this disease.
2. Identify effects of malnutrition on pulmonary status.
3. Identify effects of nutrient metabolism on pulmonary function.
4. Analyze nutrition assessment data to evaluate nutritional status and identify specific nutrition problems.

5. Determine nutrition diagnoses and write appropriate PES statements.
6. Plan interventions to increase an individual's intake of energy and protein.

Stella Bernhardt was initially diagnosed with stage 1 COPD (emphysema) five years ago. She is now admitted with increasing shortness of breath and a possible upper respiratory infection.

Bernhardt, Stella, Female, 62 y.o.
Allergies: NKA
Pt. Location: RM 704

Code: FULL
Physician: D. Bradshaw

Isolation: None
Admit Date: 1/25

Patient Summary: Acute exacerbation of COPD, increasing dyspnea, hypercapnia, r/o pneumonia

History:

Onset of disease: Initially diagnosed with stage 1 COPD (emphysema) five years ago. Medical records at last admission indicate pulmonary function tests: baseline FEV^1 = 0.7L, FVC = 1.5L, FEV^1/FVC 46%.

Medical history: No occupational exposures; history of bronchitis and upper respiratory infections during winter months for most of adult life. 4 live births; 2 miscarriages.

Surgical history: No surgeries

Medications at home: Combivent (metered-dose inhaler)—2 inhalations four times daily (each inhalation delivers 18 mcg ipratropium bromide; 130 mcg albuterol sulfate)

Tobacco use: Yes. 46 years, 1 PPD history—has not smoked for past year.

Alcohol use: No

Family history: What? CA. Who? Mother, 2 aunts died from lung cancer

Demographics:

Marital status: Married, lives with husband, aged 68, who has PMH of CAD

Years education: Completed two years of college

Language: English only

Occupation: Retired office manager for independent insurance agency

Hours of work: N/A

Ethnicity: Caucasian

Religious affiliation: Methodist

Admitting History/Physical:

Chief complaint: "I'm hardly able to do anything for myself right now. Even taking a bath or getting dressed makes me short of breath. My husband had to help me out of the shower this morning. I feel that I am gasping for air. I am coughing up a lot of phlegm that is a dark brownish-green. I am always short of breath, but I can tell when things change. I was at a church meeting with a lot of people—I might have caught something there. My husband says that I am confused in the morning. I know it is hard for me to get going in the morning. Do you think my confusion is related to my COPD?"

Vital Signs: Temp: 98.8°F   Pulse: 92   Resp rate: 22
                    BP: 130/88   Height: 5'3"   Weight: 119 lbs

Heart: Regular rate and rhythm; mild jugular distension noted
HEENT: Eyes: PERRLA, no hemorrhages
          Ears: Slight redness
          Nose: Clear
          Throat: Clear

Bernhardt, Stella, Female, 62 y.o.

Allergies: NKA	**Code:** FULL	**Isolation:** None
Pt. Location: RM 704	**Physician:** D. Bradshaw	**Admit Date:** 1/25

Genitalia: Deferred
Rectal: Not performed
Neurologic: Alert, oriented; cranial nerves intact
Extremities: 1+ bilateral pitting edema. No cyanosis or clubbing.
Skin: Warm, dry
Chest/lungs: Decreased breath sounds, percussion hyperresonant; prolonged expiration with wheezing; rhonchi throughout; using accessory muscles at rest
Abdomen: Liver, spleen palpable; nondistended, nontender, normal bowel sounds

Nursing Assessment	1/25
Abdominal appearance (concave, flat, rounded, obese, distended)	rounded
Palpation of abdomen (soft, rigid, firm, masses, tense)	soft
Bowel function (continent, incontinent, flatulence, no stool)	continent
Bowel sounds (P=present, AB=absent, hypo, hyper)	
RUQ	P
LUQ	P
RLQ	P
LLQ	P
Stool color	brown
Stool consistency	soft
Tubes/ostomies	N/A
Genitourinary	
Urinary continence	catheter
Urine source	catheter
Appearance (clear, cloudy, yellow, amber, fluorescent, hematuria, orange, blue, tea)	clear, yellow
Integumentary	
Skin color	pale
Skin temperature (DI=diaphoretic, W=warm, dry, CL=cool, CLM=clammy, CD+=cold, M=moist, H=hot)	W
Skin turgor (good, fair, poor, TENT=tenting)	fair
Skin condition (intact, EC=ecchymosis, A=abrasions, P=petechiae, R=rash, W=weeping, S=sloughing, D=dryness, EX=excoriated, T=tears, SE=subcutaneous emphysema, B=blisters, V=vesicles, N=necrosis)	intact

(Continued)

Bernhardt, Stella, Female, 62 y.o.
Allergies: NKA
Pt. Location: RM 704

Code: FULL
Physician: D. Bradshaw

Isolation: None
Admit Date: 1/25

Nursing Assessment *(Continued)*

Nursing Assessment	1/25
Mucous membranes (intact, EC=ecchymosis, A=abrasions, P=petechiae, R=rash, W=weeping, S=sloughing, D=dryness, EX=excoriated, T=tears, SE=subcutaneous emphysema, B=blisters, V=vesicles, N=necrosis)	intact
Other components of Braden score: special bed, sensory pressure, moisture, activity, friction/shear (>18 = no risk, 15–16 = low risk, 13–14 = moderate risk, ≤12 = high risk)	activity, 18

Orders:

O_2 1 L/minute via nasal cannula with humidity—keep O_2 saturation 90–91%
IVF D5 ½ NS with 20 mEq KCL @ 75 cc/hr
Solumedrol 10 mg/kg q 6 hr
Ancef 500 mg q 6 hr
Ipratropium bromide via nebulizer 2.5 mg q 30 minutes × 3 treatments then q 2 hr albuterol sulfate via nebulizer 4 mg q 30 minutes × 3 doses then 2.5 mg q 4 hr
ABGs q 6 hours
CXR—EPA/LAT
Sputum cultures and Gram stain

Nutrition:

General: Patient states that her appetite is poor: "I fill up so quickly—after just a few bites." Relates that meal preparation is difficult: "By the time I fix a meal, I am too tired to eat it." In the previous two days, she states that she has eaten very little. Increased coughing has made it very hard to eat: "I don't think food tastes as good, either. Everything has a bitter taste." Highest adult weight was 145–150 lbs (5 years ago). States that her family constantly tells her how thin she has gotten: "I haven't weighed myself for a while but I know my clothes are bigger." Dentures are present but fit loosely.

Usual dietary intake:
AM: Coffee, juice or fruit, dry cereal with small amount of milk
Lunch: Large meal of the day—meat; vegetables; rice, potato, or pasta, but patient admits she eats only very small amounts.
Dinner/evening meal: Eats very light in evening—usually soup, scrambled eggs, or sandwich. Drinks Pepsi® throughout the day (usually 3 12-oz cans).

24-hr recall: ½ c coffee with nondairy creamer, few sips of orange juice, ½ c oatmeal with 1 tsp sugar, ¾ c chicken noodle soup, 2 saltine crackers, ½ c coffee with nondairy creamer; sips of Pepsi® throughout the day and evening—estimated amount 32 oz

Bernhardt, Stella, Female, 62 y.o.
Allergies: NKA   **Code:** FULL   **Isolation:** None
Pt. Location: RM 704   **Physician:** D. Bradshaw   **Admit Date:** 1/25

Anthropometric data: Ht. 5'3", Wt. 119 lbs, UBW 145–150 lbs, last recorded weight: 139 lbs 1 year ago; MAC 19.05 cm, TSF 15 mm
Food allergies/intolerances/aversions: Avoids milk: "People say it will increase mucus production."
Previous nutrition therapy? No.
Food purchase/preparation: Self; "My daughters come and help sometimes."
Vit/min intake: None

Laboratory Results

	Ref. Range	1/25 0704	1/27 0704
Chemistry			
Sodium (mEq/L)	136–145	136	
Potassium (mEq/L)	3.5–5.5	3.7	
Chloride (mEq/L)	95–105	101	
Carbon dioxide (CO_2, mEq/L)	23–30	32 !↑	
BUN (mg/dL)	8–18	9	
Creatinine serum (mg/dL)	0.6–1.2	0.9	
Glucose (mg/dL)	70–110	92	
Phosphate, inorganic (mg/dL)	2.3–4.7	3.1	
Magnesium (mg/dL)	1.8–3	1.8	
Calcium (mg/dL)	9–11	9.1	
Bilirubin, direct (mg/dL)	<0.3	0.1	
Protein, total (g/dL)	6–8	5.8 !↓	
Albumin (g/dL)	3.5–5	3.3 !↓	
Prealbumin (mg/dL)	16–35	16	
Ammonia (NH_3, µmol/L)	9–33	25	
Alkaline phosphatase (U/L)	30–120	112	
ALT (U/L)	4–36	8	
AST (U/L)	0–35	22	
CPK (IU/L)	30–135 F 55–170 M	40	
Cholesterol (mg/dL)	120–199	145	
HDL-C (mg/dL)	>55 F, >45 M	61	
LDL (mg/dL)	<130	98	
LDL/HDL ratio	<3.22 F <3.55 M	1.61	

(Continued)

Bernhardt, Stella, Female, 62 y.o.
Allergies: NKA   **Code:** FULL   **Isolation:** None
Pt. Location: RM 704   **Physician:** D. Bradshaw   **Admit Date:** 1/25

Laboratory Results *(Continued)*

	Ref. Range	1/25 0704	1/27 0704
Triglycerides (mg/dL)	35–135 F 40–160 M	120	
HbA$_{1c}$ (%)	3.9–5.2	4.6	
Coagulation (Coag)			
PT (sec)	12.4–14.4	13.1	
Hematology			
WBC (×10^3/mm^3)	4.8–11.8	15.0 !↑	
RBC (×10^6/mm^3)	4.2–5.4 F 4.5–6.2 M	4 !↓	
Hemoglobin (Hgb, g/dL)	12–15 F 14–17 M	11.5 !↓	
Hematocrit (Hct, %)	37–47 F 40–54 M	35 !↓	
Hematology, Manual Diff			
Lymphocyte (%)	15–45	10 !↓	
Monocyte (%)	3–10	3	
Eosinophil (%)	0–6	1	
Segs (%)	0–60	83 !↑	
Arterial Blood Gases (ABGs)			
pH	7.35–7.45	7.29 !↓	7.4
pCO$_2$ (mm Hg)	35–45	50.9 !↑	40.1
SO$_2$ (%)	≥95	92 !↓	90.2 !↓
CO$_2$ content (mmol/L)	25–30	31 !↑	29.8
O$_2$ content (%)	15–22	12 !↓	18
Base excess (mEq/L)	>3		6.0
Base deficit (mEq/L)	<3	3.6 !↑	
HCO$_3^-$ (mEq/L)	24–28	29.6 !↑	24.7

Case Questions

I.   Understanding the Disease and Pathophysiology

1.   Mrs. Bernhardt was diagnosed with stage 1 emphysema/COPD five years ago. What criteria are used to classify this staging?

2.   COPD includes two distinct diagnoses. Outline the similarities and differences between emphysema and chronic bronchitis.

3.   What risk factors does Mrs. Bernhardt have for this disease?

4.   a.   Identify symptoms described in the MD's history and physical that are consistent with Mrs. Bernhardt's diagnosis. Then describe the pathophysiology that may be responsible for each symptom.

   b.   Now identify at least four features of the physician's physical examination consistent with her admitting diagnosis. Describe the pathophysiology that might be responsible for each physical finding.

5.   Mrs. Bernhardt's medical record indicates previous pulmonary function tests as follows: baseline FEV^1 = 0.7 L, FVC= 1.5 L, FEV^1/FVC 46%. Define FEV, FVC, and FEV/FVC, and indicate how they are used in the diagnosis of COPD. How can these measurements be used in treating COPD?

6.   Look at Mrs. Bernhardt's arterial blood gas values from the day she was admitted.

   a.   Why would arterial blood gases (ABGs) be drawn for this patient?

   b.   Define each of the following and interpret Mrs. Bernhardt's values:

      pH:

      pCO_2:

SO_2:

HCO_3^-:

    c. Mrs. Bernhardt was placed on oxygen therapy. What lab values tell you the therapy is working?

7. Mrs. Bernhardt has quit smoking. Shouldn't her condition now improve? Explain.

8. What is a respiratory quotient? How is this figure related to nutritional intake and respiratory status?

II.   Understanding Nutrition Therapy

9. What are the most common nutritional concerns for someone with COPD? Why is the patient diagnosed with COPD at higher risk for malnutrition?

10. Is there specific nutrition therapy prescribed for these patients?

III.   Nutrition Assessment

11. Calculate Mrs. Bernhardt's % UBW and BMI. Does either of these values indicate she is at nutritional risk? How would her 1+ bilateral pitting edema affect evaluation of her weight?

12. Calculate arm muscle area using the anthropometric data for mid-arm muscle circumference (MAC) and triceps skinfold (TSF). How would this data be interpreted?

13. Calculate Mrs. Bernhardt's energy and protein requirements. What activity and stress factors would you use? What is your rationale?

14. Using Mrs. Bernhardt's nutrition history and 24-hour recall as a reference, does she have an adequate oral intake? Explain.

15. Evaluate Mrs. Bernhardt's laboratory values. Identify those that are abnormal. Which of these may be used to assess her nutritional status?

16. Why may Mrs. Bernhardt be at risk for anemia? Do her laboratory values indicate that she is anemic?

17. What factors can you identify from her nutrition interview that probably contribute to her difficulty in eating?

IV.   Nutrition Diagnosis

18. Select two high-priority nutrition problems and complete the PES statement for each.

V.   Nutrition Intervention

19. What is the current recommendation on the appropriate mix of calories from carbohydrate, protein, and lipid for this patient?

20. For each of the PES statements you have written, establish an ideal goal (based on the signs and symptoms) and an appropriate intervention (based on the etiology).

21. What goals might you set for Mrs. Bernhardt as she is discharged and beginning pulmonary rehabilitation?

VI.   Nutrition Monitoring and Evaluation

22. You are now seeing Mrs. Bernhardt at her second visit to pulmonary rehabilitation. She provides you with the following information from her food record. Her weight is now 116 lbs. She explains adjustment to her medications and oxygen at home has been difficult, so she hasn't felt like eating very much. When you talk with her, you find she is hungriest in the morning, and often by evening she is too tired to eat. She is having no specific intolerances, but she does tell you she hasn't consumed any milk products because she thought they would cause more sputum to be produced.

Monday

Breakfast: Coffee, 1 c with 2 tbsp nondairy creamer; orange ½ c; 1 poached egg;
½ slice toast

Lunch: ¼ tuna salad sandwich (3 tbsp tuna salad on 1 slice wheat bread); coffee, 1 c
with 2 tbsp nondairy creamer

Supper: Cream of tomato soup, 1 c; ½ slice toast; ½ banana; Pepsi—approx 36 oz

Tuesday

Breakfast: Coffee, 1 c with 2 tbsp nondairy creamer; orange juice, ½ c; ½ c oatmeal with
2 tbsp brown sugar

Lunch: 1 chicken leg from Kentucky Fried Chicken; ½ c mashed potatoes; 2 tbsp
gravy; coffee, 1 c with 2 tbsp nondairy creamer

Supper: Cheese, 2 oz; 8 saltine crackers; 1 can V8 juice (6 oz); Pepsi, approx 36 oz

a.   Is she meeting her calorie and protein goals?

b.   What would you tell her regarding the use of supplements and/or milk and sputum
production?

c.   Using information from her food diary as a teaching tool, identify three interventions
you would propose for Mrs. Bernhardt to increase her calorie and protein intakes.

Bibliography

Academy of Nutrition and Dietetics. Evidence Analysis Library. Chronic Obstructive Pulmonary Disease Guideline. http://andevidencelibrary.com/topic.cfm?cat=1401. Accessed on 12/1/12.

Academy of Nutrition and Dietetics. Nutrition Care Manual. Chronic Obstructive Pulmonary Disease (COPD). http://www.nutritioncaremanual.org/topic.cfm?ncm_heading=Diseases%2FConditions&ncm_toc_id=22249. Accessed on 12/1/12.

Bergman EA, Hawk SN. Diseases of the respiratory system. In: Nelms M, Sucher K, Lacey K, Roth SL. *Nutrition Therapy and Pathophysiology*, 2nd ed. Belmont, CA: Wadsworth, Cengage Learning; 2011:648–681.

Decramer M, De Benedetto F, Del Ponte A, Marinari S. Systemic effects of COPD. *Respir Med*. 2005;99 Suppl B:S3–10.

Gronberg AM, Slinde F, Engstrom CP, Hulthen L, Larsson S. Dietary problems in patients with severe chronic obstructive pulmonary disease. *J Hum Nutr Diet*. 2005;18(6):445–52.

Hallin R, Koivisto-Hursti UK, Lindberg E, Janson C. Nutritional status, dietary energy intake and the risk of exacerbations in patients with chronic obstructive pulmonary disease (COPD). *Respir Med*. 2006 Mar;100(3):561–7.

Koehler F, Doehner W, Hoernig S, Witt C, Anker SD, John M. Anorexia in chronic obstructive pulmonary disease—association to cachexia and hormonal derangement. *Int J Cardiol*. 2007;119:83–9.

Lerario MC, Sachs A, Lazaretti-Castro M, Saraiva LG, Jardim JR. Body composition in patients with chronic obstructive pulmonary disease: Which method to use in clinical practice? *Br J Nutr*. 2006;96:86–92.

National Guideline Clearinghouse (NGC). Guideline summary NGC-7966: Chronic obstructive pulmonary disease. In: National Guideline Clearinghouse (NGC). Rockville (MD): Agency for Healthcare Research and Quality (AHRQ); 2010 Oct. Available: http://www.guideline.gov.

National Guideline Clearinghouse (NGC). Guideline synthesis: Chronic obstructive pulmonary disease (COPD): diagnosis and management of acute exacerbations. In: National Guideline Clearinghouse (NGC). Rockville (MD): Agency for Healthcare Research and Quality (AHRQ); 2001 Dec (revised 2011 Oct). Available: http://www.guideline.gov.

Nici L, Donner C, Wouters E, et al. American Thoracic Society/European Respiratory Society statement on pulmonary rehabilitation. *Am J Respir Crit Care Med*. 2006;173(12):1390–413.

Odencrants S, Ehnfors M, Grobe SJ. Living with chronic obstructive pulmonary disease (COPD): part II. RNs' experience of nursing care for patients with COPD and impaired nutritional status. *Scand J Caring Sci*. 2007;21(1):56–63.

Odencrants S, Ehnfors M, Grobe SJ. Living with chronic obstructive pulmonary disease: part I. Struggling with meal-related situations: experiences among persons with COPD. *Scand J Caring Sci*. 2005;19(3):230–9.

Sergi G, Coin A, Marin S, et al. Body composition and resting energy expenditure in elderly male patients with chronic obstructive pulmonary disease. *Respir Med*. 2006;100(11):1918–24.

Velloso M, Jardim JR. Study of energy expenditure during activities of daily living using and not using body position recommended by energy conservation techniques in patients with COPD. *Chest*. 2006;130(1):126–32.

Vermeeren MA, Creutzberg EC, Schols AM, et al. Prevalence of nutritional depletion in a large outpatient population of patients with COPD. *Respir Med*. 2006;100(8):1349–55.

Vermeeren MA, Wouters EF, Geraerts-Keeris AJ, Schols AM. Nutritional support in patients with chronic obstructive pulmonary disease during hospitalization for an acute exacerbation; A randomized controlled feasibility trial. *Clin Nutr*. 2004;23(5):1184–92.

Internet Resources

American Lung Association:
http://www.lung.org/lung-disease/copd/
Centers for Disease Control and Prevention:
http://www.cdc.gov/copd/http://www.lungusa.org/site/pp.asp?c=dvLUK9O0E&b=35020
COPD International: http://www.copd-international.com/

National Heart Lung and Blood Institute/ Learn More Breathe Better: http://www.nhlbi.nih.gov/health/public/lung/copd/index.htm
USDA Nutrient Data Laboratory: http://www.ars.usda.gov/main/site_main.htm?modecode=12-35-45-00

COPD with Respiratory Failure

Objectives

After completing this case, the student will be able to:

1. Describe the pathophysiology of chronic obstructive pulmonary disease and its relationship to acute respiratory failure.
2. Describe the role of nutrition support in treatment of patients on mechanical ventilation.
3. Outline the metabolic implications of acute respiratory failure.
4. Interpret biochemical indices for assessment of respiratory function.

5. Interpret biochemical indices for assessment of nutritional status.
6. Plan, interpret, and evaluate nutrition support for mechanically ventilated patients.

Daishi Hayoto, a 65-year-old male, is brought to the University Hospital emergency room by his wife when he experiences severe shortness of breath. Mr. Hayoto has a long-standing history of COPD.

Hayato, Daishi, Male, 65 y.o.
Allergies: Penicillin
Pt. Location: RM 405

Code: FULL
Physician: M. McFarland

Isolation: None
Admit Date: 3/26

Patient Summary: Acute respiratory distress, COPD, peripheral vascular disease with intermittent claudication

History:

Onset of disease: Patient has a long-standing history of COPD, presumably as a result of chronic tobacco use, 2 PPD for 50 years. He was in his usual state of health today with marked limitation of his exercise capacity due to dyspnea on exertion. He also notes two-pillow orthopnea, swelling in both lower extremities. Today while performing some yardwork he noted sudden onset of marked dyspnea. His wife brought him to the emergency room right away. There, a chest radiograph showed a tension pneumothorax involving the left lung. Patient also states he gets cramping in his right calf when he walks.

Medical history: Cholecystectomy 20 years ago. Total dental extraction 5 years ago. Patient describes intermittent claudication. Claims to be allergic to penicillin. Diagnosed with emphysema more than 10 years ago. Has been treated successfully with Combivent (metered dose inhaler)—2 inhalations 4 ×/d (each inhalation delivers 18 mcg ipratropium bromide; 130 mcg albuterol sulfate).

Surgical history: Cholecystectomy 20 years ago

Medications at home: Combivent, Lasix, O_2 2 L/hour via nasal cannula at night

Tobacco use: Yes; 2 PPD for 50 years

Alcohol use: Yes; 1–2 drinks 1–2 ×/week

Family history: What? Lung cancer. Who? Father

Demographics:

Marital status: Married, lives with wife, age 62, who is well; four adult children not living in the area

Years education: Bachelor's degree

Language: English and Japanese

Occupation: Retired manager of local grocery chain

Hours of work: N/A

Ethnicity: Nisei

Religious affiliation: Methodist

Admitting History/Physical:

Chief complaint: "My husband has had emphysema for many years. He was working in the yard today and got really short of breath. I called our doctor and she said to go straight to the emergency room."

Vital Signs: Temp: 98°F   Pulse: 118   Resp rate: 36
                    BP: 110/80   Height: 5'4"   Weight: 122 lbs

Heart: Normal heart sounds; no murmurs or gallops
HEENT: Within normal limits; funduscopic exam reveals AV nicking
          Eyes: Pupil reflex normal
          Ears: Slight neurosensory deficit acoustically

Hayato, Daishi, Male, 65 y.o.
Allergies: Penicillin
Pt. Location: RM 405

Code: FULL
Physician: M. McFarland

Isolation: None
Admit Date: 3/26

Nose: Unremarkable

Throat: Jugular veins appear distended. Trachea is shifted to the right. Carotids are full, symmetrical, and without bruits.

Genitalia: Unremarkable

Rectal: Prostate normal; stool hematest negative

Neurologic: DTR full and symmetric; alert and oriented × 3

Extremities: Cyanosis, 1+ pitting edema

Skin: Warm, dry to touch

Chest/lungs: Hyperresonance to percussion over the left chest anteriorly and posteriorly. Harsh inspiratory breath sounds are noted over right chest with absent sounds on the left. Using accessory muscles at rest.

Abdomen: Old surgical scar RUQ. No organomegaly or masses. BS reduced.

Circulation: R femoral bruit present. Right PT and DP pulses were absent.

Nursing Assessment	3/26
Abdominal appearance (concave, flat, rounded, obese, distended)	rounded
Palpation of abdomen (soft, rigid, firm, masses, tense)	soft
Bowel function (continent, incontinent, flatulence, no stool)	continent
Bowel sounds (P=present, AB=absent, hypo, hyper)	
RUQ	P
LUQ	P
RLQ	P
LLQ	P
Stool color	brown
Stool consistency	soft
Tubes/ostomies	N/A
Genitourinary	
Urinary continence	catheter
Urine source	catheter
Appearance (clear, cloudy, yellow, amber, fluorescent, hematuria, orange, blue, tea)	clear, yellow
Integumentary	
Skin color	pale
Skin temperature (DI=diaphoretic, W=warm, dry, CL=cool, CLM=clammy, CD+=cold, M=moist, H=hot)	W
Skin turgor (good, fair, poor, TENT=tenting)	fair

(Continued)

Hayato, Daishi, Male, 65 y.o.
Allergies: Penicillin   **Code:** FULL   **Isolation:** None
Pt. Location: RM 405   **Physician:** M. McFarland   **Admit Date:** 3/26

Nursing Assessment *(Continued)*

Nursing Assessment	3/26
Skin condition (intact, EC=ecchymosis, A=abrasions, P=petechiae, R=rash, W=weeping, S=sloughing, D=dryness, EX=excoriated, T=tears, SE=subcutaneous emphysema, B=blisters, V=vesicles, N=necrosis)	intact
Mucous membranes (intact, EC=ecchymosis, A=abrasions, P=petechiae, R=rash, W=weeping, S=sloughing, D=dryness, EX=excoriated, T=tears, SE=subcutaneous emphysema, B=blisters, V=vesicles, N=necrosis)	intact
Other components of Braden score: special bed, sensory pressure, moisture, activity, friction/shear (>18 = no risk, 15–16 = low risk, 13–14 = moderate risk, ≤12 = high risk)	activity, 18

Orders:
ABG, pulse oximetry, CBC, chemistry panel, UA
Chest X-ray, ECG Proventil 0.15 in 1.5 cc NS q 30 min × 3 followed by Proventil 0.3 cc in 3 cc normal saline q 2 hr per HHN (handheld nebulizer)
Spirogram post nebulizer Tx
IVF D_5 ½ NS at TKO Solu-Medrol 10–40 mg q 4–6 hr; high dose = 30 mg/kg q 4–6 hr (2 days max)
NPO

Nutrition:
General: Wife relates general appetite is only fair. Usually breakfast is the largest meal. Appetite has been decreased for past several weeks. She states his highest weight was 135 lbs, but feels he weighs much less than that now.
Usual dietary intake:
AM: Egg, hot cereal, bread or muffin, hot tea (with milk and sugar)
Lunch: Soup, sandwich, hot tea (with milk and sugar)
Dinner: Small amount of meat, rice, 2–3 kinds of vegetables, hot tea (with milk and sugar)

24-hr recall: 2 scrambled eggs, few bites of Cream of Wheat, sips of hot tea, bite of toast; ate nothing rest of day—sips of hot tea

Food allergies/intolerances/aversions: NKA
Previous nutrition therapy? No
Food purchase/preparation: Wife
Vit/min intake: None

Anthropometric data: Ht 5'4", Wt 122 lbs, UBW 135 lbs

Hayato, Daishi, Male, 65 y.o.
Allergies: Penicillin   **Code:** FULL   **Isolation:** None
Pt. Location: RM 405   **Physician:** M. McFarland   **Admit Date:** 3/26

Laboratory Results

	Ref. Range	3/26 1405
Chemistry		
Sodium (mEq/L)	136–145	138
Potassium (mEq/L)	3.5–5.5	3.9
Chloride (mEq/L)	95–105	101
Carbon dioxide (CO_2, mEq/L)	23–30	29
BUN (mg/dL)	8–18	11
Creatinine serum (mg/dL)	0.6–1.2	0.7
Glucose (mg/dL)	70–110	108
Phosphate, inorganic (mg/dL)	2.3–4.7	3.2
Magnesium (mg/dL)	1.8–3	1.9
Calcium (mg/dL)	9–11	9.1
Bilirubin, direct (mg/dL)	<0.3	0.8 !↑
Protein, total (g/dL)	6–8	6.1
Albumin (g/dL)	3.5–5	3.6
Prealbumin (mg/dL)	16–35	26
Ammonia (NH_3, µmol/L)	9–33	9
Alkaline phosphatase (U/L)	30–120	114
ALT (U/L)	4–36	15
AST (U/L)	0–35	22
CPK (IU/L)	30–135 F 55–170 M	152
Lactate dehydrogenase (U/L)	208–378	210
Cholesterol (mg/dL)	120–199	155
HDL-C (mg/dL)	>55 F, >45 M	32 !↓
LDL (mg/dL)	<130	142 !↑
LDL/HDL ratio	<3.22 F <3.55 M	4.44 !↑
Triglycerides (mg/dL)	35–135 F 40–160 M	155
Coagulation (Coag)		
PT (sec)	12.4–14.4	12.7

(Continued)

Hayato, Daishi, Male, 65 y.o.
Allergies: Penicillin
Pt. Location: RM 405

Code: FULL
Physician: M. McFarland

Isolation: None
Admit Date: 3/26

Laboratory Results *(Continued)*

	Ref. Range	3/26 1405
Hematology		
WBC ($\times 10^3$/mm^3)	4.8–11.8	5.6
RBC ($\times 10^6$/mm^3)	4.2–5.4 F 4.5–6.2 M	4.7
Hemoglobin (Hgb, g/dL)	12–15 F 14–17 M	13.2 !↓
Hematocrit (Hct, %)	37–47 F 40–54 M	39 !↓
Hematology, Manual Diff		
Grans (%)	3.6–79.2	52.3
Lymphocyte (%)	15–45	10 !↓
Monocyte (%)	3–10	3
Segs (%)	0–60	83 !↑

Arterial Blood Gases (ABGs)	Ref. Range	3/26 1030	3/27 0700	3/28 0700	3/30 0700
pH	7.35–7.45	7.2 !↓	7.30 !↓	7.36	7.22 !↓
pCO$_2$ (mm Hg)	35–45	65 !↑	59 !↑	50 !↑	66 !↑
CO$_2$ content (mmol/L)	25–30	35 !↑	30	29	36 !↑
pO$_2$ (mm Hg)	≥80	56 !↓	58 !↓	60 !↓	57 !↓
HCO$_3^-$ (mEq/L)	24–28	38 !↑	33 !↑	32 !↑	37 !↑

Hayato, Daishi, Male, 65 y.o.
Allergies: Penicillin
Pt. Location: RM 405

Code: FULL
Physician: M. McFarland

Isolation: None
Admit Date: 3/26

Intake/Output

Date		3/27 0701–3/28 0700			
Time		0701–1500	1501–2300	2301–0700	Daily total
IN	P.O. formula	200	200	Withheld due to high residuals	400
	P.O. flush (mL/kg/hr)	50 (0.56)	50 (0.56)		100 (0.37)
	I.V. (mL/kg/hr)	400 (0.90)	400 (0.90)	800 (1.80)	1600 (1.20)
	I.V. piggyback				
	TPN				
	Total intake (mL/kg)	650 (11.7)	650 (11.7)	800 (14.4)	2100 (37.9)
OUT	Urine (mL/kg/hr)	325 (0.73)	575 (1.30)	765 (1.72)	1665 (1.25)
	Emesis output				
	Other				
	Stool	200	100		300
	Total output (mL/kg)	525 (9.5)	675 (12.2)	765 (13.8)	1965 (35.4)
Net I/O		+125	−25	+35	+135
Net since admission (3/26)		+125	+100	+135	+135

Case Questions

I.   Understanding the Disease and Pathophysiology

1. Mr. Hayato was diagnosed with emphysema more than 10 years ago. Define *emphysema* and explain its underlying pathophysiology.

2. In the emergency room, a chest tube was inserted into the left thorax with drainage under suction. Subsequently the oropharynx was cleared. A resuscitation bag and mask were used to ventilate the patient with high-flow oxygen. Endotracheal intubation was then performed, using a laryngoscope so the trachea could be directly visualized. The patient was then ventilated with the help of a volume-cycled ventilator. Ventilation is 15 breaths/min with an FiO_2 of 100%, a positive end-expiratory pressure of 6, and a tidal volume of 700 mL. Daily chest radiographs and ABGs were used each AM to adjust the ventilator settings. Define the following terms found in the history and physical for Mr. Hayato:

   a.   Dyspnea

   b.   Orthopnea

   c.   Pneumothorax

   d.   Endotracheal intubation

   e.   Cyanosis

3. Identify features of the physician's physical examination consistent with his admitting diagnosis. Describe the pathophysiology that might be responsible for each physical finding.

II.   Understanding the Nutrition Therapy

4. What is the relationship between nutritional status and respiratory function? Define *respiratory quotient (RQ)*. What dietary factors affect RQ?

5. Do nutrition support and nutritional status play a role in enabling a patient to be weaned from a respiratory ventilator? Explain.

III. Nutrition Assessment

6. Evaluate Mr. Hayato's admitting anthropometric data.

7. Determine Mr. Hayato's energy and protein requirements using the Mifflin-St. Jeor, Ireton-Jones, and COPD predictive equations. Compare them. As Mr. Hayato's clinician, which would you set as your goal for meeting his energy needs?

8. Determine Mr. Hayato's fluid requirements.

9. Evaluate Mr. Hayato's biochemical indices relevant to nutritional status on 3/26.

IV. Nutrition Diagnosis

10. Select two high-priority nutrition problems and complete the PES statement for each.

V. Nutrition Intervention

11. A nutrition consult was completed on 3/27, and enteral feedings were initiated. Mr. Hayato was started on Isosource HN @ 25 cc/hr continuously over 24 hours.

   a. At this rate, how many kcal and grams of protein should he receive per day?

   b. Calculate his nutrition prescription utilizing this enteral formula. Include the goal rate, free water requirements, and the appropriate progression of the rate.

12. What type of formula is Isosource HN? What are the percentages of kilocalories from carbohydrate, protein, and lipid? What is the rationale for formulas that have additional nutrients added to assist with pulmonary function? List these nutrients and the proposed rationale.

VI.   Nutrition Monitoring and Evaluation

13. Examine the intake/output record. How much enteral feeding (kcal, protein) did the patient receive?

14. You read in the physician's orders that the patient experienced high gastric residual volume (GRV) and the enteral feeding was discontinued. Define *high GRV*. What is the probable cause for this patient?

15. Were any additional signs of EN intolerance documented? Do you agree with the decision to discontinue the feeding? Why or why not?

16. What options are available to improve tolerance of the tube feeding?

17. On 3/29, the enteral feeding was restarted at 25 mL/hr and then increased to 50 mL/hr after 12 hours. What were Mr. Hayato's energy and protein intakes for 3/29?

18. Examine the values documented for arterial blood gases (ABGs).

    a. On the day Mr. Hayato was intubated, his ABGs were as follows: pH 7.2, pCO_2 65, CO_2 35, pO_2 56, HCO_3^- 38. What can you determine from each of these values?

    b. On 3/28, while Mr. Hayato was on the ventilator, his ABGs were as follows: pH 7.36, pCO_2 50, CO_2 29, pO_2 60, HCO_3^- 32. What can you determine from each of these values?

    c. On 3/30, after the enteral feeding was resumed, his ABGs were as follows: pH 7.22, pCO_2 66, pO_2 57, CO_2 36, HCO_3^- 37. In addition, indirect calorimetry indicated an RQ of 0.95 and his measured energy intake was 1350 kcal. How does the patient's measured energy intake compare to your previous calculations? What does the RQ indicate?

19. The patient was weaned from the ventilator on 4/2 and discharged to home on 4/5. As Mr. Hayato is prepared for discharge, what nutritional goals might you set with him and his wife to improve his overall nutritional status?

Bibliography

Academy of Nutrition and Dietetics. Evidence Analysis Library. Chronic Obstructive Pulmonary Disease Guideline. http://andevidencelibrary.com/topic .cfm?cat=1401. Accessed on 12/1/12.

Academy of Nutrition and Dietetics. Evidence Analysis Library. Recommendations Summary. CIU: Optimizinf Enteral Nutrition Delivery. http:// andevidencelibrary.com/template.cfm?template=guide _summary&key=3256. Accessed on 1/3/13.

Academy of Nutrition and Dietetics. Nutrition Care Manual. Chronic Obstructive Pulmonary Disease (COPD). http://www.nutritioncaremanual.org/topic .cfm?ncm_heading=Diseases%2FConditions&ncm _toc_id=22249. Accessed on 12/1/12.

Altintas ND, AydinK, Türkoglu MA, Abbasoglu O, Topeli A. Effect of enteral versus parenteral nutrition on outcome of medical patients requiring mechanical ventilation. *Nutr Clin Pract*. 2011;26:322–329.

Bergman EA, Hawk SN. Diseases of the respiratory system. In: Nelms M, Sucher K, Lacey K, Roth SL. *Nutrition Therapy and Pathophysiology*. 2nd ed. Belmont, CA: Wadsworth, Cengage Learning; 2011:648–681.

Decramer M, De Benedetto F, Del Ponte A, Marinari S. Systemic effects of COPD. *Respir Med*. 2005;99 Suppl B:S3–10.

Demiling RH, DeSanti L. Effect of a catabolic state with involuntary weight loss on acute and chronic respiratory disease. Available from: http://www.medscape .com/viewprogram/1816_pnt.

Doley J, Mallampalli A, Sandberg M. Nutrition management for the patient requiring prolonged mechanical ventilation. *Nutr Clin Pract*. 2011;26:232–241.

Nahikian-Nelms M. Enteral and parenteral nutrition support. In: Nelms M, Sucher K, Lacey K, Roth SL. *Nutrition Therapy and Pathophysiology*.

2nd ed. Belmont, CA: Wadsworth, Cengage Learning; 2011:80–105.

National Guideline Clearinghouse (NGC). Guideline summary NGC-7966: Chronic obstructive pulmonary disease. In: National Guideline Clearinghouse (NGC). Rockville, MD: Agency for Healthcare Research and Quality (AHRQ); 2010 Oct. Available: http://www .guideline.gov.

National Guideline Clearinghouse (NGC). Guideline synthesis: Chronic obstructive pulmonary disease (COPD): diagnosis and management of acute exacerbations. In: National Guideline Clearinghouse (NGC). Rockville, MD: Agency for Healthcare Research and Quality (AHRQ); 2001 Dec (revised 2011 Oct). Available: http://www.guideline.gov.

Nelms MN, Habash D. Nutrition assessment: Foundation of the nutrition care process. In: Nelms M, Sucher K, Lacey K, Roth SL. *Nutrition Therapy and Pathophysiology*. 2nd ed. Belmont, CA: Wadsworth, Cengage Learning; 2011:34–65.

Nelms MN. Enteral and parenteral nutrition support. In: Nelms M, Sucher K, Lacey K, Roth SL. *Nutrition Therapy and Pathophysiology*. 2nd ed. Belmont, CA: Wadsworth, Cengage Learning; 2011:80–105.

Nelms MN. Metabolic stress and the critically ill. In: Nelms M, Sucher K, Lacey L, Roth SL. *Nutrition Therapy and Pathophysiology*. 2nd ed. Belmont, CA: Wadsworth, Cengage Learning; 2011:682–701.

Nici L, Donner C, Wouters E, et al. American Thoracic Society/European Respiratory Society statement on pulmonary rehabilitation. *Am J Respir Crit Care Med*. 2006;173(12):1390–413.

Ukleja A, Freeman KL, Gilbert K, et al. Standards for nutrition support: Adult hospitalized patients. *Nutr Clin Pract*. 2010;25:403–414.

Internet Resources

American Lung Association: http://www.lung.org/lung-disease/copd/

Centers for Disease Control and Prevention: http://www.cdc.gov/copd/

COPD International: http://www.copd-international.com/

National Heart Lung and Blood Institute—Learn More Breathe Better: http://www.nhlbi.nih.gov/health /public/lung/copd/index.htm

USDA Nutrient Data Laboratory: http://www.ars.usda.gov /main/site_main.htm?modecode=12-35-45-00

Unit Eleven

NUTRITION THERAPY FOR METABOLIC STRESS AND CRITICAL ILLNESS

The physiological response to stress, trauma, and infection has been an important area of nutrition research for the past several decades. This metabolic response is characterized by catabolism of stored nutrients to meet the increased energy requirements.

Unlike other situations in which the body faces increased energy requirements, the stress response demands a preferential use of glucose for fuel. Because glycogen stores are quickly depleted, the body turns to lean body mass for glucose produced via gluconeogenesis.

Under the influence of counterregulatory hormones such as glucagon, epinephrine, norepinephrine, and cortisol, as well as cytokines such as interleukin and tumor necrosis factor, the body shifts from its normal balanced state or anabolism to catabolism. All sources of fuel metabolism are affected by the stress response and the subsequent control of counterregulatory hormones. Despite increased lipolysis, the efficiency for the use of fatty acids and glycerol as sources of fuel is reduced.

The body's inability to keep up with the rate of protein catabolism results in loss of skeletal muscle and high urinary losses of nitrogen. The liver's rate of gluconeogenesis is increased, and hyperglycemia is common. In addition, many tissues—especially skeletal tissue—develop insulin resistance, which contributes to the hyperglycemic state.

Cases 28, 29, 30, and 31 focus on the application of evidenced-based guidelines to guide the decision making for nutrition support. Nutrition support can serve as an integral component of the medical care and can impact the metabolic response. With increasing rates of obesity in the United States, the necessity of providing nutrition support in the morbidly obese patient (as in Case 31) is an increasingly common challenge for the registered dietitian.

Case 28 allows you to assess a patient with a closed head injury from a motor vehicle accident. Closed head injuries are an excellent example of the post-traumatic, hypermetabolic state. Determining nutritional needs, prescribing appropriate nutrition support, and monitoring daily progress are all addressed in this pediatric case. Other conditions resulting in metabolic stress include trauma, open wounds, and sepsis. These situations also demand close attention to nutrition support to minimize complications of protein-calorie malnutrition and to optimize recovery through medical nutrition therapy. You can easily transfer the same concepts for nutrition assessment and support to other individual cases you may encounter.

Pediatric Traumatic Brain Injury: Metabolic Stress with Nutrition Support

Objectives

After completing this case, the student will be able to:

1. Describe and discuss the pathophysiology of a closed head injury.
2. Describe the metabolic response to stress and trauma.
3. Determine nutrient, fluid, and electrolyte requirements for a pediatric patient.
4. Evaluate nutrition assessment data to determine nutritional status and identify specific nutrition problems.
5. Determine nutrition diagnoses and write appropriate PES statements.

6. Design an enteral nutrition support plan.
7. Determine steps to monitor and evaluate enteral nutrition support and the transition to an oral diet.

Chelsea Montgomery is an 8-year-old female admitted through the emergency room after being injured as a restrained front-seat passenger in a motor vehicle accident. She is transferred to the pediatric intensive care unit with a traumatic brain injury.

Montgomery, Chelsea, Female, 8 y.o.

Allergies: NKA	**Code:** FULL	**Isolation:** None
Pt. Location: Bed #8 PICU	**Physician:** E. Mantio	**Admit Date:** 4/22

Patient Summary: 8-year-old female admitted through the ER after being injured as a restrained front-seat passenger in a motor vehicle accident. She is transferred to the pediatric intensive care unit with a traumatic brain injury.

History:
Full-term infant weighing 9 lbs 1 oz delivered via caesarean. Healthy except for severe myopia—second-grade student; participates in a variety of activities including gymnastics, softball, and Girl Scouts.

Medical history: None

Surgical history: None

Medications at home: None

Tobacco use: No

Alcohol use: No

Family history: What? CAD; Who? Paternal grandfather; diabetes—older brother

Demographics:
Lives with parents and three siblings—ages 10, 14, 16

Language: English only

Parents' occupations: Father works for local insurance company—mother is assistant in physician's office.

Ethnicity: Caucasian

Religious affiliation: Episcopalian

Admitting History/Physical:
General appearance: 8-year-old female child alternating between crying and unconsciousness

Vital Signs:	Temp: 97	Pulse: 100	Resp rate: 27
	BP: 138/90	Height: 52"	Weight: 61 lbs

Heart: RRR, nl S1-S2, tachycardia, no murmur

HEENT: Eyes: Pupils 4 mm reactive; no battle/raccoon signs

   Ears: WNL

   Nose: WNL

   Throat: WNL

Genitalia: +rectal tone, heme negative

Neurologic: GCS = 10 E4V2M4. Obtundation and L-sided hemiparesis. No verbal responses. Withdrawal and moaning when touched.

Extremities: DTR symmetric; 2 cm laceration on R knee

Skin: Warm and dry

Chest/lungs: Breath sounds bilaterally

Peripheral vascular: No edema

Abdomen: Soft; bowel sounds diminished, linear mark in LUQ, +guarding throughout.

Dx: Traumatic brain injury secondary to MVA

Montgomery, Chelsea, Female, 8 y.o.

Allergies: NKA

Pt. Location: Bed #8 PICU

Code: FULL

Physician: E. Mantio

Isolation: None

Admit Date: 4/22

Orders:

Admit to PICU; $D_5 0.9NS$ with 10 mEq KCl @ 65 mL/hr; Zantac 25 mg every 6 hours; Tylenol 450 mg every 6 hours; Zofran 2 mg IV every 6 hours. NPO—O_2 to keep sat >95%; I/O; foley to gravity.

Department of Radiology—CT Report

Date: 4/22

Patient: Chelsea Montgomery

Physician: Elizabeth Mantio, MD

Two areas of increased density in L frontal lobe near vertex and possibly left central modality

Victoria Roundtree, MD, Dept. of Radiology

Department of Radiology—MRI Report

Date: 4/27

Patient: Chelsea Montgomery

Physician: Elizabeth Mantio, MD

MRI showed areas of hemorrhagic edema in deep white matter of L frontal lobe anteriorly. Additionally, heme and edema found in the splenium of corpus callosum. 3.4 cm × 4.2 cm × 1.0 cm representing areas of shearing injury.

James Morgan, MD, Dept. of Radiology

Nutrition:

History: Parents indicate that patient had normal growth and appetite PTA.

Usual dietary intake:

AM:	Cereal, milk, juice, toast
Lunch:	At school cafeteria
Snack before or after school activity:	Granola bar, juice box, crackers
PM:	Meat, pasta or potatoes, rolls or bread. Likes only green beans, corn, salad with Ranch dressing. Likes all fruit.

Food allergies/intolerances/aversions: None

Previous nutrition therapy? No

Food purchase/preparation: Parents

Vit/min intake: Gummy vitamin—Target brand.

Excerpt from nutrition consult 4/22: Recommend enteral feeding Pediasure 1.5 @ 10 cc/hr. Increase by 10 cc every 6 hours until goal rate of 57 cc/hr continuous drip. Free water 200 mL four times daily to meet fluid requirements when IV is discontinued.

Department of Speech Pathology

RE: Interpretation of FEES/swallowing evaluation

Date: 5/2

Montgomery, Chelsea, Female, 8 y.o.
Allergies: NKA   **Code:** FULL   **Isolation:** None
Pt. Location: Bed #8 PICU   **Physician:** E. Mantio   **Admit Date:** 4/22

Patient: Chelsea Montgomery
Physician: Elizabeth Mantio, MD
Patient accepted macaroni and cheese with appropriate tongue lateralization and chewing skills but choked after 5–7 ice chips. Oral skills appropriate. Showed significant signs of fatigue and decreased cooperation after a few swallows, which therefore inhibited PO feeding. No evidence of penetration or aspiration.
Cindy Davie, MS, SLP

Nursing Assessment	5/2
Abdominal appearance (concave, flat, rounded, obese, distended)	flat
Palpation of abdomen (soft, rigid, firm, masses, tense)	soft
Bowel function (continent, incontinent, flatulence, no stool)	incontinent
Bowel sounds (P=present, AB=absent, hypo, hyper)	
RUQ	P, hypo
LUQ	P, hypo
RLQ	P, hypo
LLQ	P, hypo
Stool color	brown
Stool consistency	soft
Tubes/ostomies	NA
Genitourinary	
Urinary continence	catheter
Urine source	catheter
Appearance (clear, cloudy, yellow, amber, fluorescent, hematuria, orange, blue, tea)	cloudy, hematuria
Integumentary	
Skin color	pale
Skin temperature (DI=diaphoretic, W=warm, dry, CL=cool, CLM=clammy, CD+=cold, M=moist, H=hot)	CLM, DI
Skin turgor (good, fair, poor, TENT=tenting)	good
Skin condition (intact, EC=ecchymosis, A=abrasions, P=petechiae, R=rash, W=weeping, S=sloughing, D=dryness, EX=excoriated, T=tears, SE=subcutaneous emphysema, B=blisters, V=vesicles, N=necrosis)	D, A
Mucous membranes (intact, EC=ecchymosis, A=abrasions, P=petechiae, R=rash, W=weeping, S=sloughing, D=dryness, EX=excoriated, T=tears, SE=subcutaneous emphysema, B=blisters, V=vesicles, N=necrosis)	intact, A
Other components of Braden score: special bed, sensory pressure, moisture, activity, friction/shear (>18 = no risk, 15–16 = low risk, 13–14 = moderate risk, ≤12 = high risk)	13

Montgomery, Chelsea, Female, 8 y.o.
Allergies: NKA   **Code:** FULL   **Isolation:** None
Pt. Location: Bed #8 PICU   **Physician:** E. Mantio   **Admit Date:** 4/22

MD Progress Note Hospital Day 12:

5/3 0840

Subjective: Chelsea Montgomery's previous 24 hours reviewed
Vitals: Temp: 99.1, Pulse: 82, Resp rate: 25, BP: 122/78, Wt. 23 kg
Urine Output: 970 mL

Physical Exam:

General: Latest MRI showed areas of hemorrhagic edema in deep white matter of L frontal lobe anteriorly. Additionally heme and edema found in the splenium of corpus callosum.
3.4 cm × 4.2 cm × 1.0 cm representing areas of shearing injury. Continues to arouse easily—automatic speech of No-No-No. One level commands followed. Oriented to parents, but not place or time. Failed Speech/Swallowing evaluation.
HEENT: WNL
    Neck: WNL
    Heart: WNL
    Lungs: Clear to auscultation
    Abdomen: Soft with + BS

Assessment/Plan: Continue PT, OT, and current medical care. Continue enteral feeding—consult for transfer to rehab unit.

        E. Mantio, MD

Intake/Output

Date		5/2 0701–5/3 0700			
Time		0701–1500	1501–2300	2301–0700	Daily total
IN	Tube feeding water flush	50	50	50	150
	Tube feeding Pediasure 1.5	342	228	456	1026
	I.V. (mL/kg/hr)				
	I.V. piggyback				
	Total intake	392	278	506	1176
	(mL/kg)	(17.0)	(12.1)	(22.0)	(51.1)
OUT	Urine	250	300	420	970
	(mL/kg/hr)	(1.4)	(1.6)	(2.3)	(1.8)
	Emesis output				
	Other				
	Stool	×1		×2	
	Total output	250	300	420	970
	(mL/kg)	(10.9)	(13.0)	(18.3)	(42.2)
Net I/O		+142	−22	+86	+206
Net since admission (4/22)		−464	−486	−400	−400

Montgomery, Chelsea, Female, 8 y.o.
Allergies: NKA   **Code:** FULL   **Isolation:** None
Pt. Location: Bed #8 PICU   **Physician:** E. Mantio   **Admit Date:** 4/22

Laboratory Results

	Ref. Range	4/22 (day 1)	5/2 (day 11)
Chemistry			
Sodium (mEq/L)	136–145	142	139
Potassium (mEq/L)	3.5–5.5	3.9	3.6
Chloride (mEq/L)	95–105	101	105
Carbon dioxide (CO_2, mEq/L)	23–30	28	29
BUN (mg/dL)	8–18	8	10
Creatinine serum (mg/dL)	0.6–1.2	0.7	0.6
BUN/Crea ratio	10.0–20.0	12.8	16.6
Glucose (mg/dL)	70–110	145 !↑	109
Phosphate, inorganic (mg/dL)	2.3–4.7	3.8	3.3
Magnesium (mg/dL)	1.8–3	2.0	1.9
Calcium (mg/dL)	9–11	9.1	9.3
Bilirubin total (mg/dL)	≤1.5	1.6 !↑	1.1
Bilirubin, direct (mg/dL)	<0.3	0.8 !↑	0.1
Protein, total (g/dL)	6–8	6.8	5.1 !↓
Albumin (g/dL)	3.5–5	4.9	2.8 !↓
Prealbumin (mg/dL)	16–35	33	17
Ammonia (NH_3, μmol/L)	9–33	20	14
Alkaline phosphatase (U/L)	30–120	122 !↑	138 !↑
ALT (U/L)	4–36	30	34
AST (U/L)	0–35	22	25
CPK (U/L)	30–135 F 55–170 M	100	110
Lactate (mmol/L)	<1.00	2.1 !↑	0.9
Fibrinogen (mg/dL)	160–450	472 !↑	460 !↑
C-reactive protein (mEq/L)	0.3–2.3	2.5 !↑	3.2 !↑
Coagulation (Coag)			
PT (sec)	12.4–14.4	13.8	12.6
INR	0.9–1.1	1.0	0.92
PTT (sec)	24–34	28	25
Hematology			
WBC ($\times 10^3$/mm^3)	4.8–11.8	5.6	10.2

(Continued)

Montgomery, Chelsea, Female, 8 y.o.

Allergies: NKA

Pt. Location: Bed #8 PICU

Code: FULL

Physician: E. Mantio

Isolation: None

Admit Date: 4/22

Laboratory Results *(Continued)*

	Ref. Range	4/22 (day 1)	5/2 (day 11)
RBC ($\times 10^6$/mm^3)	4.2–5.4 F 4.5–6.2 M	5.2	4.2
Hemoglobin (Hgb, g/dL)	12–15 F 14–17 M	12.1	11.5 ‼↓
Hematocrit (Hct, %)	37–47 F 40–54 M	38	35 ‼↓

Case Questions

I.   Understanding the Disease and Pathophysiology

1.   Define *traumatic brain injury*. What is the Glasgow coma scale? What was Chelsea's initial GCS score? What findings from the physical exam are consistent with this score?

2.   Read the radiology reports and the MD progress note dated 5/3. What causes edema and bleeding after a traumatic brain injury? What general functions occur in the frontal lobe? How might Chelsea's injury affect her in the long term?

3.   Describe the inflammatory response that occurs in metabolic stress. Explain the effects of this response on carbohydrate, protein, and lipid metabolism.

II.   Understanding the Nutrition Therapy

4.   Based on evidence-based guidelines, what is the proposed role of nutrition support in Chelsea's medical care?

5.   Are there specific nutrients that are recommended to support the care of an individual with a TBI?

6.   Chelsea is 8 years old. What specific concerns should the RD have for planning the nutrition care of a pediatric patient?

III.   Nutrition Assessment

7.   Assess Chelsea's admitting height and weight. Provide the rationale for the reference standards that you have used.

8.   Determine Chelsea's admission requirements for the following:
   • Fluid
   • Calories
   • Protein
   • Vitamins
   • Minerals
   • Electrolytes

9. Chelsea was to receive Pediasure 1.5 at a goal rate of 57 cc/hr. How much energy and protein does this provide? Show your calculations. Does it meet her needs that you determined in question #8?

10. Using the intake/output record for 5/2, answer the following:

    a. What was the total volume of her feeding for 5/2?

    b. What was the nutritional value of her feeding for that day? Calculate the total energy and protein.

    c. What percentage of her needs was met? What percentage of her prescribed feeding did she actually receive? What factors may interfere with the patient receiving her prescribed nutrition support? What steps can be taken to ensure that the patient is receiving her prescribed enteral feeding in full?

11. Assess Chelsea's laboratory values at admission and on day 11. Please explain your interpretation of each abnormal lab.

12. What information in the MD progress note (written on day 12) provides data you can use to plan Chelsea's nutrition support? Assess Chelsea's current nutritional status on day 12 of her admission. Evaluate her current hydration status, enteral feeding, and any additional information you have available to assess her current condition.

13. On 5/2, a 24-hour urine sample was collected for nitrogen balance. Her total urine urea nitrogen was 12 g.

    a. Using the intake/output information for that day, calculate her nitrogen balance. How would you assess this information? Explain your response in the context of her potential hypermetabolism.

    b. Are there any factors that may affect the accuracy of this test?

    c. The intern taking care of Chelsea pages you when he reads your note regarding her negative nitrogen balance. He asks whether he should change the enteral formula to one higher in nitrogen. Explain your response to him.

IV.   Nutrition Diagnosis

14. Select two nutrition problems and complete the PES statement for each.

V.   Nutrition Intervention

15. For each of the PES statements that you have written, establish an ideal goal (based on the signs and symptoms) and an appropriate intervention (based on the etiology).

16. Write your follow-up nutrition note for 5/3.

VI.   Nutrition Monitoring and Evaluation

17. Chelsea has worked with an occupational therapist, a speech therapist, and a physical therapist. Summarize the training that each of these professionals receives and describe their expected roles in Chelsea's rehabilitation.

18. The speech pathologist saw Chelsea for a swallowing evaluation. What is a FEES? What factors in the speech pathologist's report indicate the continued need for enteral feeding?

19. As Chelsea's recovery proceeds, she begins a PO mechanical soft diet. Her kcalorie counts are as follows:

5/14: oatmeal ¼ c; brown sugar 2 tbsp; whole milk 1 c; 240 cc Carnation Instant Breakfast (CIB) prepared with 2% milk; mashed potatoes 1 c; gravy 2 tbsp

5/15: Cheerios 1 c; whole milk 1 c; 240 cc CIB prepared with 2% milk; grilled cheese sandwich (2 slices bread, 1 oz American cheese, 1 tsp margarine); Jell-O 1 c; 240 cc CIB prepared with 2% milk

    a. Calculate her daily kcal and protein intakes and the average for these 2 days of kcalorie counts.

    b. What recommendations would you make regarding her enteral feeding at this time?

Bibliography

Boselli M, Awuilani R, Baiardi P, et al. Supplementation of essential amino acids may reduce the occurrence of infections in rehabilitation patients with brain injury. *Nutr Clin Prac.* 2012;27:99–113.

Cook AM, Peppard Am, Magnuson B. Nutrition considerations in traumatic brain injury. *Nutr Clin Prac.* 2008;23:608–620.

Havalad S, Quaid MA, Sapiega V. Energy expenditure in children with severe head injury: Lack of agreement between measured and estimated energy expenditure. *Nutr Clin Pract.* 2006;21:175–181.

Mehta NM, Compher C, ASPEN. ASPEN Clinical guidelines: Nutrition support of the critically ill child. *J Parenter Enteral Nutr.* 2009;33:260–276.

Mehta NM, McAleer D, Hamilton S, Naples E, Leavitt K, Mitchell P, Duggan C. Challenges to optimal enteral nutrition in a multidisciplinary pediatric intensive care unit. *J Parenter Enteral Nutr.* 2010;34:38–45.

Ruf K, Magnuson B, Jatton J, Cook AM. Nutrition in neurologic impairment. In: *The ASPEN Adult Nutrition Support Core Curriculum.* 2nd ed. Washington, DC: American Society for Parenteral and Enteral Nutrition; 2012.

Skillman HE, Wischmeyer PE. Nutrition therapy in critically ill infants and children. *J Parenter Enteral Nutr.* 2008;32:520–534.

Internet Resources

Centers for Disease Control and Prevention: Injury Prevention and Control: Traumatic Brain Injury: http://www.cdc.gov/traumaticbraininjury/

National Institute of Neurological Disorders and Stroke: Traumatic Brain Injury: http://www.ninds.nih.gov/disorders/tbi/tbi.htm

Pediatric Nutrition Care Manual: www.nutritioncaremanual.org

Metabolic Stress and Trauma: Open Abdomen

Deborah Cohen, DCN, MMSc, RD
University of New Mexico

Objectives

After completing the case, the student will be able to:

1. Apply the knowledge of the pathophysiology of trauma and metabolic stress in order to provide nutrition support for the critically ill patient.
2. Identify the basic components of indirect calorimetry
3. State specific indications for the use of indirect calorimetry in critically ill patients.
4. Interpret the respiratory quotient.
5. Compare different predictive equations that are appropriate for use in the critically ill population.
6. Use current evidence-based guidelines to evaluate and plan nutrition support for the critically ill.

Juan Perez is a 29-year-old male admitted to the surgical intensive care unit with a gunshot wound to the abdomen. He experienced gastric, duodenal, and jejunal injuries, liver laceration, and a left pleural effusion.

Perez, Juan, Male, 29 y.o.
Allergies: NKA
Pt. Location: Bed #2 SICU

Code: FULL
Physician: D. Kuhls, MD

Isolation: Contact
Admit Date: 3/22

Patient Summary: The patient was brought into the emergency room by a friend after he was shot in the abdomen. He was vomiting blood, and complained of severe back and "stomach" pain. He was able to respond to a few questions initially but stated the "pain is too bad for me to think." He denied being allergic to any medications or having any chronic medical problems.

History:
Medical history: Unremarkable
Surgical history: Unknown
Medications at home: Unknown
Tobacco use: Yes
Alcohol use: Unknown

Demographics:
Marital status: Single; lives with his brother, his brother's wife, and their two children, ages 2 and 4
Language: Spanish/English
Occupation: Convenience store clerk
Ethnicity: Hispanic
Religious affiliation: Catholic

Admitting History/Physical:
General appearance: Mildly obese 29-year-old Hispanic male on mechanical ventilation

Vital Signs:	Temp: 102.5	Pulse: 135	Resp rate: 20
	BP: 115/65	Height: 70"	Weight: 102.7 kg

Heart: Noncontributory
HEENT: NG tube in place for decompression
Genitalia: Deferred
Neurologic: Sedated
Extremities: 4+ bilateral edema
Skin: Warm, moist
Chest/lungs: Lungs clear to auscultation and percussion
Peripheral vascular: Pulses full—no bruits
Abdomen: Abdominal distention, wound VAC in place, three tubes draining peritoneal fluid, hypoactive BS present in all regions. Liver percusses 8 cm at the midclavicular line, one fingerbreadth below the right costal margin.

Dx: Abdominal GSW

Tx plan: Emergent OR

Perez, Juan, Male, 29 y.o.
Allergies: NKA
Pt. Location: Bed #2 SICU

Code: FULL
Physician: D. Kuhls, MD

Isolation: Contact
Admit Date: 3/22

MD Progress Note:
Hospital Day #7
3/29 0840
Admission History: Juan Perez is a 29-yo male who suffered a GSW to the abdomen. In the ER a FAST (focused assessment with sonography for trauma) scan was performed and found to be positive in the ED and he was transferred emergently to the OR. In the OR, he underwent exploratory laparotomy, gastric repair, control of liver hemorrhage, and resection of proximal jejunum, leaving his GI tract in discontinuity. Following the OR, he was transferred to the SICU intubated and hemodynamically stable.

Hospital Day #2: Returned to surgery to remove packs and to reestablish bowel continuity. An abdominal vacuum-assisted closure (VAC) device was placed. Three Jackson Pratt drains left in place.

Hospital Day #3: Returned to surgery for anastomotic leak—gastrojejunostomy tube inserted through the patient's stomach, with the jejunal limb shortened in order to provide antegrade intraluminal drainage, as well as retrograde jejunostomy tube for drainage.

Hospital Day #7: Returned to surgery for abdominal washout, insertion of a distally placed jejunostomy tube for feeding, and a VAC change.

Subjective: Juan Perez previous 24 hours reviewed.
Vitals: Temp: 99.1 Pulse: 82 Resp rate: 15 BP: 122/78 Wt. 109 kg
O_2 Sat (%): 93%
I/O last 3 completed shifts: In: 5472 (IV 1800, TPN 3312, IV Piggyback 360)
                                               Out: 4584 (Urine 2889, Other 1695)

Physical Exam
General/Constitutional: NAD, lying in bed, sedated
HEENT: Normocephalic, PERRL
Cardiac: RRR
Pulmonary/Chest: No wheezes or crackles
Abdominal: VAC in place, jejunostomy feeding tube in place, moderate distention with appropriate tenderness
Extremities: WNL
Periperal Vascular Exam: Peripheral pulses throughout
Neurological: Moves all extremities spontaneously—sedated
Skin: Warm, dry
IV Fluid: D5 0.45 @75 mL/h – KVO
Nutrition: NPO, TPN
Glycemic Control: Sliding scale insulin
Mobility: As tolerates per PT treatment plan

....T Franks, MD—Surgical Resident
....D. Kuhls, MD—Trauma Attending

Perez, Juan, Male, 29 y.o.
Allergies: NKA   **Code:** FULL   **Isolation:** Contact
Pt. Location: Bed #2 SICU   **Physician:** D. Kuhls, MD   **Admit Date:** 3/22

Nutrition Consult—Follow-up: 3/29 Hospital Day #7

A̲: 29-yo male s/p GSW to the abdomen with emergent exploratory laparatomy; s/p gastric repair, control of liver hemorrhage and resection of proximal jejunum—jejunostomy feeding tube in place.
Ht. 70" Adm Wt: 102.7 kg Current wt: 109 kg IBW: 75.45 kg
Metabolic cart measurement: REE 3657 kcal RQ 0.76
Current nutrition support: Dextrose:140 CAA: 60 FAT/L:20 Goal Rate: 135 mL/hr
Propofol @35 mL/hr providing additional 924 kcal for total of:
3888 kcal (51 kcal/kg IBW/day) and 194 g protein (g/kg IBW/day)
Labs and medications reviewed.

D̲: Increased energy expenditure related to open abdomen and posttrauma status as evidenced by metabolic cart measurement of 3657 kcal expenditure.
Altered GI function related to open abdomen and GSW with subsequent surgery as evidenced by lack of bowel sounds, no stool output, and continued wound VAC for open abdomen.

I̲: Current metabolic cart measurement does not indicate overfeeding so will recommend to continue current TPN. Recommend to initiate trickle feeds of Pivot at 5 mL/hr.

M/E̲: Will continue to monitor electrolytes, nutritional parameters, and metabolic tolerance of nutrition support daily. Would recommend to delay nitrogen balance study as it will most likely be inaccurate due to large amounts of protein losses via abdominal wounds.

.... N. Ridgway, MS, RD, CNSC

Nursing Assessment	3/29
Abdominal appearance (concave, flat, rounded, obese, distended)	distended—wound VAC in place
Palpation of abdomen (soft, rigid, firm, masses, tense)	
Bowel function (continent, incontinent, flatulence, no stool)	no stool
Bowel sounds (P=present, AB=absent, hypo, hyper)	
RUQ	AB
LUQ	AB
RLQ	AB
LLQ	AB
Stool color	N/A
Stool consistency	N/A
Tubes/ostomies	J-P drains—wound VAC
Genitourinary	
Urinary continence	catheter
Urine source	catheter
Appearance (clear, cloudy, yellow, amber, fluorescent, hematuria, orange, blue, tea)	cloudy, pale yellow

Perez, Juan, Male, 29 y.o.
Allergies: NKA
Pt. Location: Bed #2 SICU

Code: FULL
Physician: D. Kuhls, MD

Isolation: Contact
Admit Date: 3/22

Nursing Assessment *(Continued)*

Nursing Assessment	3/29
Integumentary	
Skin color	light brown
Skin temperature (DI=diaphoretic, W=warm, dry, CL=cool, CLM=clammy, CD+=cold, M=moist, H=hot)	DI
Skin turgor (good, fair, poor, TEN I =tenting)	fair
Skin condition (intact, EC=ecchymosis, A=abrasions, P=petechiae, R=rash, W=weeping, S=sloughing, D=dryness, EX=excoriated, T=tears, SE=subcutaneous emphysema, B=blisters, V=vesicles, N=necrosis)	A, D
Mucous membranes (intact, EC=ecchymosis, A=abrasions, P=petechiae, R=rash, W=weeping, S=sloughing, D=dryness, EX=excoriated, T=tears, SE=subcutaneous emphysema, B=blisters, V=vesicles, N=necrosis)	intact
Other components of Braden score: special bed, sensory pressure, moisture, activity, friction/shear (>18 = no risk, 15–16 = low risk, 13–14 = moderate risk, ≤12 = high risk)	10

Intake/Output

Date		3/29 0701–3/30 0700			
Time		0701–1500	1501–2300	2301–0700	Daily total
IN	P.O.	0	0	0	0
	I.V.	600	600	600	1800
	(mL/kg/hr)	(0.69)	(0.69)	(0.69)	(0.69)
	I.V. piggyback	120	120	120	360
	TPN	1104	1104	1104	3312
	Total intake	1824	1824	1824	5472
	(mL/kg)	(16.7)	(16.7)	(16.7)	(50.2)
OUT	Urine	868	790	1231	2889
	(mL/kg/hr)	(0.99)	(0.90)	(1.41)	(1.10)
	Emesis output				
	Other (JP drains)	525	555	615	1695
	Stool				
	Total output	1393	1345	1846	4584
	(mL/kg)	(12.78)	(12.34)	(16.93)	(42.05)
Net I/O		+431	+479	−22	+888
Net since admission (3/22)		+3264	+3743	+3721	+3721

Perez, Juan, Male, 29 y.o.
Allergies: NKA
Pt. Location: Bed #2 SICU

Code: FULL
Physician: D. Kuhls, MD

Isolation: Contact
Admit Date: 3/22

Laboratory Results

	Ref. Range	3/29 1522	4/1 1809
Chemistry			
Sodium (mEq/L)	136–145	146 !↑	140
Potassium (mEq/L)	3.5–5.5	4.0	3.7
Chloride (mEq/L)	95–105	99	99
Carbon dioxide (CO_2, mEq/L)	23–30	25	26
BUN (mg/dL)	8–18	23 !↑	25 !↑
Creatinine serum (mg/dL)	0.6–1.2	1.4 !↑	1.6 !↑
BUN/Crea ratio	10.0–20.0	16.42	15.6
Uric acid (mg/dL)	2.8–8.8 F 4.0–9.0 M	8.9	
Glucose (mg/dL)	70–110	164 !↑	140 !↑
Phosphate, inorganic (mg/dL)	2.3–4.7	2.2 !↓	2.4
Magnesium (mg/dL)	1.8–3	1.9	1.5 !↓
Calcium (mg/dL)	9–11	7.1	
Osmolality (mmol/kg/H_2O)	285–295	309.3 !↑	296.7 !↑
Bilirubin total (mg/dL)	≤1.5	0.9	
Bilirubin, direct (mg/dL)	<0.3	0.15	
Protein, total (g/dL)	6–8	5.2 !↓	5.1 !↓
Albumin (g/dL)	3.5–5	1.4 !↓	1.9 !↓
Prealbumin (mg/dL)	16–35	3.0 !↓	5.0 !↓
Ammonia (NH_3, µmol/L)	9–33	11	
Alkaline phosphatase (U/L)	30–120	540 !↑	
ALT (U/L)	4–36	435 !↑	
AST (U/L)	0–35	190 !↑	
CPK (U/L)	30–135 F 55–170 M	182 !↑	
Lactate dehydrogenase (U/L)	208–378	750 !↑	
C-reactive protein (mg/dL)	<1.0	245 !↑	220 !↑
Cholesterol (mg/dL)	120–199	180	
HDL-C (mg/dL)	>55 F, >45 M	40 !↓	
VLDL (mg/dL)	7–32	110 !↑	

Perez, Juan, Male, 29 y.o.
Allergies: NKA
Pt. Location: Bed #2 SICU

Code: FULL
Physician: D. Kuhls, MD

Isolation: Contact
Admit Date: 3/22

Laboratory Results *(Continued)*

	Ref. Range	3/29 1522	4/1 1809
LDL (mg/dL)	<130	140 !↑	
LDL/HDL ratio	<3.22 F <3.55 M	3.5	
Triglycerides (mg/dL)	35–135 F 40–160 M	274 !↑	265 !↑
HbA$_{1C}$ (%)	3.9–5.2	7.1 !↑	
Coagulation (Coag)			
PT (sec)	12.4–14.4	9 !↓	
INR	0.9–1.1	0.6 !↓	
PTT (sec)	24–34	21 !↓	
Hematology			
WBC ($\times 10^3$/mm^3)	4.8–11.8	15.2 !↑	
RBC ($\times 10^6$/mm^3)	4.2–5.4 F 4.5–6.2 M	3.2 !↓	
Hemoglobin (Hgb, g/dL)	12–15 F 14–1/ M	14	
Hematocrit (Hct, %)	37–47 F 40–54 M	35 !↓	
Mean cell volume (μm^3)	80–96	82	
Mean cell Hgb (pg)	26–32	27	
Mean cell Hgb content (g/dL)	31.5–36	33	
RBC distribution (%)	11.6–16.5	12	
Platelet count ($\times 10^3$/mm^3)	140–440	180	
Urinalysis			
Collection method	—	catheter	
Color	—	pale yellow	
Appearance	—	cloudy	
Specific gravity	1.003–1.030	1.045 !↑	
pH	5–7	5.1	
Protein (mg/dL)	Neg	+ !↑	
Glucose (mg/dL)	Neg	+ !↑	
Ketones	Neg	+ !↑	
Blood	Neg	Neg	

(Continued)

Perez, Juan, Male, 29 y.o.
Allergies: NKA   **Code:** FULL   **Isolation:** Contact
Pt. Location: Bed #2 SICU   **Physician:** D. Kuhls, MD   **Admit Date:** 3/22

Laboratory Results *(Continued)*

	Ref. Range	3/29 1522	4/1 1809
Bilirubin	Neg	Neg	
Nitrites	Neg	Neg	
Urobilinogen (EU/dL)	<1.1	<0.1	
Leukocyte esterase	Neg	Neg	
Prot chk	Neg	Neg	
WBCs (/HPF)	0–5	0	
RBCs (/HPF)	0–5	0	
Bact	0	5 !↑	
Mucus	0	5 !↑	
Crys	0	0	
Casts (/LPF)	0	0	
Yeast	0	2 !↑	

Case Questions

I.   Understanding the Disease and Pathophysiology

1.   The patient has suffered a gunshot wound to the abdomen. This has resulted in an open abdomen. Define *open abdomen*. The medical record describes the use of a wound "VAC." Describe this procedure and its connection to the diagnosis for open abdomen.

2.   The patient underwent gastric resection and repair, control of liver hemorrhage, and resection of proximal jejunum, leaving his GI tract in discontinuity. Describe the potential effects of surgery on this patient's ability to meet his nutritional needs.

3.   The metabolic stress response to trauma has been described as a progression through three phases: the ebb phase, the flow phase, and finally the recovery or resolution. Define each of these and determine how they may correspond to this patient's hospital course.

4.   Acute-phase proteins are often used as a marker of the stress response. What is an acute-phase protein? What is the role of C-reactive protein in the nutritional assessment of critically ill trauma patients? What other acute-phase proteins may be followed to assess the inflammatory stress response?

II.   Understanding the Nutrition Therapy

5.   Metabolic stress and trauma significantly affect nutritional requirements. Describe the changes in nutrient metabolism that occur in metabolic stress. Specifically address energy requirements and changes in carbohydrate, protein, and lipid metabolism.

6.   Are there specific nutrients that should be considered when designing nutrition support for a trauma patient? Explain the rationale and current recommendations regarding glutamine, arginine, and omega-3 fatty acids for this patient population.

7.   Using current evidence-based guidelines, explain the decision-making process that would be applied in determining the route for nutrition support for the trauma patient.

III.   Nutrition Assessment

8.   Calculate and interpret the patient's BMI.

9. What factors make assessing his actual weight difficult on a daily basis?

10. Calculate energy and protein requirements for Mr. Perez. Use at least two methods (including the Penn State) to estimate his energy needs. Explain your rationale for using each one. For the Penn State calculation, the minute ventilation is 3.5 L/minute and the maximum temperature is 39.2.

11. What does indirect calorimetry measure?

12. Compare the estimated energy needs calculated using the predictive equations with each other and with those obtained by indirect calorimetry measurements.

13. Interpret the RQ value. What does it indicate?

14. What factors contribute to the elevated energy expenditure in this patient?

15. Mr. Perez was prescribed parenteral nutrition. Determine how many kilocalories and grams of protein are provided with his prescription. Read the nutrition consult follow-up and the I/O record. What was the total volume of PN provided that day?

16. Compare this nutrition support to his measured energy requirements obtained by the metabolic cart on day 4. Based on the metabolic cart results, what changes would you recommend be made to the TPN regimen, if any? What are the limitations that prevent the health care team from making significant changes to the nutrition support regimen?

17. The patient was also receiving propofol. What is this, and why should it be included in an assessment of his nutritional intake? How much energy did it provide?

18. The RD recommended that trickle feeds be initiated. What is this and what is the rationale? The RD recommended the formula Pivot 1.5 for these trickle feeds. What type of formula is this, and what would be the rationale for choosing this formula?

19. List abnormal biochemical values for 3/29, describe why they might be abnormal, and explain any nutrition-related implications.

20. Current guidelines recommend using a nitrogen balance study to assess the adequacy of nutrition support.

    a. According to the Powell (2012) article (see bibliography below), what adjustments should be made to assess for nitrogen losses through fistulas, drains, or wound output?

    b. A 24-hour nitrogen collection is completed for Mr. Perez with results of UUN 42 g. Calculate his nitrogen balance.

IV.   Nutrition Diagnosis

21. Identify the nutrition diagnosis you would use in your follow-up note. Complete the PES statement.

V.   Nutrition Intervention

22. For the PES statement that you have written, establish an ideal goal (based on the signs and symptoms) and an appropriate intervention (based on the etiology).

VI.   Nutrition Monitoring and Evaluation

23. What are the standard recommendations for monitoring the nutritional status of a patient receiving nutrition support?

24. Hyperglycemia was noted in the laboratory results. Why is hyperglycemia of concern in the critically ill patient? How was this handled for this patient?

25. What would be the standard guidelines and subsequent recommendations to begin weaning TPN and increasing enteral feeds?

Bibliography

Bankhead R, Boullata J, Brantley S, et al. Enteral nutrition practice recommendations. *J Parenter Enteral Nutr.* 2009;33(2):122–167.

Byrnes MC, Reicks P, Irwin E. Early enteral nutrition can be successfully implemented in trauma patients with an "open abdomen." *Am J Surg.* 2010;199:359–363.

Cohen D, Kuhls D. Energy expenditure and open abdomen following trauma. *Top Clin Nutr.* 2009;24:122–129.

Diaz JJ, et al. The management of the open abdomen in trauma and emergency general surgery: Part 1—damage control. *J Trauma.* 2010;68:1425–1438.

Friese RS. The open abdomen: Definitions, management principles and nutrition support considerations. *Nutr Clin Prac.* 2012;27:492–498.

McClave SA, Martindale RG, Vanek VW, et al. Guidelines for the provision and assessment of nutrition support therapy in the adult critically ill patient: Society of Critical Care Medicine and American Society for Parenteral and Enteral Nutrition. *J Parenter Enteral Nutr.* 2009;33:277–316.

McDonnell M, Umpierrez G. Insulin therapy for the management of hyperglycemia in hospitalized patients. *Endocrinology & Metabolism Clinics of North America* [serial online]. March 2012;41(1):175–201. Available from: CINAHL Plus with Full Text, Ipswich, MA. Accessed October 26, 2012.

Nahikian-Nelms ML. Metabolic stress and the critically ill. In: Nelms M, Sucher K, Lacey K, Roth SL. *Nutrition Therapy and Pathophysiology.* 2nd ed. Belmont, CA: Wadsworth, Cengage Learning; 2011:682–701.

Nahikian-Nelms ML, Habash D. Nutrition assessment: Foundation of the nutrition care process. In: Nelms M, Sucher K, Lacey K, Roth SL. *Nutrition Therapy and Pathophysiology.* 2nd ed. Belmont, CA: Wadsworth, Cengage Learning, 2011:34–65.

Powell NJ, Collier B. Nutrition and the open abdomen. *Nutr Clin Pract.* 2012;27:499–506.

Internet Resources

ADA Evidence Analysis Library:
http://www.adaevidencelibrary.com

Nutrition Care Manual:
http://www.nutritioncaremanual.org

Case 30

Nutrition Support for Burn Injury

Sheela Thomas, MS, RD, LD, CNSC
Wexner Medical Center at The Ohio State University

Objectives

After completing this case, the student will be able to:

1. Understand the metabolic response to burn injury and the related nutritional considerations.
2. Understand the nutritional implications of the following: total body surface area burn; inhalation injury; circumferential burns; fluid resuscitation; age and comorbid conditions in burn injury.
3. Demonstrate the ability to assess the nutritional needs of patients with burn injury.
4. Evaluate the current literature regarding nutrition support for patients with burn injury.

5. Apply evidence-based guidelines for nutrition support in the critically ill burn patient.
6. Develop a nutrition care plan—with appropriate measurable goals, interventions, and strategies for monitoring and evaluation—that addresses the nutrition diagnoses for this case.

Mr. Angelo is a 65-year-old male who has been admitted to the surgical intensive care unit for treatment of serious burns estimated to cover 40% of his body as well as suspected smoke inhalation injury.

Angelo, Joe, male, 65 y.o.

Allergies: Tylenol	**Code:** FULL	**Isolation:** None
Pt. Location: SICU Bed 36	**Physician:** L. Martin	**Admit Date:** 9/9

Patient Summary: Mr. Angelo is a 65-year-old male admitted as a level 2 trauma with 40% total body surface area burns after being involved in a trailer fire. He is admitted to the surgical intensive care unit for management of burn injury.

History:

Onset of disease: Patient is unclear about what occurred, and his story changed several times during assessment. Patient lost his job recently and was coming from Atlanta to move in with his parents. He was driving behind an RV in a caravan when the RV caught fire. Apparently he was in the front cab of the RV trying to put out an engine fire when his clothes caught on fire. He jumped out of the car and rolled on the ground to put out the flames. At one time he stated that he jumped into a ravine, but later he stated this was not the case. He received 1650 cc of normal saline en route to hospital. The burn involves the face, bilateral upper extremity, bilateral lower extremity circumferentially, scrotum, back, and buttocks. The ENT service evaluated the patient and performed a nasopharyngolaryngoscopy. Findings included laryngeal edema and soot on the vocal cords bilaterally. Recommendation is to intubate for airway protection due to edema and soot on the vocal cords. Patient does have occasional wheezing and some patchy infiltrates on chest X-ray that could be related to smoke inhalation. Pt was started on fluid resuscitation per Parkland formula using lactated Ringer's (LR) @ 610 mL/hr.

Medical history: Diabetes, HTN, GERD
Surgical history: s/p cholecystectomy 30 years ago
Medications at home: None
Tobacco use: Smokes 1 PPD for >30 yrs
Alcohol use: 2–3 beers daily and a case on Saturday and Sunday
Family history: Father: HTN; mother: anxiety disorder, HTN; brother: healthy

Demographics:

Years education: 11
Language: English only
Occupation: Unemployed
Household members: Lives alone
Ethnicity: Caucasian
Religious affiliation: Unknown

Primary Assessment per EMS:

Airway: Intact
Breathing: Clear
Circulation 2+ carotid and radial pulses, 2+ femoral, 1+ DP pulses diminished
Glasgow coma score: 14

Admitting History/Physical:

General appearance: Alert, cooperative, mild distress, appears stated age. The wounds appear to have ruptured blisters and devitalized skin. The patient's ROM to affected area is diminished in range with pain.

Angelo, Joe, male, 65 y.o.
Allergies: Tylenol
Pt. Location: SICU Bed 36

Code: FULL
Physician: L. Martin

Isolation: None
Admit Date: 9/9

Vital Signs: Temp: 100   Pulse: 120   Resp rate: 22
                   BP: 140/93   Height: 72"   Weight: 71.2 kg   SpO$_2$: 98%

HEENT: Head/Face: Non-rebreather mask in place. Burns involving entire face, singed eyebrows, hair, and facial hair.
         Eyes: PERRLA
         Ears: Clear
         Nose: Soot noted in nares and oropharynx
         Throat: Dry mucous membranes
Neck: C-collar in place
Lungs: Clear to auscultation bilaterally
Heart: Tachycardia, regular rhythm; S1, S2 normal, no murmur, click, rub, or gallop
Abdomen: Soft, skin tender. Bowel sounds normal. No masses, no organomegaly, partial thickness and 1st degree burns near umbilicus.
Upper extremities: Burns noted R bicep, forearm, hand, left bicep and hand, mostly second degree. Skin sloughing and devitalized tissue.
Lower extremities: Mostly full thickness burns noted to bilateral lower extremities circumferentially
Back: Second degree burns in mid and left back
Genitourinary: Erythema and blistering at head of penis and scrotum
Peripheral vascular: Pulses:
Right pulses: FEM: present 2+, POP: present 2+, DP: present 1+, PT: present 1+
Left pulses: FEM: present 2+, POP: present 2+, DP: present 1+, PT present 1+

Nursing Assessment	9/9
Abdominal appearance (concave, flat, rounded, obese, distended)	distended
Palpation of abdomen (soft, rigid, firm, masses, tense)	soft
Bowel function (continent, incontinent, flatulence, no stool)	continent
Bowel sounds (P=present, AB=absent, hypo, hyper)	
RUQ	P, hypo
LUQ	P, hypo
RLQ	P, hypo
LLQ	P, hypo
Stool color	none
Stool consistency	
Tubes/ostomies	N/A
Genitourinary	
Urinary continence	catheter
Urine source	catheter
Appearance (clear, cloudy, yellow, amber, fluorescent, hematuria, orange, blue, tea)	yellow

(Continued)

Angelo, Joe, male, 65 y.o.
Allergies: Tylenol
Pt. Location: SICU Bed 36

Code: FULL
Physician: L. Martin

Isolation: None
Admit Date: 9/9

Nursing Assessment *(Continued)*

Nursing Assessment	9/9
Integumentary	
Skin color	beefy red to pale
Skin temperature (DI=diaphoretic, W=warm, dry, CL=cool, CLM=clammy, CD+=cold, M=moist, H=hot)	CL, M
Skin turgor (good, fair, poor, TENT=tenting)	poor
Skin condition (intact, EC=ecchymosis, A=abrasions, P=petechiae, R=rash, W=weeping, S=sloughing, D=dryness, EX=excoriated, T=tears, SE=subcutaneous emphysema, B=blisters, V=vesicles, N=necrosis)	W, S, B, N
Mucous membranes (intact, EC=ecchymosis, A=abrasions, P=petechiae, R=rash, W=weeping, S=sloughing, D=dryness, EX=excoriated, T=tears, SE=subcutaneous emphysema, B=blisters, V=vesicles, N=necrosis)	B, D
Other components of Braden score: special bed, sensory pressure, moisture, activity, friction/shear (>18=no risk, 15–16=low risk, 13–14=moderate risk, ≤12=high risk)	10

Admission Orders:
Laboratory: C-reactive protein now and routine every Monday morning
CBC, EDIF, platelet routine every morning
Chem 7, IP, Mg, Ca routine every morning
Hemoglobin A_{1C} routine one time
Hepatic function panel routine, every Monday morning
Prealbumin now and routine every Monday morning
Ionized calcium routine every morning
PT, IR, PTT routine every morning
ABGs routine every morning

Radiology:
CT head, neck, abdomen
KUB—NG placement and enteral feeding tube placement verification
Chest X-ray for CVC placement and ET tube

Vital Signs: Routine, every 1 hour
I & O recorded every hour
NG tube to low intermittent suction
Oral care per ventilator protocol: Oral mouth swab every 4 hours and PRN. Teeth and gum brushing every 12 hours. Supraglottic oral suctioning every 8 hours and prior to manipulation of the ETT.

Notify burn MD during first 48 hours: If urinary output is less than 0.5 mL/kg/hr or greater than 1.0 mL/kg/hr. Notify for any fluid bolus orders for urinary output.

Angelo, Joe, male, 65 y.o.

Allergies: Tylenol	**Code:** FULL	**Isolation:** None
Pt. Location: SICU Bed 36	**Physician:** L. Martin	**Admit Date:** 9/9

Notify house officer:
If urinary output is less than 0.5 mL/kg/hr or greater than 1.0 mL/kg/hr.
Systolic blood pressure greater than 180, less than 90.
Diastolic blood pressure greater than 110, less than 60.
Heart rate greater than 120, less than 60.
Temperature greater than 102.
Respiratory rate greater than 28, less than 10.
Oxygen saturation less than 92%.
Absent peripheral pulses or decreased in Q1H circ checks.
If repleted with any electrolyte more than twice per day.

Diet: NPO with EN. Impact with Glutamine @ 20 mL/hr, advance 20 mL/hr every 4 hours to 60 mL/hr.
Final goal rate per RD.

Activity:
Position any burned extremity elevated
Position HOB 30 degrees or greater
Position with no pillow under head if neck burns present

Scheduled and PRN Medications:
Ascorbic acid 500 mg every 12 hours
Chlorhexidine 0.12% oral solution 15 mL every 12 hours
Famotidine tablet 20 mg every 12 hours
Heparin injection 5,000 units every 8 hours
Insulin regular injection every 6 hours
Multivitamin tablet 1 Tab daily
Zinc sulfate 220 mg daily
Methadone 5 mg every 8 hours
Oxandrolone 10 mg every 12 hours
Senna tablet 8.6 mg daily
Docusate oral liquid 100 mg every 12 hours
Silver sulfadiazine 1% cream topical application daily
Acetaminophen 650 mg oral every 4 hours as needed
Midazolam HCl (Versed) 100 mg in sodium chloride 0.9% 100 mL IV infusion, initiate infusion at 1 mg/hr
Hydromorphone (Dilaudid) injection 0.5–1 mg, intravenous every 3 hours as needed
Fentanyl (Sublimaze) injection 50–100 mcg intravenous every 15 minutes as needed
Propofol (Diprivan) 10 mg/mL premix infusion, start at 25 mcg/kg/min intravenous continuous
Thiamin 100 mg × 3 days
Folate 1 mg × 3 days

Dressing change:
Hydromorphone (Dilaudid) injection 0.5–1.5 mg, IV give 10 min prior to dressing change and every
15 min prn if pain score is greater than 4 out of 10 during dressing change.

Angelo, Joe, male, 65 y.o.
Allergies: Tylenol
Pt. Location: SICU Bed 36

Code: FULL
Physician: L. Martin

Isolation: None
Admit Date: 9/9

Ketamine 50 mg injection, administer as slow IV push over greater than 1 minute immediately prior to dressing change. If needed, may administer 50 mg IV push every 15 minutes for prn pain score greater than 4 out of 10 during dressing change up to 200 mg maximum.
Midazolam (Versed) 2–5 mg, intravenous. Administer as directed, for anxiety, pre- and intraprocedure pain management. Give 10 minutes prior to dressing change and may repeat prn during dressing change if duration greater than 60 minutes.
Oxycodone (Roxicodone) 5–10 mg, per NG tube every 4 hours as needed for moderate pain.

IVF: LR @ 610 mL/hr × first 8 hours and decrease to 305 mL/hr × 16 hours

Nutrition: NPO with TF Impact with Glutamine @ 60 mL/hr.
History: Not following any specific diet. Stable weight for past 6 months. Has not been monitoring blood glucose levels for about a year.

MD Progress Note:
9/11 0500
Subjective: 65-yo male who presented as level 1 trauma with 40% total body surface area burns. Intubated on arrival to SICU for airway protection. Plan per burn team to do bronchoscopy at 11:30 today. Patient with significant respiratory acidosis. S/P escharotomy of bilateral lower extremities overnight per trauma team. Hypotensive overnight. Received 4 L of fluids.

Principal Problem:
Burn involving 40% body surface area
Active problems:
Respiratory failure
Acute pain due to injury
Oliguria
Malnutrition
Vitals: Temp: 100.2 Pulse: 104 Resp. rate: 18 BP: 87/59 O_2 Sat (%) 100%

Hemodynamic/Invasive Device Data (24 hours):
Arterial Line (1) Monitoring
Arterial Line (1) BP: 87/59 mm Hg
Arterial Line (1) MAP: 71 mm Hg

Ventilation/Oxygen Therapy (24 hours):
Oxygen therapy
O_2 Sat (%): 100%
Oxygen therapy O_2 device: Ventilator (mechanical ventilation)
FIO_2%: 40%

Angelo, Joe, male, 65 y.o.
Allergies: Tylenol   **Code:** FULL   **Isolation:** None
Pt. Location: SICU Bed 36   **Physician:** L. Martin   **Admit Date:** 9/9

Ventilator settings and monitoring (adult)
Ventilator type: 840
Adult/Pediatric Modes SIMV-BiPhasic: SIMV VC
Set/target tidal volume: 550
Set rate: 18
RR total (breaths/min): 18
PEEP: 5
Pressure support: 10
Peak inspiratory pressure (cmH$_2$O): 18
I:E ratio: 1:2.30

Fluid Management (24 hours):
I/O last 3 completed shifts
In: 16425 mL
Out: 1696 mL
Urine output: 1295 mL (18 mL/kg)

Physical Exam:
General appearance: Intubated, sedated
Head: Burns involving entire face, singed eyebrows, hair, and facial hair
Back: Partial thickness burns over lower back and buttocks
Lungs: Clear to auscultation bilaterally
Heart: Tachycardia, regular rhythm; S1, S2 normal, no murmur, click, rub, or gallop
Abdomen: Soft, non-tender. Bowel sounds normal. No masses, no organomegaly, partial thickness
and 1st degree burns near umbilicus.
Male genitalia: Abnormal findings: blistering over scrotum and head of penis
Extremities: Partial thickness burns to bilateral upper extremities and full thickness circumferential
burns to lower extremities. S/P escharotomy of bilateral lower extremities.

Assessment/Plan:
40% TBSA burn: Managed per burn team. Continue daily dressing changes. OR today for debridement and split thickness skin grafting.
Respiratory failure: Intubated 9/9 for airway protection. Bronchoscopy at 11:30 today.
Pain: Versed gtt, increase methadone to 10 mg every 8 hours. Dilaudid and fentanyl prn. Wean propofol to off possibly by the end of the day. Currently at 25 mL/hr.
Hyperkalemia: Secondary to metabolic, respiratory acidosis. Improving. Last K+5.9. Continue to resuscitate with LR.
Protein-calorie malnutrition: Advance TF to goal rate per nutrition.
Acute kidney injury: Continue fluid resuscitation.

.... L. Martin MD

Angelo, Joe, male, 65 y.o.
Allergies: Tylenol
Pt. Location: SICU Bed 36

Code: FULL
Physician: L. Martin

Isolation: None
Admit Date: 9/9

Intake/Output

Date		9/09 0700–9/10 0659			
Time		0700–1459	1500–2259	2300–0659	Daily total
IN	P.O.	**0**	**0**	**0**	**0**
	I.V.	**4880**	**2440**	**2440**	**9760**
	(mL/kg/hr)	(8.6)	(4.3)	(4.3)	(5.71)
	IV bolus	**1000**	**1000**	**4000**	**6000**
	Enteral feeding	**0**	**228**	**337**	**565**
	IV piggyback	**0**	**100**	**0**	**100**
	Total intake	**5880**	**3768**	**6777**	**16425**
	(mL/kg)	(82.6)	(52.9)	(95.2)	(230.7)
OUT	Urine	**665**	**325**	**305**	**1295**
	(mL/kg/hr)	(1.2)	(0.6)	(0.5)	(0.76)
	Emesis output				
	NG Tube	**0**	**300**	**100**	**400**
	Stool	**0**	**0**	**×1**	**×1**
	Total output	**665**	**625**	**405**	**1695**
	(mL/kg)	(9.3)	(8.8)	(5.7)	(23.8)
Net I/O		**+5215**	**+3143**	**+6372**	**+14730**
Net since admission (9/09)		**+5215**	**+8358**	**+14730**	**+14730**

Laboratory Results

	Ref. Range	9/9 0200
Chemistry		
Sodium (mEq/L)	136–145	137
Potassium (mEq/L)	3.5–5.5	5.9 !↑
Chloride (mEq/L)	95–105	113 !↑
Carbon dioxide (CO_2, mEq/L)	23–30	20 !↓
BUN (mg/dL)	8–18	13
Creatinine serum (mg/dL)	0.6–1.2	1.26 !↑
BUN/Crea ratio	10.0–20.0	10.3
Glucose (mg/dL)	70–110	211 !↑
Phosphate, inorganic (mg/dL)	2.3–4.7	3.4
Magnesium (mg/dL)	1.8–3	1.5 !↓
Calcium (mg/dL)	9–11	6.9 !↓

Angelo, Joe, male, 65 y.o.

Allergies: Tylenol	
Pt. Location: SICU Bed 36	

Code: FULL

Physician: L. Martin

Isolation: None

Admit Date: 9/9

Laboratory Results *(Continued)*

	Ref. Range	9/9 0200
Osmolality (mmol/kg/H$_2$O)	285–295	295
Bilirubin total (mg/dL)	≤1.5	1.2
Bilirubin, direct (mg/dL)	<0.3	0.2
Protein, total (g/dL)	6–8	4.7 !↓
Albumin (g/dL)	3.5–5	2.1 !↓
Prealbumin (mg/dL)	16–35	12 !↓
Alkaline phosphatase (U/L)	30–120	70
ALT (U/L)	4–36	21
AST (U/L)	0–35	44 !↑
C-reactive protein (mg/dL)	<1	12 !↑
Coagulation (Coag)		
PT (sec)	12.4–14.4	13.7
PTT (sec)	24–34	26
Hematology		
WBC (×10^3/mm^3)	4.8–11.8	18.1 !↑
RBC (×10^6/mm^3)	4.2–5.4 F 4.5–6.2 M	5.97
Hemoglobin (Hgb, g/dL)	12–15 F 14–17 M	18.7 !↑
Hematocrit (Hct, %)	37–47 F 40–54 M	54.4 !↑
Arterial Blood Gases (ABGs)		
pH	7.35–7.45	7.31 !↓
pCO$_2$ (mm Hg)	35–45	39.8
SO$_2$ (%)	≥95	99.8
pO$_2$ (mm Hg)	≥80	106.5
HCO$_3^-$ (mEq/L)	24–28	19.6 !↓

Case Questions

I. **Understanding the Diagnosis and Pathophysiology**

1. Describe how burn wounds are classified. Identify and describe Mr. Angelo's burn injuries.

2. Explain the "rule of nines" used in assessment of burn injury.

3. Mr. Angelo's fluid resuscitation order was: *LR @ 610 mL/hr × first 8 hours and decrease to 305 mL/hr × 16 hours*. What is the primary goal of fluid resuscitation? Briefly explain the Parkland formula. What common intravenous fluid is used in burn patients for fluid resuscitation? What are the components of this solution?

4. What is inhalation injury? How can it affect patient management?

5. Burns are often described as one of the most metabolically stressful injuries. Discuss the effects of a burn on metabolism and how this will affect nutritional requirements.

6. List all medications that Mr. Angelo is receiving. Identify the action of each medication and any drug–nutrient interactions that you should monitor.

II. **Understanding the Nutrition Therapy**

7. Using evidence-based guidelines, describe the potential benefits of early enteral nutrition in burn patients.

8. What are the common criteria used to assess readiness for the initiation of enteral nutrition in burn patients?

9. What are the specialized nutrient recommendations for the enteral nutrition formula administered to burn and trauma patients per ASPEN/SCCM guidelines?

10. What additional micronutrients will need supplementation in burn therapy? What dosages are recommended?

III.   Nutrition Assessment

11. Using Mr. Angelo's height and admit weight, calculate IBW, % IBW, BMI, and BSA.

12. Energy requirements can be estimated using a variety of equations. The Xie and Zawacki equations are frequently used. Estimate Mr. Angelo's energy needs using these equations. How many kcal/kg does he require based on these equations?

13. Determine Mr. Angelo's protein requirements. Provide the rationale for your estimate.

14. The MD's progress note indicates that the patient is experiencing acute kidney injury. What is this? If the patient's renal function continues to deteriorate and he needs continuous renal replacement therapy, what changes will you make to your current nutritional regimen and why?

15. This patient is receiving the medication propofol. Using the information that you listed in question #6, what changes will you make to your nutritional regimen and how will you assess tolerance to this medication?

IV.   Nutrition Diagnosis

16. Identify at least 2 of the most pertinent nutrition problems and the corresponding nutrition diagnoses.

17. Write your PES statement for each nutrition problem.

V.   Nutrition Intervention

18. The patient is receiving enteral feeding using Impact with Glutamine @ 60 mL/hr. Determine the energy and protein provided by this prescription. Provide guidelines to meet the patient's calculated needs using the Xie equation.

19. By using the information on the intake/output record, determine the energy and protein provided during this time period. Compare the energy and protein provided by the enteral feeding to your estimation of Mr. Angelo's needs.

20. One of the residents on the medical team asks you if he should stop the enteral feeding because the patient's blood pressure has been unstable. What recommendations can you make to the patient's critical care team regarding tube feeding and hemodynamic status?

VI.   Nutrition Monitoring and Evaluation

21. List factors that you would monitor to assess the tolerance to and adequacy of nutrition support.

22. What is the best method to assess calorie needs in critically ill patients? What are the factors that need to be considered before the test is ordered?

23. Write an ADIME note that provides your nutrition assessment and enteral feeding recommendations and/or evaluation of the current enteral feeding orders.

Bibliography

Bankhead R, Boullata J, Brantley S, et al. Enteral nutrition practice recommendations. *J Parenter Enteral Nutr.* 2009;33:122–167.

Chan MM, Chan GM. Nutritional therapy for burns in children and adults. *Nutrition.* 2009;25: 261–269.

Cioffi G. What's New in Burns and Metabolism. *J Am Coll Surg.* 2001;192:241–253.

Coen JR, Carpenter AM, Shupp JW, et al. The results of a national survey regarding nutritional care of obese burn patients. *J Burn Care & Res.* 2011;32:561–565.

Endorf FW, Ahrenholz D. Burn management. *Curr Op in Crit Care.* 2011;17:601–605.

Graves C, Saffle J, Cochran A. Actual burn nutrition care practices: An update. *J Burn Care Res.* 2009; 30:77–82.

Hall B. Care for the patient with burns in the trauma rehabilitation setting. *Critical Care Nursing Quarterly.* 2012;3:272–280.

Holt B, Graves C, Faraklas I, Cochran A. Compliance with nutrition support guidelines in acutely burned patients. *Burns.* 2012;38:645–649.

Kurmis R, Parker A, Greenwood J. The use of immunonutrition in burn injury care: Where are we? *J Burn Care & Res.* 2010;31:677–691.

Latenser BA. Critical care of the burn patient: The first 48 hours. *Crit Care Med.* 2009;37:2819–2826.

Masters B, Aarabi S, Sidhwa F, Wood F. High-carbohydrate, high-protein, low-fat versus low-carbohydrate, high-protein, high-fat enteral feeds for burns. *Cochrane Database of Systematic Reviews.* 2012;1:Article No.:CD006122.DOI:10.1002/14651858. CD006122. pub3.

Mayes T, Gottschlich MM, Warden GD. Clinical nutrition protocols for continuous quality improvements in the outcomes of patients with burns. *J Burn Care Rehabil.* 1997;18:365–368.

McClave SA, et al. Guidelines for the provision and assessment of nutrition support therapy in the adult critically ill patient: Society of Critical Care Medicine (SCCM) and American Society for Parenteral and Enteral Nutrition (ASPEN). *J Parenter Enteral Nutr.* 2009;33:277–316.

Moore CL, Schmidt PM. A burn progressive care unit: Customized care from admission through discharge. *Perioperative Nursing Clinics.* 2012;7:99–105.

Mendonça MN, Gragnani A, Masako F. Burns, metabolism and nutritional requirements. *Nutr Hosp.* 2011;26:692–700.

Nahikian-Nelms ML. Metabolic stress and the critically ill. In: Nelms M, Sucher K, Lacey K, Roth SL. *Nutrition Therapy and Pathophysiology.* 2nd ed. Belmont, CA: Wadsworth, Cengage Learning; 2011:682–701.

Nahikian-Nelms ML, Habash D. Nutrition assessment: Foundation of the nutrition care process. In: Nelms M, Sucher K, Lacey K, Roth SL. *Nutrition Therapy and Pathophysiology.* 2nd ed. Belmont, CA: Wadsworth, Cengage Learning; 2011:34–65.

Rodriguez NA, Jeschke MG, Williams FN, Kamolz LP, Herndon DN. Nutrition in burns: Galveston contributions. *JPEN.* 2011;35:704–14.

Sullivan J. Nutrition and metabolic support in severe burn injury. *Support Line.* 2010;32:3–13.

Vanek VW, et al. A.S.P.E.N. Position Paper: Parenteral Nutrition Glutamine Supplementation. *Nutrition in Clinical Practice.* 2011;26:479–494.

Internet Resources

ADA Evidence Analysis Library:
http://www.adaevidencelibrary.com

American Burn Association: http://www.ameriburn.org

Nutrition Care Manual: http://www.nutritioncaremanual.org

UpToDate: http://www.uptodate.com/contents /overview-of-nutritional-support-for-moderate-to -severe-burn-patients

Case 31

Nutrition Support in Sepsis and Morbid Obesity

Objectives

After completing this case, the student will be able to:

1. Identify the current surgical procedures used for bariatric surgery.
2. Identify the physiological consequences of bariatric surgery.
3. Describe the pathophysiology of sepsis.
4. Apply evidence-based guidelines for provision of nutrition support in the morbidly obese patient.
5. Identify specific nutrients that may assist with treatment of sepsis.
6. Describe refeeding syndrome and the current recommendations for its prevention.
7. Interpret nutrition assessment data to assist with the design of measurable goals, interventions, and strategies for monitoring and evaluation that address the nutrition diagnoses for the patient.

Chris McKinley is a 37-year-old obese male admitted to the MICU from the ER with probable sepsis 4 months after undergoing a Roux-en-Y gastric bypass.

McKinley, Chris, Male, 37 y.o.
Allergies: NKA
Pt. Location: MICU Bed #5

Code: FULL
Physician: P Walker

Isolation: None
Admit Date: 2/23

Patient Summary: Mr. McKinley has suffered from type 2 diabetes mellitus, hyperlipidemia, hypertension, and osteoarthritis over the previous 10 years. Mr. McKinley has weighed over 250 lbs since age 15 with steady weight gain since that time. He had attempted to lose weight numerous times but the most weight he ever lost was 75 lbs, which he regained over the following two-year period. He had recently reached his highest weight of 425 lbs, lost 24 lbs prior to his planned bariatric surgery through the preoperative nutrition education program, and then had the Roux-en-Y gastric bypass surgery 4 months ago. He has done well at home with a total weight loss of approximately 100 lbs to date. Now, however, he is admitted to the MICU from the ER with probable sepsis.

History:
Onset of disease: Experienced flu-like symptoms over previous 48 hours and then became acutely short of breath—brought to ER at that time.
Medical history: Type 2 diabetes mellitus, hypertension, hyperlipidemia, osteoarthritis
Surgical history: s/p Roux-en-Y gastric bypass surgery November 1; R total knee replacement 3 years previous
Medications at home: Lovastatin 60 mg/day (previously on Lantus and metformin—off diabetes medications for 2 months)
Tobacco use: None
Alcohol use: Socially, 2–3 beers per week—has not had alcohol since surgery
Family history: Father: Type 2 DM, CAD, Htn, COPD; Mother: Type 2 DM, CAD, osteoporosis

Demographics:
Marital status: Single
Number of children: 0
Years education: Associate's degree
Language: English only
Occupation: Office manager for real estate office
Hours of work: 8–5 daily—sometimes on weekends
Household members: Lives with roommate
Ethnicity: Caucasian
Religious affiliation: None stated

Admitting History/Physical:
Chief complaint: On mechanical ventilation
General appearance: Obese white male

| **Vital Signs:** | Temp: 102.5 | Pulse: 98 | Resp rate: 23 |
| | BP: 135/90 | Height: 5'10" | Weight: 325 lbs |

McKinley, Chris, Male, 37 y.o.

Allergies: NKA   **Code:** FULL   **Isolation:** None
Pt. Location: MICU Bed #5   **Physician:** P Walker   **Admit Date:** 2/23

Heart: Elevated rate, regular rhythm, normal heart; diminished distal pulses. Exam reveals no gallop and no friction rub.
HEENT: Head: WNL
          Eyes: PERRLA
          Ears: Clear
          Nose: WNL
          Throat: Dry mucous membranes without exudates or lesions
Genitalia: Normally developed 37-year-old male
Extremities: Ecchymosis, abrasions, petechiae on lower extremities, 2+ pitting edema
Skin: Warm, dry to touch
Chest/lungs: Respirations rapid with rales
Peripheral vascular: Diminished pulses bilaterally
Abdomen: Obese, rash present under skinfolds

Nursing Assessment	2/23
Abdominal appearance (concave, flat, rounded, obese, distended)	obese
Palpation of abdomen (soft, rigid, firm, masses, tense)	soft
Bowel function (continent, incontinent, flatulence, no stool)	continent
Bowel sounds (P5present, AB5absent, hypo, hyper)	
RUQ	P
LUQ	P
RLQ	P
LLQ	P
Stool color	light brown
Stool consistency	formed
Tubes/ostomies	NA
Genitourinary	
Urinary continence	NA
Urine source	catheter
Appearance (clear, cloudy, yellow, amber, fluorescent, hematuria, orange, blue, tea)	clear, yellow
Integumentary	
Skin color	pale
Skin temperature (DI=diaphoretic, W=warm, dry, CL=cool, CLM=clammy, CD+=cold, M=moist, H=hot)	W, M
Skin turgor (good, fair, poor, TENT=tenting)	good

(Continued)

McKinley, Chris, Male, 37 y.o.
Allergies: NKA   **Code:** FULL   **Isolation:** None
Pt. Location: MICU Bed #5   **Physician:** P Walker   **Admit Date:** 2/23

Nursing Assessment *(Continued)*

Nursing Assessment	2/23
Skin condition (intact, EC=ecchymosis, A=abrasions, P=petechiae, R=rash, W=weeping, S=sloughing, D=dryness, EX=excoriated, T=tears, SE=subcutaneous emphysema, B=blisters, V=vesicles, N=necrosis)	EC, A, R
Mucous membranes (intact, EC=ecchymosis, A=abrasions, P=petechiae, R=rash, W=weeping, S=sloughing, D=dryness, EX=excoriated, T=tears, SE=subcutaneous emphysema, B=blisters, V=vesicles, N=necrosis)	intact
Other components of Braden score: special bed, sensory pressure, moisture, activity, friction/shear (>18 = no risk, 15–16 = low risk, 13–14 = moderate risk, ≤12 = high risk)	13

Orders:

Initiate Sepsis Bundle Orders: Central line placement; arterial line placement—arterial line care per protocol
If SBP <90 or MAP <65: norepinephrine 4 mcg/min; dopamine 5 mcg/kg/min; epinephrine 1 mcg/min
Normal Saline 500 mL bolus until CVP 8–12, then continue at 150 mL/hr
ECG, urine culture, U/A with microscopic; blood culture 1st and 2nd peripheral site
Serum lactate
Basic metabolic panel
Hepatic function panel, CBC, EDIF, platelets.
Insert peripheral IV and maintain venous access
Vancomycin 2 g in sodium chloride IVPB
Piperacillin-tazobactam (Zosyn) 4.5 g in dextrose 100 mL IVPB
Height, Weight, Intake/Output
Insert feeding tube—nutrition consult

Nutrition:

Meal type: NPO
Fluid requirement: 1800–2000 mL

MD Progress Note:

2/24
Subjective: Chris McKinley's previous 24 hours reviewed

Vitals: Temp: 102.4, Pulse: 88, Resp rate: 30, BP:108/58
Urine Output: 3270 mL, Glu 220

McKinley, Chris, Male, 37 y.o.
Allergies: NKA
Pt. Location: MICU Bed #5

Code: FULL
Physician: P Walker

Isolation: None
Admit Date: 2/23

Physical Exam:

HEENT: Obese neck, no adenopathy, no JVD appreciated, RIJ CVC in place

Neck: WNL

Heart: Regular rate, regular rhythm, no M/R/G appreciated

Lungs: Coarse breath sounds bilaterally with scattered rhonchi R>L; no wheezes or crackles

Abdomen: Morbidly obese, soft, non-distended, no organomegaly, bowel sounds present

Extremities: Good radial pulses bilaterally. Ecchymosis, abrasions, petechiae on lower extremities, 2+ pitting edema.

Skin: WNL

Neurologic: Intubated, sedated, pupils equal and reactive to light

Assessment/Plan: 37-yo male transferred from ER with severe sepsis, pneumonia. Maintain current mechanical ventilation, cultures pending but will continue vancomycin and Zosyn. Sedation with Versed and fentanyl. Initiate enteral feeding per nutrition consult.

.... P. Walker, MD

Intake/Output

Date		2/23 0701 – 2/24 0700			
Time		0701–1500	1501–2300	2301–0700	Daily total
IN	P.O.	0	0	0	0
	I.V.	1550	1200	1200	3950
	(mL/kg/hr)	(1.31)	(1.01)	(1.01)	(1.11)
	I.V. piggyback	250	250	250	750
	TPN	0	0	0	0
	Total intake	1800	1450	1450	4700
	(mL/kg)	(12.2)	(9.81)	(9.81)	(31.8)
OUT	Urine	1320	1000	950	3270
	(mL/kg/hr)	(1.12)	(0.85)	(0.80)	(0.92)
	Emesis output				
	Other				
	Stool				
	Total output	1320	1000	950	3270
	(mL/kg)	(8.93)	(6.77)	(6.43)	(22.1)
Net I/O		+480	+450	+500	+1430
Net since admission (2/23)		+480	+930	+1430	+1430

McKinley, Chris, Male, 37 y.o.
Allergies: NKA
Pt. Location: MICU Bed #5

Code: FULL
Physician: P Walker

Isolation: None
Admit Date: 2/23

Laboratory Results

	Ref. Range	2/23
Chemistry		
Sodium (mEq/L)	136–145	136
Potassium (mEq/L)	3.5–5.5	5.8 !↑
Chloride (mEq/L)	95–105	99
Carbon dioxide (CO_2, mEq/L)	23–30	31 !↑
BUN (mg/dL)	8–18	15
Creatinine serum (mg/dL)	0.6–1.2	0.9
BUN/Crea ratio	10.0–20.0	16.7
Glucose (mg/dL)	70–110	185 !↑
Phosphate, inorganic (mg/dL)	2.3–4.7	2.2 !↓
Magnesium (mg/dL)	1.8–3	1.8
Calcium (mg/dL)	9–11	9.5
Osmolality (mmol/kg/H_2O)	285–295	289
Bilirubin total (mg/dL)	≤1.5	0.8
Bilirubin, direct (mg/dL)	<0.3	0.7 !↑
Protein, total (g/dL)	6–8	5.8 !↓
Albumin (g/dL)	3.5–5	1.9 !↓
Prealbumin (mg/dL)	16–35	11 !↓
Ammonia (NH_3, μmol/L)	9–33	35 !↑
Alkaline phosphatase (U/L)	30–120	118
ALT (U/L)	4–36	37 !↑
AST (U/L)	0–35	38 !↑
CPK (U/L)	30–135 F 55–170 M	220 !↑
C-reactive protein (mg/dL)	<1.00	5.8 !↑
Fibrinogen (mg/dL)	160–450	525 !↑
Lactate (mEq/L)	0.3–2.3	4.2 !↑
Cholesterol (mg/dL)	120–199	320 !↑
HDL-C (mg/dL)	>55 F, >45 M	32 !↓
VLDL (mg/dL)	7–32	45 !↑
LDL (mg/dL)	<130	232 !↑
LDL/HDL ratio	<3.22 F <3.55 M	7.5 !↑

McKinley, Chris, Male, 37 y.o.
Allergies: NKA
Pt. Location: MICU Bed #5

Code: FULL
Physician: P Walker

Isolation: None
Admit Date: 2/23

Laboratory Results *(Continued)*

	Ref. Range	2/23
Triglycerides (mg/dL)	35–135 F 40–160 M	245 !↑
T_4 (µg/dL)	4–12	6.1
T_3 (µg/dL)	75–98	82
Amylase (U/L)	25–125	26
Lipase (U/L)	10–140	11
HbA$_{1C}$ (%)	3.9–5.2	6.8 !↑
Hematology		
WBC ($\times 10^3$/mm³)	4.8–11.8	23.5 !↑
RBC ($\times 10^6$/mm³)	4.2–5.4 F 4.5–6.2 M	5.5
Hemoglobin (Hgb, g/dL)	12–15 F 14–17 M	12.5 !↓
Hematocrit (Hct, %)	37–47 F 40–54 M	38 !↓
Platelet count ($\times 10^3$/mm³)	140–440	210
Transferrin (mg/dL)	250–380 F 215–365 M	385 !↑
Ferritin (mg/mL)	20–120 F 20–300 M	14 !↓
Vitamin B$_{12}$ (ng/dL)	24.4–100	25
Folate (ng/dL)	5–25	15
Urinalysis		
Collection method	—	catheter
Color	—	yellow
Appearance	—	clear
Specific gravity	1.003–1.030	1.004
pH	5–7	6.9
Protein (mg/dL)	Neg	+ !↑
Glucose (mg/dL)	Neg	+ !↑
Ketones	Neg	+ !↑
Blood	Neg	Neg
Bilirubin	Neg	Neg
Nitrites	Neg	Neg
Urobilinogen (EU/dL)	<1.1	Neg

(Continued)

McKinley, Chris, Male, 37 y.o.
Allergies: NKA
Pt. Location: MICU Bed #5

Code: FULL
Physician: P Walker

Isolation: None
Admit Date: 2/23

Laboratory Results *(Continued)*

	Ref. Range	2/23
Leukocyte esterase	Neg	Neg
Prot chk	Neg	+ !↑
WBCs (/HPF)	0–5	2
RBCs (/HPF)	0–5	0
Bact	0	+ !↑
Mucus	0	0
Crys	0	0
Casts (/LPF)	0	0
Yeast	0	0

Case Questions

I.   Understanding the Pathophysiology

1.   Mr. McKinley's admission orders indicate he is being treated for probable sepsis and SIRS. Define these conditions.

2.   Describe the metabolic alterations that occur as a result of sepsis and the systemic inflammatory response. Using the medical record information, identify the specific criteria that are consistent with the diagnosis of sepsis.

3.   Mr. McKinley had a Roux-en-Y gastric bypass 4 months ago and has lost approximately 100 lbs. Describe this procedure. Identify the most probable nutritional concerns associated with this rapid weight loss.

II.   Understanding the Nutrition Therapy

4.   Using evidenced-based guidelines, determine whether Mr. McKinley should receive nutrition support. Explain the rationale for your decisions.

5.   How will Mr. McKinley's bariatric surgery affect your recommendations for nutrition support?

6.   Define *refeeding syndrome*. How will Mr. McKinley's recent 100-lb weight loss affect your nutrition support recommendations?

III.   Nutrition Assessment

7.   Assess Mr. McKinley's height and weight. Calculate his BMI and % usual body weight.

8.   After reading the physician's history and physical, identify any signs or symptoms that are most likely a consequence of Mr. McKinley's admitting critical illness.

9.   Identify any abnormal biochemical indices and discuss the probable underlying etiology.

10.   Assess Mr. McKinley's current hydration status using the first 24 hours of I/O and the nursing assessment.

11. Determine Mr. McKinley's energy and protein requirements. Explain the rationale for the method you used to calculate these requirements.

IV. Nutrition Diagnosis

12. Identify the pertinent nutrition problems and the corresponding nutrition diagnoses.

13. Are you able to diagnose Mr. McKinley using the etiology-based malnutrition criteria? If so, describe the information you used to make this diagnosis.

V. Nutrition Intervention

14. Outline the nutrition support regimen you would recommend for Mr. McKinley. This should include formula choice (and rationale) and rate initiation and advancement.

VI. Nutrition Monitoring and Evaluation

15. Identify the steps you would take to monitor Mr. McKinley's nutritional status in the intensive care unit.

16. What factors may affect his tolerance to enteral feeding?

17. Write a note for your initial inpatient nutrition assessment with nutrition support recommendations.

Bibliography

Academy of Nutrition and Dietetics. Nutrition Care Manual. Bariatric Surgery. Available at: http://nutrition-caremanual.org/content.cfm?ncm_content_id=79396. Accessed March 20, 2012.

Arabi YM, Tamim HM, Dhar GS, et al. Permissive underfeeding and intensive insulin therapy in critically ill patients: A randomized controlled trial. *Am J Clin Nutr.* 2011;93:569–577.

Davis CJ, Sowa D, Keim KS, et al. The use of prealbumin and C-reactive protein for monitoring nutrition support in adult patients receiving enteral nutrition in an urban medical center. *J Parenter Enteral Nutr.* 2012;36:197–204.

Dickerson RN. Optimal caloric intake for critically ill patients: First, do no harm. *Nutr Clin Pract.* 2001;26:48–54.

Gariballa S. Refeeding syndrome: A potentially fatal condition but remains underdiagnosed and undertreated. *Nutrition.* 2008;24:604–606.

Jecjeebhoy KN. Permissive underfeeding of the critically ill patient. *Nutr Clin Pract.* 2004;19:477–480.

Kastrup M, Spies C. Less is more? Is permissive underfeeding in crtically ill patients necessary? *Am J Clin Nutr.* 2011;94:957–958.

Kushner RF, Drover JW. Current strategies of critical care assessment and therapy of the obese patient (hypocaloric feeding): What are we doing and what do we need to do? *J Parenter Enteral Nutr.* 2011;35(5 Suppl):36S–43S.

Laferrère B. Diabetes remission after bariatric surgery: Is it just the incretins? *International Journal of Obesity.* 2011;35:S22–S25.

Miller KR, Kiraly L, Martindale RG. Critical Care Sepsis. In: ASPEN Adult Nutrition Support Core Curriculum. 2nd ed. 2012:377–391.

Miller KR, Laszlo NK, Lowen CC, et al. "CAN WE FEED?" A mnemonic to merge nutrition and intensive care assessment of the critically ill patient. *J Parenter Enteral Nutr.* 2011;35:643–660.

Nelms M. Metabolic stress and the critically ill. In: Nelms M, Sucher K, Lacey K, Roth SL. *Nutrition Therapy and Pathophysiology.* 2nd ed. Belmont, CA: Wadsworth, Cengage Learning; 2011:682–701.

Owais AR, Bumby RF, MacFie J. Review article: Permissive underfeeding in short-term nutritional support. *Aliment Pharmacol Ther.* 2010;32:628–636.

Singer P, Anbar R, Cohen J, et al. The tight calorie control study (TICACOS): A prospective, randomized, controlled pilot study of nutritional support in critically ill patients. *Intensive Care Med.* 2011;37:601–609.

Tresley J, Sheean PM. Refeeding syndrome: Recognition is the key to prevention and management. *J Am Diet Assoc.* 2008;108:2105–2108.

Turner KL, Moore FA, Martindale R. Nutrition support for the acute lung injury/adult respiratory distress patient: A review. *Nutr Clin Pract.* 2011;26:14–25.

Ukleja A, Freeman KL, Gilbert K, et al. Standards for nutrition support: Adult hospitalized patients. *Nutr Clin Pract.* 2010;25:403–414.

Internet Resources

American Society for Metabolic and Bariatric Surgery: http://ASMBS.org

Nutrition Care Manual: http://www.nutritioncaremanual.org

Sepsis Alliance: http://www.sepsisalliance.org/

Surviving Sepsis Campaign: http://www.survivingsepsis.org/Pages/default.aspx

U.S. National Library of Medicine: http://www.ncbi.nlm.nih.gov/pubmedhealth/PMH0001687/

Unit Twelve
NUTRITION THERAPY FOR NEOPLASTIC DISEASE

The layperson often uses *cancer* as a name for one disease. The term *cancer*, or *neoplasm*, actually describes any condition in which cells proliferate at a rapid rate and in an unrestrained manner. Each type of cancer is a different disease with different origins and responses to therapy. It is difficult to generalize about the role of nutrition in cancer treatment, because each diagnosis is truly an individual case. However, it is obvious to any clinician participating in the care of cancer patients that nutrition problems are common.

More than 80 percent of patients with cancer experience some degree of malnutrition. Nutrition problems may be some of the first symptoms the patient recognizes. Unexplained weight loss, changes in ability to taste, or a decrease in appetite are often present at diagnosis. The malignancy itself may affect not only energy requirements, but also the metabolism of nutrients.

As the patient begins therapy for a malignancy—surgery, radiation therapy, chemotherapy, immunotherapy, or bone marrow transplant—treatment side effects occur that can affect nutritional status. And, as with many medical conditions, cancer patients also face significant psychosocial issues. Can nutrition make a difference? Adequate nutrition helps prevent surgical complications, meet increased energy and protein requirements, and repair and rebuild tissues, which cancer therapies often damage. Furthermore, good nutrition allows increased tolerance of therapy and helps maintain the patient's quality of life.

All these factors must be considered when planning nutritional and medical care. The cases in this section allow you to plan nutritional care for some of the most common problems during cancer diagnosis and therapy. In addition, the cases in Unit Twelve let you practice nutrition support and tackle psychosocial issues in complementary and alternative medicine.

Acute Lymphoblastic Leukemia Treated with Hematopoietic Cell Transplantation

Kimberlee Orben MS, RD, LD, CSO
The Ohio State University Comprehensive Cancer Center – Arthur G. James
Cancer Hospital and Richard J. Solove Research Institute – Wexner Medical Center

Objectives

After completing this case, the student will be able to:

1. Describe the role of blast cells in acute lymphoblastic leukemia.
2. Explain the rationale for using high-dose chemotherapy, total body irradiation, and hematopoietic cell transplantation in the treatment of hematologic malignancies.
3. Identify nutrition risk factors in patients undergoing a hematopoietic cell transplant.
4. Explain the mechanism of graft versus host disease (GVHD) and the related nutrition implications.
5. Apply the current recommendations regarding nutrition support in patients receiving hematopoietic cell transplants.
6. Develop a nutrition care plan—with appropriate measurable goals, interventions, and strategies for monitoring and evaluation—that addresses the nutrition diagnoses for this case.

Scott Bear is 28-year-old white male with acute lymphoblastic leukemia (ALL) in chronic remission 2. He is being admitted to the transplant unit for a 10/10 HLA matched unrelated donor transplant (MUD) with myeloablative cyclophosphamide and total body irradiation (TBI).

Bear, Scott, Male, 28 y.o.
Allergies: NKA   **Code:** FULL   **Isolation:** None
Pt. Location: BMT   **Physician:** S. Smoot   **Admit Date:** 11/7

Patient Summary: Patient is a 28-year-old white male with acute lymphoblastic leukemia (ALL) admitted to the transplant unit for a 10/10 HLA matched unrelated donor transplant (MUD) with myeloablative cyclophosphamide and total body irradiation (TBI). He will receive TBI twice daily on days −6, −5, and −4 and cyclophosphamide daily on days −3 and −2. After transplant, he will receive 4 doses of methotrexate for graft-versus-host disease (GVHD) prophylaxis on days +1, +3, +6, and +11.

Oncology History:

Pt was initially diagnosed with ALL in 2008; however, it is unknown what his cytogenetics or treatment regimen was at that time. He was doing well until spring of 2012 when he was admitted to a local ED with fever, fatigue, dyspnea, and red spots on his skin. After further evaluation, he was diagnosed with a relapse of Pre-B cell ALL. The patient was started on HyperCVAD, and received 5 cycles of HyperCVAD prior to admission. On admission the patient is in chronic remission 2 (CR2).
Medical history: ALL, obesity, anxiety, depression, osteomyelitis, back pain
Surgical history: Mediport placement, mandible reconstruction 2006
Medications at home: Morphine, temazepam
Tobacco use: ½ pack daily, states he "quit this morning"
Alcohol use: 1 drink daily, beer or alcohol
Family history: Mother—diabetes; father—hypertension

Demographics:

Marital status: Married
Years education: 12
Language: English only
Occupation: Server at local restaurant
Hours of work: 8- to 12-hour days
Household members: Wife, 2 sons (ages 4,6), 1 daughter (age 2)
Ethnicity: Caucasian
Religious affiliation: Christian

Admitting History/Physical:

Chief complaint: "I'm here to get some new cells."
General appearance: Comfortable, not in acute distress, well-nourished

Vital Signs: Temp: 98.1   Pulse: 80   Resp rate: 16
                    BP: 126/78   Height: 5'9"   Weight: 198 lbs

Heart: RRR, unremarkable
HEENT: Head: WNL
          Eyes: PERRLA

Bear, Scott, Male, 28 y.o.
Allergies: NKA
Pt. Location: BMT

Code: FULL
Physician: S. Smoot

Isolation: None
Admit Date: 11/7

Ears: Clear
Nose: Dry mucous membranes
Throat: Moist mucous membranes
Genitalia: WNL
Neurologic: WNL
Extremities: Normal range of motion in all four extremities. No cyanosis, clubbing, or peripheral edema.
Skin: Warm and dry, not diaphoretic
Chest/lungs: Lungs are clear to auscultation bilaterally. No wheezes, rhonchi, or rales noted.
Peripheral vascular: WNL
Abdomen: Abdomen with normoactive bowel sounds in all four quadrants. Soft, nontender, nondistended. No organomegaly.

Nursing Assessment	11/7 (day –7)	11/30 (day +16)
Abdominal appearance (concave, flat, rounded, obese, distended)	rounded	rounded
Palpation of abdomen (soft, rigid, firm, masses, tense)	soft	firm, tense
Bowel function (continent, incontinent, flatulence, no stool)	continent	continent
Bowel sounds (P=present, AB=absent, hypo, hyper)		
RUQ	P	P, hyper
LUQ	P	P, hyper
RLQ	P	P, hyper
LLQ	P	P, hyper
Stool color	none	green
Stool consistency	NA	liquid
Tubes/ostomies	NA	NA
Genitourinary		
Urinary continence	continent	continent
Urine source	clean catch	clean catch
Appearance (clear, cloudy, yellow, amber, fluorescent, hematuria, orange, blue, tea)	clear	orange
Integumentary		
Skin color	WNL	WNL
Skin temperature (DI=diaphoretic, W=warm, dry, CL=cool, CLM=clammy, CD+=cold, M=moist, H=hot)	WNL	CLM

(Continued)

Bear, Scott, Male, 28 y.o.
Allergies: NKA
Pt. Location: BMT

Code: FULL
Physician: S. Smoot

Isolation: None
Admit Date: 11/7

Nursing Assessment *(Continued)*

Nursing Assessment	11/7 (day –7)	11/30 (day +16)
Skin turgor (good, fair, poor, TENT=tenting)	good	fair
Skin condition (intact, EC=ecchymosis, A=abrasions, P=petechiae, R=rash, W=weeping, S=sloughing, D=dryness, EX=excoriated, T=tears, SE=subcutaneous emphysema, B=blisters, V=vesicles, N=necrosis)	intact	R
Mucous membranes (intact, EC=ecchymosis, A=abrasions, P=petechiae, R=rash, W=weeping, S=sloughing, D=dryness, EX=excoriated, T=tears, SE=subcutaneous emphysema, B=blisters, V=vesicles, N=necrosis)	intact	B, S
Edema (1+=trace, 2+=mild, 3+=moderate, 4+=severe, P=pitting)		
Right upper extremity	none	1+
Left upper extremity	none	1+
Right lower extremity	none	2+
Left lower extremity	none	2+
Other components of Braden score: special bed, sensory pressure, moisture, activity, friction/shear (>18 = no risk, 15–16 = low risk, 13–14 = moderate risk, ≤12 = high risk)	21	13

Admission Orders:
Laboratory: ALT/AST, Bilirubin—Total and Direct, CMV, PT, IRN, PTT—on admit and every Mon/Thurs; Calcium, Phosphate—on admit and every Monday; CBC, EDif, Platelet—on admit and every 2 days; Chem 7, Mg—on admit and daily; prealb on admit
Radiology: None
Vital Signs:
Every 4 hrs
I & O recorded every 8 hrs
Diet: Regular, low bacterial
Activity: As tolerated
IVF: None

Scheduled Medications on Admission:
Lorazepam 0.5–1 mg po every 8 hrs prn
Docusate 100 mg po, twice daily prn
Oxycodone 5 mg po every 4 hrs prn
Senna 8.6mg daily prn

Bear, Scott, Male, 28 y.o.
Allergies: NKA
Pt. Location: BMT

Code: FULL
Physician: S. Smoot

Isolation: None
Admit Date: 11/7

Nutrition:

Meal type (prior to admission): Regular
History: Mr. Bear states that his usual body weight was always approximately 230 lbs. Prior to diagnosis and with initial chemotherapy treatment in 2008, pt lost approximately 50 lbs (weight 180 lbs). Over the next year, pt regained 20 lbs of his lost weight and has maintained weight at approximately 200 lbs since then. 48-hour food recall shows intake prior to admission good and likely meeting ≥ 100% of calorie and protein needs. No food allergies.

MD Progress Note:

11/8/12 0640
Subjective: Scott Bear's previous 24 hours reviewed
Vitals: Temp: 98.2, Pulse: 80, Resp rate: 16, BP: 126/78
Daily wt: 198 lbs
Urine Output: 2100 mL

Physical Exam:
General: Well developed, alert, and oriented. Aware of plan for transplant.
HEENT: WNL
Neck: WNL
Heart: WNL
Lungs: Clear to auscultation
Abdomen: Soft, nontender. Normal BS×4. No complaints of N/V/D.
Assessment/Plan: Pt without complaints at this time. Educated on transplant process.
Dx: ALL; plan for MUD transplant with myeloablative Cy/TBI day–6
Plan: Plan to start TBI today. Nutrition Consult for initial transplant assessment.
.... S. Smoot, MD

MD Progress Note:

11/30/12 0700
Subjective: Scott Bear's previous 24 hours reviewed
Vitals: Temp: 98.4, Pulse: 83, Resp rate: 20, BP: 130/82
Daily wt: 205 lbs
Urine output past 24 hrs: 1205 mL
Stool output past 24 hrs: 3500 mL

Physical Exam:
General: Macropapular rash on palms and trunk
HEENT: Grade 2 mucositis in mouth
Neck: WNL
Heart: WNL

Bear, Scott, Male, 28 y.o.
Allergies: NKA
Pt. Location: BMT

Code: FULL
Physician: S. Smoot

Isolation: None
Admit Date: 11/7

Lungs: Clear to auscultation
Abdomen: Firm, tense. Hyperactive BS×4.
Dx: ALL s/p URD transplant with myeloablative Cy/TBI day+16

Assessment/Plan:

Pt reports feeling fatigued. Complains of mouth pain and mild discomfort from rash. Diarrhea continues, per pt 12–16 bowel movements yesterday. 3.5 L liquid stool past 24 hours plus 520 mL urine and stool mix. Occasional nausea, controlled with anti-emetics. Pt on tacrolimus for GVHD prophylaxis, 4 planned doses of methotrexate received. PO intake has been decreasing; now NPO due to high-volume diarrhea and plan for GI tests.

Plan:

Pt day +16 s/p URD transplant for ALL with myeloablative cyclophosphamide/TBI
Counts engrafting. WBC 2.9, remains on valacyclovir and fluconazole for ID prophylaxis. Tacrolimus for GVHD prophylaxis, received 4 doses of methotrexate.
Diarrhea
C-diff PCR negative from today. Stool studies negative. CS to GI for biopsies for likely GVHD. Will start methylprednisolone 2 mg/kg/day. After bowel prep, may consider Imodium, Lomotil, or opium tincture to slow stools. Increase IV fluids to 125 mL/hr.
Skin
Macropapular rash on palms and trunk. Skin bx pending.
Mucositis
Grade 2, slowly improving. Pt remains on morphine PCA for pain, hydromorphone prn. Continue mouth care.
Nausea
Stable. Continue scheduled lorazepam and scopolamine, prn ondansetron and Phenergan.
Nutrition
Consult nutrition for nutrition support.

.... S. Smoot, MD

Laboratory Results

	Ref. Range	11/7 (day −7)	11/30 (day +16)
Chemistry			
Sodium (mEq/L)	136–145	136	144
Potassium (mEq/L)	3.5–5.5	3.8	4
Chloride (mEq/L)	95–105	97	100
Carbon dioxide (CO_2, mEq/L)	23–30	28	30

Bear, Scott, Male, 28 y.o.
Allergies: NKA
Pt. Location: BMT

Code: FULL
Physician: S. Smoot

Isolation: None
Admit Date: 11/7

Laboratory Results *(Continued)*

	Ref. Range	11/7 (day –7)	11/30 (day +16)
BUN (mg/dL)	8–18	13	23 !↑
Creatinine serum (mg/dL)	0.6–1.2	1	1
BUN/Crea ratio	10.0–20.0	13	23 !↑
Glucose (mg/dL)	70–110	78	104
Phosphate, inorganic (mg/dL)	2.3–4.7	3.4	2.7
Magnesium (mg/dL)	1.8–3	1.9	1.7 !↓
Calcium (mg/dL)	9–11	10.2	9
Osmolality (mmol/kg/H_2O)	285–295	290	290
Bilirubin total (mg/dL)	≤1.5	1.2	1.4
Bilirubin, direct (mg/dL)	<0.3	0.1	0.2
Protein, total (g/dL)	6–8	6.2	6.3
Albumin (g/dL)	3.5–5	3.5	2 !↓
Prealbumin (mg/dL)	16–35	24	<1 !↓
ALT (U/L)	4–36	20	30
AST (U/L)	0–35	8	25
Coagulation (Coag)			
PT (sec)	12.4–14.4	13.5	
INR	0.9–1.1	1	
PTT (sec)	24–34	33	
Hematology			
WBC ($\times 10^3$/mm^3)	4.8–11.8	4.5 !↓	2.9 !↓
RBC ($\times 10^6$/mm^3)	4.2–5.4 F 4.5–6.2 M	2.82 !↓	2.33 !↓
Hemoglobin (Hgb, g/dL)	12–15 F 14–17 M	9.8 !↓	7.9 !↓
Hematocrit (Hct, %)	37–47 F 40–54 M	29.1 !↓	23.7 !↓
Mean cell Hgb (pg)	26–32	20.4 !↓	16.5 !↓
RBC distribution (%)	11.6–16.5	7.6 !↓	8.9 !↓
Hematology, Manual Diff			
Neutrophil (%)	50–70	48.3 !↓	56
Lymphocyte (%)	15–45	33.1	11.3 !↓

(Continued)

Bear, Scott, Male, 28 y.o.
Allergies: NKA
Pt. Location: BMT

Code: FULL
Physician: S. Smoot

Isolation: None
Admit Date: 11/7

Laboratory Results *(Continued)*

	Ref. Range	11/7 (day −7)	11/30 (day +16)
Monocyte (%)	3–10	16.5 !↑	22.6 !↑
Eosinophil (%)	0–6	0.7	0
Basophil (%)	0–2	0.7	0
Blasts (%)	3–10	0 !↓	0 !↓
Segs (%)	0–60	2.17	1.4
Urinalysis			
Collection method	—	clean catch	
Color	—	yellow	
Appearance	—	clear	
Specific gravity	1.003–1.030	1.02	
pH	5–7	5.5	
Protein (mg/dL)	Neg	Neg	
Glucose (mg/dL)	Neg	Neg	
Ketones	Neg	Neg	
Blood	Neg	Neg	
Bilirubin	Neg	Neg	
Nitrites	Neg	Neg	
Urobilinogen (EU/dL)	<1.1	0.2	
Leukocyte esterase	Neg	Neg	

Case Questions

I. Understanding the Diagnosis and Pathophysiology

1. What is acute lymphoblastic leukemia (ALL)? Describe the role of blast cells in this disease.

2. What are the signs and symptoms of ALL? Which of these symptoms did Mr. Bear present with when he was diagnosed with his relapse of ALL?

3. What are the primary goals of high-dose chemotherapy and/or total body radiation in an allogeneic transplant? What are common side effects related to these?

4. Describe the primary difference among an autologous transplant, an allogeneic transplant, and a syngeneic transplant.

5. Why do you think Mr. Bear is receiving an allogeneic transplant as opposed to an autologous transplant?

6. What is the difference between a mycloablative transplant and a non-myeloablative transplant (also referred to as reduced-intensity conditioning [RIC])? In what patient populations would a non-myeloablative transplant be a good treatment option?

7. From what three sources can stem cells be collected for a transplant?

8. Explain the terms *neutropenic* and *engraftment* and their relationship to white blood cell count. What health risks are increased for a neutropenic patient?

9. Describe graft-versus-host disease (GVHD). Name the cells that the new immune system views as foreign and attacks.

II. Understanding the Nutrition Therapy

10. Historically, bone marrow transplant patients were prescribed a low-bacterial diet. What are the current recommendations for the use of this diet during periods of immunosuppression?

11. What are A.S.P.E.N.'s recommendations for nutrition support in hematopoietic cell transplant patients?

12. Describe a situation in which TPN is appropriate in hematopoietic cell transplant patients and a situation in which enteral nutrition should be used.

13. State the long-term nutritional implications associated with the chronic use of corticosteroids for the treatment of GVHD. Mr. Bear will be discharged on steroids. What should you encourage him to eat and what supplement may he need to take?

III.   Nutrition Assessment

14. Assess Mr. Bear's height and weight. Calculate his BMI and % IBW.

15. Determine Mr. Bear's energy, protein, and fluid requirements.

16. On admission, did Mr. Bear present with any nutrition risk factors?

17. Prior to completing your nutrition assessment for his admission, you go and talk with Mr. Bear. Please list 2 additional questions you would ask him and give a reason why you chose each question.

18. Look at the MD note, labs, and RN documentation from 11/30. What nutrition risk factors has the patient developed? For each, name a reason why he may have developed this risk factor.

19. Review the MD note from 11/30. List the names and actions (benefits for this patient) of all the medications that are mentioned in the doctor's note. List the major nutrition side effects for the following: methotrexate, methylprednisolone, morphine/hydromorphone, and scopolamine patch.

20. The MD note from 11/30 mentions that Mr. Bear's mucositis is improving. Why do you think it is starting to improve?

IV.   Nutrition Diagnosis

21. Identify the pertinent nutrition problems and the corresponding nutrition diagnoses for Mr. Bear upon his admission on 11/7.

22. Write your PES statement for each high-priority nutrition problem that you have identified for Mr. Bear upon his admission on 11/7.

V.   Nutrition Intervention

23. Design a parenteral nutrition support regimen for Mr. Bear, based on his labs from 11/30. Calculate the dextrose, lipids, and amino acids for his parenteral solution.

24. Mr. Bear and his family have asked to meet with you; they want to know why he needs TPN. How would you explain his need for TPN?

25. Mr. Bear's gut biopsy came back positive for GVHD. Once his diarrhea has resolved, how would you recommend that his diet be advanced?

VI.   Nutrition Monitoring and Evaluation

26. What labs would you want to make sure are ordered now that TPN is being started? Do you think measuring nitrogen balance would be helpful for this patient at this time? Why or why not?

27. Mr. Bear's blood glucose has increased, and now ranges from 280–410 mg/dL. What could be contributing to this sudden increase? What changes could you make to the TPN to improve his blood sugar values?

Bibliography

August, DA, Huhmann, MB, & the American Society for Parenteral and Enteral Nutrition Board of Directors. A.S.P.E.N. clinical guidelines: Nutrition support therapy during adult anticancer treatment and in hematopoietic cell transplantation. *Journal of Parenteral and Enteral Nutrition.* 2009;33:472–500.

Cohen DA. Neoplastic disease. In: Nelms M, Sucher KP, Lacey K, Roth SL. *Nutrition Therapy and Pathophysiology.* 2nd ed. Belmont, CA: Wadsworth, Cengage Learning; 2011:702–734.

Cohen J, Maurice L. Adequacy of nutritional support in pediatric blood and marrow transplantation. *Journal of Pediatric Oncology Nursing.* 2010;27:40–47.

Farrington M, Cullen L, Dawson C. Assessment of oral mucositis in adult and pediatric oncology patients: An evidence-based approach. *ORL-Head & Neck Nursing.* 2010;28:8–15.

Martin-Salces M, de Paz R, Canales MA, Mesejo A, Hernandez-Navarro F. Nutritional recommendations in hematopoietic stem cell transplantation. *Nutrition.* 2008;24:769–775.

Murray SM, Pindoria S. Nutrition support for bone marrow transplant patients. *Cochrane Database of Systematic Reviews.* 2009;1:CD002920.

Van Dalen EC, Mank A, Leclercq E, et al. Low bacterial diet versus control diet to prevent infection in cancer patients treated with chemotherapy causing episodes of neutropenia. *Cochrane Database of Systematic Reviews.* 2012;9:CD006247.

Internet Resources

American Cancer Society: http://www.cancer.org/acs /groups/cid/documents/webcontent/003215-pdf.pdf

Leukemia & Lymphoma Society: http://www.lls.org /content/nationalcontent/resourcecenter /freeeducationmaterials/leukemia/pdf/all.pdf

Nutrition Care Manual: http://www.nutritioncaremanual.org

Esophageal Cancer Treated with Surgery and Radiation[1]

Objectives

After completing this case, the student will be able to:

1. Identify and explain common metabolic and nutritional problems associated with malignancy.
2. Explain complications of medical treatment for cancer and the potential nutritional consequences.
3. Apply understanding of nutrition support in the treatment of and recovery from malignancy.
4. Analyze nutrition assessment data to evaluate nutritional status and identify specific nutrition problems.
5. Determine nutrition diagnoses and write appropriate PES statements.
6. Evaluate the adequacy of an enteral feeding regimen for a cancer patient.

Nick Seyer is a 58-year-old man who, after suffering from recurrent heartburn for over a year, seeks medical attention. He presents to his physician with difficulty swallowing and a significant unexplained weight loss.

---

[1] Adapted with permission from: Whitman M. Esophageal Cancer. Available from: Virtual Health Care Team' School of Health-Professions University of Missouri–Columbia. http://www.vhct.org/index.htm.

Seyer, Nick, Male, 58 y.o.
Allergies: IVP dye, seafood
Pt. Location: RM 832

Code: FULL
Physician: H. Brown

Isolation: None
Admit Date: 9/5

Patient Summary: After undergoing chest X-ray, endoscopy with brushings and biopsy, and CT scan, Mr. Seyer was diagnosed with Stage IIB (T1, N1, M0) adenocarcinoma of the esophagus.

History:

Onset of disease: Dysphagia × 3-4 months; odynophagia × 5-6 months
Medical history: Patient describes significant heartburn for the previous year. He has been taking TUMS, Alka-Seltzer, and Pepcid consistently for the past year. He has noted weight loss of over 30 lbs. in the last several months. He states that he just has not been able to eat because of the pain and heartburn. Now, difficulty swallowing foods—especially anything with texture—has brought him to his physician. Patient also describes a recurrent cough at night.
Medications at home: TUMS®, Alka-Seltzer®, and Pepcid®
Tobacco use: Yes. 2 ppd; wife also smokes.
Alcohol use: Yes. 1–2 drinks 1–2 × /week.
Family history: What? Liver cancer Who? Mother—died age 58

Demographics:

Marital status: Wife, age 52; son, age 18; two other sons are away at college—ages 19 and 22
Years education: Some college
Language: English only
Occupation: Contractor
Hours of work: Variable but usually 5–6 days per week—starts as early as 6:30 and works often until after 6 pm
Ethnicity: Caucasian
Religious affiliation: Catholic

Admitting History/Physical:

Chief complaint: Heartburn for "a long time" and difficulty swallowing during the past 4 or 5 months. Occasionally food seems to "hang up" in his throat. He points to the upper portion of his neck, directly beneath his chin.

Vital Signs: Temp: 98.3°F   Pulse: 88   Resp rate: 13
                    BP: 132/92   Height: 6'3"   Weight: 198 lbs

Heart: Unremarkable
HEENT: Within normal limits; funduscopic exam reveals AV nicking
        Eyes: Sunken; sclera clear without evidence of tears
        Ears: Clear
        Nose: Dry mucous membranes
        Throat: Dry mucous membranes, no inflammation
Genitalia: Unremarkable
Rectal: Prostate normal; stool hematest negative

Seyer, Nick, Male, 58 y.o.
Allergies: IVP dye, seafood   **Code:** FULL   **Isolation:** None
Pt. Location: RM 832   **Physician:** H. Brown   **Admit Date:** 9/5

Neurologic: Alert, oriented × 3
Extremities: Joints appear prominent with evidence of some muscle wasting. No edema.
Skin: Warm, dry
Chest/lungs: Clear to auscultation and percussion
Abdomen: Epigastric tenderness on palpation

Nursing Assessment	9/5
Abdominal appearance (concave, flat, rounded, obese, distended)	rounded
Palpation of abdomen (soft, rigid, firm, masses, tense)	tense
Bowel function (continent, incontinent, flatulence, no stool)	continent
Bowel sounds (P=present, AB=absent, hypo, hyper)	
RUQ	P
LUQ	P
RLQ	P
LLQ	P
Stool color	brown
Stool consistency	soft
Tubes/ostomies	N/A
Genitourinary	
Urinary continence	catheter
Urine source	catheter
Appearance (clear, cloudy, yellow, amber, fluorescent, hematuria, orange, blue, tea)	clear, yellow
Integumentary	
Skin color	pale
Skin temperature (DI=diaphoretic, W=warm, dry, CL=cool, CLM=clammy, CD+=cold, M=moist, H=hot)	W
Skin turgor (good, fair, poor, TENT=tenting)	fair
Skin condition (intact, EC=ecchymosis, A=abrasions, P=petechiae, R=rash, W=weeping, S=sloughing, D=dryness, EX=excoriated, T=tears, SE=subcutaneous emphysema, B=blisters, V=vesicles, N=necrosis)	D
Mucous membranes (intact, EC=ecchymosis, A=abrasions, P=petechiae, R=rash, W=weeping, S=sloughing, D=dryness, EX=excoriated, T=tears, SE=subcutaneous emphysema, B=blisters, V=vesicles, N=necrosis)	D
Other components of Braden score: special bed, sensory pressure, moisture, activity, friction/shear (>18=no risk, 15–16=low risk, 13–14=moderate risk, ≤12=high risk)	activity, 16

Seyer, Nick, Male, 58 y.o.
Allergies: IVP dye, seafood
Pt. Location: RM 832

Code: FULL
Physician: H. Brown

Isolation: None
Admit Date: 9/5

Orders:
Surgery consult to evaluate for surgical resection
Evaluate for pre- and postoperative external beam radiation therapy

Nutrition:
General: Prior to admission has noted decreased appetite, feeling full all the time, and regurgitation of some foods. He notes pain upon swallowing as well as rather constant heartburn.

Usual dietary intake:
AM: Used to eat eggs, bacon, toast every morning but has not eaten this for at least the past month. Most recently has had just coffee and cereal.
Lunch: Previously, ate cold lunch packed for the work site. Included sandwich, cold meat or other leftovers from previous dinner, fruit, cookies, and tea.
Dinner: All meats, pasta or rice, 2–3 vegetables, 1–2 beers.

24-hour recall:
AM: 1 packet of instant oatmeal; sips of coffee
Lunch: 6 oz tomato soup with 2–4 crackers
Dinner: Macaroni and cheese—homemade, ½ c
Bedtime: 1 scoop of chocolate ice cream

Food allergies/intolerances/aversions: Seafood
Previous nutrition therapy? No
Food purchase/preparation: Wife
Vit/min intake: None

Laboratory Results

	Ref. Range	9/5 0832	9/11 0832
Chemistry			
Sodium (mEq/L)	136–145	137	136
Potassium (mEq/L)	3.5–5.5	3.8	3.6
Chloride (mEq/L)	95–105	101	99
Carbon dioxide (CO_2, mEq/L)	23–30	26	25
BUN (mg/dL)	8–18	9	10
Creatinine serum (mg/dL)	0.6–1.2	0.7	0.9
Glucose (mg/dL)	70–110	71	108
Phosphate, inorganic (mg/dL)	2.3–4.7	3.2	
Magnesium (mg/dL)	1.8–3	1.8	1.8
Calcium (mg/dL)	9–11	9.1	9.4

Seyer, Nick, Male, 58 y.o.
Allergies: IVP dye, seafood
Pt. Location: RM 832

Code: FULL
Physician: H. Brown

Isolation: None
Admit Date: 9/5

Laboratory Results *(Continued)*

	Ref. Range	9/5 0832	9/11 0832
Bilirubin, direct (mg/dL)	<0.3	0.2	0.3 !↑
Protein, total (g/dL)	6–8	5.7 !↓	5.7 !↓
Albumin (g/dL)	3.5–5	3.1 !↓	3.0 !↓
Prealbumin (mg/dL)	16–35	15 !↓	12 !↓
Ammonia (NH$_3$, μmol/L)	9–33	11	21
Alkaline phosphatase (U/L)	30–120	101	99
ALT (U/L)	4–36	21	33
AST (U/L)	0–35	32	27
CPK (IU/L)	30–135 F 55–170 M	162	145
Lactate dehydrogenase (U/L)	208–378	300	290
Cholesterol (mg/dL)	120–199	180	170
HDL-C (mg/dL)	>55 F, >45 M	47	
LDL (mg/dL)	<130	129	
LDL/HDL ratio	<3.22 F <3.55 M	2.74	
Triglycerides (mg/dL)	35–135 F 40–160 M	158	
Coagulation (Coag)			
PT (sec)	12.4–14.4	12 !↓	12.8
Hematology			
WBC (×10^3/mm^3)	4.8–11.8	5.2	6.9
RBC (×10^6/mm^3)	4.2–5.4 F 4.5–6.2 M	4.2 !↓	4.3 !↓
Hemoglobin (Hgb, g/dL)	12–15 F 14–17 M	13.5 !↓	13.9 !↓
Hematocrit (Hct, %)	37–47 F 40–54 M	38 !↓	38 !↓
Mean cell volume (μm^3)	80–96	90	86
Mean cell Hgb (pg)	26–32	32.4 !↑	32.3 !↑
Mean cell Hgb content (g/dL)	31.5–36	35.5	36.5 !↑
Platelet count (×10^3/mm^3)	140–440	250	232
Ferritin (mg/mL)	20–120 F 20–300 M	220	208

(Continued)

Seyer, Nick, Male, 58 y.o.
Allergies: IVP dye, seafood
Pt. Location: RM 832

Code: FULL
Physician: H. Brown

Isolation: None
Admit Date: 9/5

Laboratory Results *(Continued)*

	Ref. Range	9/5 0832	9/11 0832
Hematology, Manual Diff			
Grans (%)	3.6–79.2	75	65
Lymphocyte (%)	15–45	25	35
Monocyte (%)	3–10	4	5
Eosinophil (%)	0–6	0.5	0
Segs (%)	0–60	55	60
Bands (%)	0–10	4	3

Intake/Output

Date		9/11 0701–9/12 0700			
Time		0701–1500	1501–2300	2301–0700	Daily total
IN	P.O. formula	**600**	**535**	**600**	**1735**
	P.O. flush	**50**	**50**	**50**	**150**
	(mL/kg/hr)	(0.90)	(0.81)	(0.90)	(0.87)
	I.V.	**800**	**800**	**800**	**2400**
	(mL/kg/hr)	(1.11)	(1.11)	(1.11)	(1.11)
	I.V. piggyback				
	TPN				
	Total intake	**1450**	**1385**	**1450**	**4285**
	(mL/kg)	(16.11)	(15.39)	(16.11)	(47.61)
OUT	Urine	**1100**	**1700**	**900**	**3700**
	(mL/kg/hr)	(1.53)	(2.36)	(1.25)	(1.71)
	Emesis output				
	Other				
	Stool			**300**	**300**
	Total output	**1100**	**1700**	**1200**	**4000**
	(mL/kg)	(12.22)	(18.89)	(13.33)	(44.44)
Net I/O		**+350**	**−315**	**+250**	**+285**
Net since admission (9/5)		**+400**	**+85**	**+335**	**+335**

Case Questions

I. Understanding the Disease and Pathophysiology

1. Mr. Seyer has been diagnosed with adenocarcinoma of the esophagus. What does the term *adenocarcinoma* mean?

2. What are the two most common types of esophageal cancer? What are the risk factors for development of this malignancy? Does Mr. Seyer's medical record indicate that he has any of these risk factors?

3. Mr. Seyer's cancer was described as Stage IIB (T1, N1, M0). Explain this terminology used to describe staging for malignancies.

4. Cancer is generally treated with a combination of therapies. These can include surgical resection, radiation therapy, chemotherapy, and immunotherapy. The type of malignancy and staging of the disease will, in part, determine the types of therapies that are prescribed. Define and describe each of these therapies. Briefly describe the mechanism for each. In general, how do they act to treat a malignancy?

5. Mr. Seyer had a transhiatal esophagectomy on 9/7. Describe this surgical procedure. How may this procedure affect his digestion and absorption?

II. Understanding the Nutrition Therapy

6. Many cancer patients experience changes in nutritional status. Briefly describe the potential effect of cancer on nutritional status.

7. Both surgery and radiation affect nutritional status. Describe potential nutritional and metabolic effects of these treatments.

III. Nutrition Assessment

8. Calculate and evaluate Mr. Seyer's % UBW and BMI.

9. Summarize your findings regarding his weight status. Classify the severity of his weight loss. What factors may have contributed to his weight loss? Explain.

10. What does research tell us about the relationship between significant weight loss and prognosis in cancer patients?

11. Estimate Mr. Seyer's energy and protein requirements based on his current weight.

12. Estimate Mr. Seyer's fluid requirements based on his current weight.

13. What factors noted in Mr. Seyer's history and physical may indicate problems with eating prior to admission?

14. Mr. Seyer is currently receiving enteral nutrition, specifically Isosource HN at 75 mL/hr.

    a. Calculate the amount of energy and protein that will be provided at this rate.

    b. Next, by assessing the information on the intake/output record, determine the actual amount of enteral nutrition he received on September 11.

    c. Compare this to his estimated nutrient requirements.

    d. Compare fluids required to fluids received. Is he meeting his fluid requirements? How did you determine this? Why would you evaluate his output when assessing his fluid intake?

15. What type of formula is Isosource HN? One of the residents taking care of Mr. Seyer asks about a formula with a higher concentration of omega-3-fatty acids, antioxidants, arginine, and glutamine that could promote healing after surgery. What does the evidence indicate regarding nutritional needs for cancer patients and, in particular, nutrients to promote postoperative wound healing? What formulas may meet this profile? List them and discuss why you chose them.

16. Are any clinical signs of malnutrition noted in the patient's admission history and physical?

17. Review the patient's chemistries upon admission. Identify any that are abnormal and describe their clinical significance for this patient, including the likely reason for each abnormality and its nutritional implications.

18. Mr. Seyer has been diagnosed with a life-threatening illness. What is the definition of *terminal illness*?

19. The literature describes how a patient and his family may experience varying levels of emotional response to a terminal illness. These may include anger, denial, depression, and acceptance. How may this affect the patient's nutritional intake? How would you handle these components in your nutritional care? What questions might you have for Mr. Seyer or his family? List three.

IV.   Nutrition Diagnosis

20. Select two high-priority nutrition problems after Mr. Seyer's surgery and complete the PES statement for each.

V.   Nutrition Intervention

21. For each of the PES statements you have written, establish an ideal goal (based on the signs and symptoms) and an appropriate intervention (based on the etiology).

22. Does his current nutrition support meet his estimated nutritional needs? If not, determine the recommended changes. Discuss any areas of deficiency and ideas for implementing a new plan.

23. How may these interventions (from #21) change as he progresses postoperatively? Discuss how Mr. Seyer may transition from enteral feeding to an oral diet.

VI.   Nutrition Monitoring and Evaluation

24. List the factors you should monitor for Mr. Seyer while he is receiving enteral nutrition therapy.

25. Mr. Seyer will receive radiation therapy as an outpatient. In question #7, you identified potential nutritional complications with radiation therapy. Choose one of these nutritional complications and describe the nutrition intervention that would be appropriate.

26. Identify major assessment indices you would use to monitor his nutritional status once he begins therapy.

Bibliography

Academy of Nutrition and Dietetics Nutrition Care Manual. Esophageal Cancer. Available from http://nutritioncaremanual.org/topic.cfm?ncm _heading=Nutrition%20Care&ncm_toc_id=145164. Accessed 1/6/13.

Academy of Nutrition and Dietetics Evidence Analysis Library. Oncology evidence-based nutrition practice guidelines. Available from http://andevidencelibrary .com/topic.cfm?cat=2819. Accessed 1/6/13.

Aiko S, Yoshizumi Y, Tsuwano S, Shimanouchi M, Sugiura Y, Maehara T. The effects of immediate enteral feeding with a formula containing high levels of omega-3 fatty acids in patients after surgery for esophageal cancer. *J Parenter Enteral Nutr.* 2005 May-Jun;29(3):141–7.

Churma SA, Horrell CJ. Esophageal and gastric cancers. In: Kogut VJ & Luthringer SL. *Nutritional Issues in Cancer Care.* Pittsburgh, PA: Oncology Nursing Society; 2005:45–63.

Cohen DA. Neoplastic disease. In: Nelms M, Sucher K, Lacey K, Roth SL. *Nutrition Therapy and Pathophysiology.* 2nd ed. Belmont, CA: Wadsworth, Cengage Learning; 2011:702–734.

Fessler T, Havrilla C. Nutrition support for esophageal cancer patients—strategies for meeting the challenges while improving patient care. *Today's Dietitian.* 2012;14:1–4.

Miyata H, Yano M, Yasuda T. Randomized study of clinical effect of enteral nutrition support during neoadjuvant chemotherapy on chemotherapy-related toxicity in patients with esophageal cancer. *Clin Nutr.* 2012;31:330–6.

National Cancer Institute. Loss, Grief, and Bereavement. Available from: http://www.cancer.gov/cancertopics /pdq/supportivecare/bereavement/HealthProfessional /page1. Accessed 1/6/13.

Internet Resources

American Cancer Society: http://www.cancer.org/index

Cancer Support Community: http://www.cancersupport-community.org/MainMenu/About-Cancer

/Understanding-Cancer/General-Information .html?gclid=CKflmbWV1bQCFYw-MgodRl4AhA

National Cancer Institute: https://resresources.nci.nih.gov/

NUTRITION THERAPY FOR HIV/AIDS

The case in this section involves acquired immuno-deficiency syndrome (AIDS). Nutritional concerns for the AIDS patient have shifted as the treatment of this disease has improved. Certainly, viral load, opportunistic infections, and the presence of wasting syndrome all can increase energy expenditure and shift substrate metabolism. Long-term consequences of antiretroviral therapy and other issues for HIV and AIDS are included in this case. Drug–nutrient interactions, biochemical indices of viral load, and appropriate nutrition education are all crucial aspects of medical nutrition therapy for the patient with HIV and AIDS.

Case 34

AIDS

Objectives

After completing this case, the student will be able to:

1. Identify and explain common nutritional problems associated with HIV and AIDS.
2. Identify nutritional risk factors for the patient with acquired immunodeficiency syndrome (AIDS).
3. Analyze nutrition assessment data to evaluate nutritional status and identify specific nutrition problems.
4. Determine nutrition diagnoses and write appropriate PES statements
5. Develop a nutrition care plan—with appropriate measurable goals, interventions, and strategies for monitoring and evaluation—that addresses the nutrition diagnoses of this case.

6. Specify potential drug–nutrient interactions and appropriate interventions.
7. Describe key components of nutrition education for the patient with AIDS.
8. Evaluate risks and current recommendations for nutritional supplementation.

Mr. Terry Long, a 32-year-old African-American male, is admitted with probable pneumonia and progression to AIDS. Mr. Long was diagnosed as HIV-positive four years ago and has not received any previous treatment.

Long, Terry, Male, 32 y.o.
Allergies: NKA
Pt. Location: RM 704

Code: FULL
Physician: A. Fremont

Isolation: None
Admit Date: 10/17

Patient Summary: AIDS—Stage 3 with oral thrush; no clinical evidence of pneumonia; HAART regimen initiated with Atripla (efavirenz, tenofovir DF, emtricitabine).

History:

Onset of disease: Seropositive for HIV-1 confirmed by ELISA and Western Blot 4 years previously
Medical history: Etiology of contraction not known, but was employed in high-risk environment. Admits to intercourse with multiple partners but denies same-sex intercourse.
Surgical history: Tonsillectomy age 6; appendectomy age 18, which required a blood transfusion due to complications.
Medications at home: Multivitamin, vitamin E (1500 IU), vitamin C (500 mg twice daily), ginseng (500 mg daily), milk thistle (200 mg twice daily), *Echinacea* (3 capsules every day), St. John's wort (300 mg daily)
Tobacco use: No—quit 5 years ago
Alcohol use: Yes—2–3 drinks 3–4 ×/week
Family history: What? CAD, HTN Who? Father

Demographics:

Marital status: Single, lives with father age 69, mother age 66, both well
Years education: Bachelor's degree
Language: English only
Occupation: Currently on disability but previously worked as nurse in dialysis clinic
Hours of work: N/A
Ethnicity: African-American
Religious affiliation: AME (African Methodist Episcopal)

Admitting History/Physical:

Chief complaint: "I was diagnosed with HIV four years ago when I was living in St. Louis. I just recently moved back home because I am not able to work right now. I have not been treated before but I am fairly sure I will need to be. I am exhausted all the time—I have a really sore mouth and throat. I have lost a lot of weight. I think I've just been denying that I may have AIDS. But a lot of people I know are doing OK on drugs, so I came to this new physician. The case manager at the Health Department set it up for me. Dr. Fremont thinks I may have pneumonia as well, so she admitted me for a full workup."

Vital Signs: Temp: 98.6°F   Pulse: 92   Resp rate: 18
             BP: 120/84   Height: 6'1"   Weight: 151 lbs

Heart: Regular rate and rhythm—normal heart sounds
HEENT: Eyes: PERRLA
        Ears: Unremarkable
        Nose: Mucosa pink without drainage
        Throat: Erythematous with white, patchy exudate

Long, Terry, Male, 32 y.o.
Allergies: NKA
Pt. Location: RM 704

Code: FULL
Physician: A. Fremont

Isolation: None
Admit Date: 10/17

Genitalia: WNL
Rectal: Rectal exam normal. Stool: heme negative.
Neurologic: Oriented × 3, no focal motor or sensory deficits, cranial nerves intact, DTR +2 in all groups
Extremities: Good pulses, no edema
Skin: Warm, dry, with flaky patches
Chest/lungs: Rhonchi in lower left lung
Abdomen: Nondistended, nontender, hyperactive bowel sounds

Nursing Assessment	10/17
Abdominal appearance (concave, flat, rounded, obese, distended)	flat
Palpation of abdomen (soft, rigid, firm, masses, tense)	soft
Bowel function (continent, incontinent, flatulence, no stool)	continent
Bowel sounds (P=present, AB=absent, hypo, hyper)	
RUQ	hyper
LUQ	hyper
RLQ	hyper
LLQ	hyper
Stool color	brown
Stool consistency	soft
Tubes/ostomies	N/A
Genitourinary	
Urinary continence	clean catch
Urine source	clean catch
Appearance (clear, cloudy, yellow, amber, fluorescent, hematuria, orange, blue, tea)	clear, yellow
Integumentary	
Skin color	light brown
Skin temperature (DI=diaphoretic, W=warm, dry, CL=cool, CLM=clammy, CD+=cold, M=moist, H=hot)	W
Skin turgor (good, fair, poor, TENT=tenting)	fair
Skin condition (intact, EC=ecchymosis, A=abrasions, P=petechiae, R=rash, W=weeping, S=sloughing, D=dryness, EX=excoriated, T=tears, SE=subcutaneous emphysema, B=blisters, V=vesicles, N=necrosis)	intact, D
Mucous membranes (intact, EC=ecchymosis, A=abrasions, P=petechiae, R=rash, W=weeping, S=sloughing, D=dryness, EX=excoriated, T=tears, SE=subcutaneous emphysema, B=blisters, V=vesicles, N=necrosis)	intact

(Continued)

Long, Terry, Male, 32 y.o.
Allergies: NKA   **Code:** FULL   **Isolation:** None
Pt. Location: RM 704   **Physician:** A. Fremont   **Admit Date:** 10/17

Nursing Assessment *(Continued)*

Nursing Assessment	10/17
Other components of Braden score: special bed, sensory pressure, moisture, activity, friction/shear (>18=no risk, 15–16=low risk, 13–14=moderate risk, ≤12=high risk)	activity, 15

Orders:
CXR
WBC with diff
CD4 and viral load
$D_5$1/2 NS @ 100 cc/hr
Fluconazole IV

Nutrition:
Usual dietary intake:
General: Patient describes appetite as OK but not normal. "I have always been a picky eater. There are a lot of foods that I don't like. But in the last few days it is the sores in my mouth and throat that have made the biggest difference. It hurts pretty badly, and I can hardly even drink. I have been reading about nutrition and HIV on the Internet—I've been trying to do some research. That's when I started taking more supplements. I thought if I wasn't eating like I should that I could at least take supplements. They are expensive, though, so I don't have them every day like I probably should. My highest weight ever was about 175 lbs, which was during college almost 10 years ago. But I have never been this thin as an adult."

Usual dietary intake (before mouth sores):
Breakfast/lunch: ("I usually don't get up before noon because I stay up really late.") Cold cereal 1–2 c, ½ c whole milk
Supper: Meat—2 pork chops or other type of meat (except beef), mashed potatoes—1 c, rice or pasta, tea or soda
Snacks: Pizza, candy bar, or cookies with tea or soda. Drinks 2–3 beers 3–4 times a week.

24-hr recall: Sips of apple juice, pudding 1 c, rice and gravy 1 c, iced tea with sugar—sips throughout the day

Anthropometric data: MAC 10"; TSF 7 mm; body fat 12.5%, Ht 6'1", Wt 151 lbs, UBW 160–165 lbs

Food allergies/intolerances/aversions: Can only tolerate small amount of milk at a time; does not like beef, coffee, or vegetables (except salad)
Previous nutrition therapy? No
Food purchase/preparation: Parent(s), self
Vit/min intake: Multivitamin 1 daily, vitamin E 1500 IU daily, vitamin C 500 mg twice daily, ginseng 500 mg daily, milk thistle 200 mg twice daily, *Echinacea* 3 capsules daily (88.5 mg per capsule), St. John's wort 300 mg daily

Long, Terry, Male, 32 y.o.
Allergies: NKA
Pt. Location: RM 704

Code: FULL
Physician: A. Fremont

Isolation: None
Admit Date: 10/17

Laboratory Results

	Ref. Range	10/17 1522
Chemistry		
Sodium (mEq/L)	136–145	142
Potassium (mEq/L)	3.5–5.5	3.6
Chloride (mEq/L)	95–105	101
Carbon dioxide (CO_2, mEq/L)	23–30	27
BUN (mg/dL)	8–18	11
Creatinine serum (mg/dL)	0.6–1.2	0.8
BUN/Crea ratio	10.0–20.0	13.75
Uric acid (mg/dL)	2.8–8.8 F 4.0–9.0 M	5.2
Glucose (mg/dL)	70–110	75
Phosphate, inorganic (mg/dL)	2.3–4.7	3.2
Magnesium (mg/dL)	1.8–3	1.8
Calcium (mg/dL)	9–11	9.1
Osmolality (mmol/kg/H_2O)	285–295	292
Bilirubin, direct (mg/dL)	<0.3	0.2
Protein, total (g/dL)	6–8	6.0
Albumin (g/dL)	3.5–5	3.6
Prealbumin (mg/dL)	16–35	15 !↓
Ammonia (NH$_3$, μmol/L)	9–33	18
Alkaline phosphatase (U/L)	30–120	102
ALT (U/L)	4–36	12
AST (U/L)	0–35	17
CPK (U/L)	30–135 F 55–170 M	110
Lactate dehydrogenase (U/L)	208–378	710 !↑
Cholesterol (mg/dL)	120–199	150
HDL-C (mg/dL)	>55 F, >45 M	42 !↓
LDL (mg/dL)	<130	114
LDL/HDL ratio	<3.22 F <3.55 M	2.71
Triglycerides (mg/dL)	35–135 F 40–160 M	78
Coagulation (Coag)		
PT (sec)	12.4–14.4	12.9

(Continued)

Long, Terry, Male, 32 y.o.
Allergies: NKA
Pt. Location: RM 704

Code: FULL
Physician: A. Fremont

Isolation: None
Admit Date: 10/17

Laboratory Results *(Continued)*

	Ref. Range	10/17 1522
Hematology		
WBC ($\times 10^3$/mm³)	4.8–11.8	8.5
RBC ($\times 10^6$/mm³)	4.2–5.4 F 4.5–6.2 M	5.2
Hemoglobin (Hgb, g/dL)	12–15 F 14–17 M	14.2
Hematocrit (Hct, %)	37–47 F 40–54 M	40
Mean cell volume (μm³)	80–96	96
Mean cell Hgb (pg)	26–32	34.2 !↑
Mean cell Hgb content (g/dL)	31.5–36	35.5
RBC distribution (%)	11.6–16.5	16.3
Platelet count ($\times 10^3$/mm³)	140–440	220
Transferrin (mg/dL)	250–380 F 215–365 M	201 !↓
Hematology, Manual Diff		
Lymphocyte (%)	15–45	3 !↓
Monocyte (%)	3–10	12 !↑
Eosinophil (%)	0–6	3
Segs (%)	0–60	51
Bands (%)	0–10	4
HIV-1 RNA Quant (copies/mL)	<75	29000 !↑
Total T cells (mm³)	812–2,318	240 !↓
T-helper cells (mm³)	589–1,505	153 !↓
T-suppressor cells (mm³)	325–997	102 !↓

Case Questions

I. Understanding the Disease and Pathophysiology

1. How is HIV transmitted? Based on Mr. Long's history and physical, what risk factors for contracting HIV might he have had?

2. The history and physical indicate that he is seropositive. What does that mean? The Western Blot and ELISA confirmed he was seropositive. Describe these tests.

3. Mr. Long says he learned he was HIV-positive four years ago. Why has he only now become symptomatic?

4. What is thrush, and why might Mr. Long have this condition?

5. After this admission, Mr. Long was diagnosed with HIV Infection, Stage 3 (AIDS). What information from his medical record confirms this diagnosis?

II. Understanding the Nutrition Therapy

6. What are the common nutritional complications of HIV and AIDS?

7. Are there specific recommendations for energy, protein, vitamin, and mineral intakes for someone with AIDS?

III. Nutrition Assessment

8. Evaluate the patient's anthropometric information.

   a. Calculate % UBW and BMI.

   b. Compare the TSF to population standards. What does this comparison mean? Is this a viable comparison? Explain.

   c. Using MAC and TSF, calculate upper arm muscle area. What can you infer from this calculation?

   d. Mr. Long's body fat percentage is 12.5%. What does this mean? Compare this to standards.

9. Determine Mr. Long's energy and protein requirements.

10. Evaluate Mr. Long's dietary information. What tools could you use to evaluate his dietary intake?

11. Is Mr. Long consuming adequate amounts of food? Does his history indicate that he is having difficulty eating? Explain.

12. Mr. Long states that he consumes alcohol several times a week. Are there any contraindications for alcohol consumption for him?

13. Using this patient's laboratory values, identify those labs used to monitor his disease status. What do these specifically measure, and how would you interpret them for him? Explain how the virus affects these laboratory values.

14. What laboratory values can be used to evaluate nutritional status? Do any of Mr. Long's values indicate nutritional risk?

15. Mr. Long was started on three medications that he will be discharged on.

   a. Identify these medications and the purpose of each.

   b. Are there any specific drug–nutrient interactions to be concerned about? Explain.

16. Mr. Long is taking several vitamin and herbal supplements. What would you tell Mr. Long about these supplements? Do they pose any risks?

IV.   Nutrition Diagnosis
17. Select two high-priority nutrition problems and complete the PES statement for each.

V.   Nutrition Intervention
18. Identify 3 interventions you would recommend for improving Mr. Long's tolerance of food until his oral thrush has subsided.

19. Describe at least two areas of nutrition education that you would want to ensure that Mr. Long receives. Explain your rationale for these choices.

20. Patients with AIDS are at increased risk for infection. What nutritional practices would you teach Mr. Long to help him prevent illness related to food or water intake?

21. Why is exercise important as a component of the nutrition care plan? What general recommendations could you give to Mr. Long regarding physical activity?

VI.   Nutrition Monitoring and Evaluation
22. One of the more recent complications for AIDS and prescribed HAART is the development of lipodystrophy. Define this condition and describe the most common signs and symptoms.

23. How would the clinician monitor Mr. Long for the development of this disorder?

Bibliography

Academy of Nutrition and Dietetics Evidence Analysis Library. HIV/AIDS evidence-based nutrition practice guidelines. Available from http://andevidencelibrary.com/topic.cfm?cat=4248. Accessed 1/9/13.

Academy of Nutrition and Dietetics Nutrition Care Manual. HIV/AIDS. Available from http://www.nutritioncaremanual.org/topic.cfm?ncm_heading=Diseases%2FConditions&ncm_toc_id=20149. Accessed 1/9/13.

American Dietetic Association. Position of the American Dietetic Association: Nutrition intervention and human immunodeficiency virus infection. *J Am Diet Assoc.* 2010;110;1105–1119.

Batterham MJ, Morgan-Jones J, Greenop P, Garsia R, Gold J, Caterson I. Calculating energy requirements for men with HIV/AIDS in the era of highly active antiretroviral therapy. *Eur J Clin Nutr.* 2003;57:209–217.

Batterham MJ. Investigating heterogeneity in studies of resting energy expenditure in persons with HIV/AIDS: A meta-analysis. *Am J Clin Nutr.* 2005;81:702–713.

Carr A, Law M, HIV Lipodystrophy Case Definition Study Group. An objective lipodystrophy severity grading scale derived from the lipodystrophy case definition score. *J Acquir Immune Defic Syndr.* 2003;33(5):571–576.

Centers for Disease Control and Prevention. Revised surveillance case definitions for HIV infection among adults, adolescents, and children ages <18 months and for HIV infection and AIDS among children ages 18 months to <13 years — United States, 2008. *Morbidity and Mortality Weekly Review. Recommendations and Reports.* December 5, 2008/57(RR10);1–8.

Drug Facts and Comparisons. *Drug Facts and Comparisons 2004.* 58th ed. 2003.

Engleson ES. HIV lipodystrophy diagnosis and management. Body composition and metabolic alterations: diagnosis and management. *AIDS Reader.* 2003;13 (4 suppl):S10–S14.

Gardner CF. HIV and AIDS. In: Nelms M, Sucher K, Lacey K, Roth SL. *Nutrition Therapy and Pathophysiology.* 2nd ed. Belmont, CA: Wadsworth, Cengage Learning; 2011:735–770.

Hadigan C. Dietary habits and their association with metabolic abnormalities in human immunodeficiency virus-related lipodystrophy. *Clin Infect Dis.* 2003;37(suppl 2):S101–S104.

Hayes C, Elliot E, Krales E, Downer G. Food and water safety for persons infected with human immunodeficiency virus. *Clin Infect Dis.* 2003;36:S106–109.

Mwamburi DM, Wilson IB, Jacobson DL, et al. Understanding the role of HIV load in determining weight change in the era of highly active antiretroviral therapy. *Clin Infect Dis.* 2005;40(1):167–173.

Pronsky ZM, Meyer SA, Fields-Gardner C. *HIV Medications-Food Medication Interaction Guide.* 2nd ed. Birchrunville, PA: Food Medication Interactions; 2001.

Scevola D, Di Matteo A, Lanzarini P, et al. Effect of exercise and strength training on cardiovascular status in HIV-infected patients receiving highly active antiretroviral therapy. *AIDS.* 2003;17(suppl 1):S123–S129.

Suttmann U, Holtmannspotter M, Ockenga J, Gallati H, Deicher H, Selberg O. Tumor necrosis factor, interleukin-6, and epinephrine are associated with hypermetabolism in AIDS patients with acute opportunistic infections. *Ann Nutr Metab.* 2000;44:43–53.

Internet Resources

AIDS.gov: http://aids.gov/hiv-aids-basics/index.html

The AIDS InfoNet: http://www.aidsinfonet.org

Centers for Disease Control and Prevention: http://www.cdc.gov/hiv/

USDA Nutrient Data Laboratory: http://ndb.nal.usda.gov/ndb/foods/list

U.S. Pharmacist: http://www.uspharmacist.com/

COMMON MEDICAL ABBREVIATIONS

AAL	anterior axillary line		BS	bowel sounds, breath sounds, or blood sugar
ac	before meals		BSA	body surface area
ACTH	adrenocorticotropic hormone		BUN	blood urea nitrogen
AD	Alzheimer's disease		c	cup
ad lib	as desired (ad libitum)		c	with
ADA	American Diabetes Association		C	centigrade
ADL	activities of daily living		C.C.E.	clubbing, cyanosis, or edema
AGA	antigliadin antibody		c/o	complains of
AIDS	acquired immunodeficiency syndrome		CA	cancer; carcinoma
ALP (Alk phos)	alkaline phosphatase		CABG	coronary artery bypass graft
ALS	amyotrophic lateral sclerosis		CAD	coronary artery disease
ALT	alanine aminotransferase		CAPD	continuous ambulatory peritoneal dialysis
ANC	absolute neutrophil count			
ANCA	antineutrophil cytoplasmic antibody		cath	catheter, catheterize
			CAVH	continuous arteriovenous hemofiltration
AND	Academy of Nutrition and Dietetics			
AP	anterior posterior		CBC	complete blood count
ARDS	adult respiratory distress syndrome		cc	cubic centimeter
ARF	acute renal failure, acute respiratory failure		CCK	cholecystokinin
			CCU	coronary care unit
ASA	acetylsalicylic acid, aspirin		CDAI	Crohn's disease activity index
ASCA	antisaccharomyces antibody		CDC	Centers for Disease Control and Prevention
ASHD	arteriosclerotic heart disease			
AV	arteriovenous		CHD	coronary heart disease
BANDS	neutrophils		CHF	congestive heart failure
BCAA	branched-chain amino acids		CHI	closed head injury
BE	barium enema		CHO	carbohydrate
BEE	basal energy expenditure		CHOL	cholesterol
BG	blood glucose		CKD	chronic kidney disease
bid	twice a day (bis in die)		cm	centimeter
bili	bilirubin		CNS	central nervous system
BM	bowel movement		COPD	chronic obstructive pulmonary disease
BMI	body mass index			
BMR	basal metabolic rate		CPK	creatinine phosphokinase
BMT	bone marrow transplant		Cr	creatinine
BP (B/P)	blood pressure		CR	complete remission
BPD	bronchopulmonary dysplasia		CSF	cerebrospinal fluid
BPH	benign prostate hypertrophy		CT	computed tomography
bpm	beats per minute, breaths per minute		CVA	cerebrovascular accident
			CVD	cardiovascular disease

Note: Abbreviations can vary from institution to institution. Although the student will find many of the accepted variations listed in this appendix, other references may be needed to supplement this list.

CVP	central venous pressure		HbA_{1c}	glycosylated hemoglobin
CXR	chest X-ray		HBV	hepatitis B virus
d/c	discharge		HC	head circumference
D/C	discontinue		Hct	hematocrit
D_5NS	dextrose, 5% in normal saline		HCV	hepatitis C virus
D_5W	dextrose, 5% in water		HDL	high-density lipoprotein
DASH	Dietary Approaches to Stop Hypertension		HEENT	head, eyes, ears, nose, throat
DBW	desirable body weight		Hg	mercury
DCCT	Diabetes Control and Complications Trial		Hgb	hemoglobin
			HHNS	hyperosmolar hyperglycemic nonketotic (syndrome)
DKA	diabetic ketoacidosis		HIV	human immunodeficiency virus
dL	deciliter		HLA	human leukocyte antigen
DM	diabetes mellitus		HOB	head of bed
DRI	Dietary Reference Intake		HR	heart rate
DTR	deep tendon reflex		HS or h.s.	hours of sleep
DTs	delirium tremens		HTN	hypertension
DVT	deep vein thrombosis		Hx	history
Dx	diagnosis		I & O (I/O)	intake and output
e.g.	for example		i.e.	that is
ECF	extracellular fluid		IBD	inflammatory bowel disease
ECG/EKG	electrocardiogram		IBS	irritable bowel syndrome
EEG	electroencephalogram		IBW	ideal body weight
EGD	esophagogastroduodenoscopy		ICF	intracranial fluid
ELISA	enzyme-linked immunosorbent assay		ICP	intracranial pressure
			ICS	intercostal space
EMA	antiendomysial antibody		ICU	intensive care unit
EMG	electromyography		IGT	impaired glucose tolerance
EOMI	extra-ocular muscles intact		IM	intramuscularly
ER	emergency room		inc	incontinent
ERT	estrogen replacement therapy		IV	intravenous
ESR	erythrocyte sedimentation rate		J	joule
F	Fahrenheit		K	potassium
FBG	fasting blood glucose		kcal	kilocalorie
FBS	fasting blood sugar		KCl	potassium chloride
FDA	Food and Drug Administration		kg	kilogram
FEF	forced mid-expiratory flow		KS	Kaposi's sarcoma
FEV	forced-expiratory volume		KUB	kidney, ureter, bladder
FFA	free fatty acid		L	liter
FH	family history		lb	pounds
FTT	failure to thrive		LBM	lean body mass
FUO	fever of unknown origin		LCT	long-chain triglyceride
FVC	forced vital capacity		LDH	lactic dehydrogenase
FX	fracture		LES	lower esophageal sphincter
g	gram		LFT	liver function test
g/dL	grams per deciliter		LIGS	low intermittent gastric suction
GB	gallbladder		LLD	left lateral decubitus position
GERD	gastroesophageal reflux disease		LLQ	lower left quadrant
GFR	glomerular filtration rate		LMP	last menstrual period
GI	gastrointestinal		LOC	level of consciousness
GM-CSF	granulocyte/macrophage colony stimulating factor		LP	lumbar puncture
			LUQ	left upper quadrant
GTF	glucose tolerance factor		lytes	electrolytes
GTT	glucose tolerance test		MAC	midarm circumference
GVHD	graft versus host disease		MAMC	midarm muscle circumference
h	hour		MAOI	monoamine oxidase inhibitor
H & P (HPI)	history and physical		MCHC	mean corpuscular hemoglobin concentration
HAV	hepatitis A virus			

MCL	midclavicular line
MCT	medium-chain triglyceride
MCV	mean corpuscular volume
mEq	milliequivalent
mg	milligram
Mg	magnesium
MI	myocardial infarction
mm	millimeter
mmHg	millimeters of mercury
MNT	medical nutrition therapy
MOM	Milk of Magnesia
mOsm	milliosmol
MR	mitral regurgitation
MRI	magnetic resonance imaging
MVA	motor vehicle accident
MVI	multiple vitamin infusion
N	nitrogen
N/V	nausea and vomiting
NG	nasogastric
NH_3	ammonia
NICU	neurointensive care unit, neonatal intensive care unit
NKA	no known allergies
NKDA	no known drug allergies
NPH	neutral protamine Hagedorn insulin
NPO	nothing by mouth
NSAID	nonsteroidal antiinflammatory drug
NTG	nitroglycerin
O_2	oxygen
OA	osteoarthritis
OC	oral contraceptive
OHA	oral hypoglycemic agent
OR	operating room
ORIF	open reduction internal fixation
OT	occupational therapist
OTC	over the counter
$paCO_2$	partial pressure of dissolved carbon dioxide in arterial blood
paO_2	partial pressure of dissolved oxygen in arterial blood
pc	after meals
PCM	protein-calorie malnutrition
PD	Parkinson's disease
PE	pulmonary embolus
PED	percutaneous endoscopic duodenostomy
PEEP	positive end-expiratory pressure
PEG	percutaneous endoscopic gastrostomy
PEM	protein-energy malnutrition
PERRLA	pupils equal, round, and reactive to light and accommodation
pH	hydrogen ion concentration
PKU	phenylketonuria
PMI	point of maximum impulse
PMN	polymorphonuclear
PN	parenteral nutrition
PO	by mouth (per os)
PPD	packs per day

PPN	peripheral parenteral nutrition
prn	may be repeated as necessary (pro re nata)
pt	patient
PT	patient, physical therapy, prothrombin time
PTA	prior to admission
PTT	partial thromboplastin time
PUD	peptic ulcer disease
PVC	premature ventricular contraction
PVD	peripheral vascular disease
R/O	rule out
RA	rheumatoid arthritis
RBC	red blood cell
RBW	reference body weight
RD	registered dietitian
RDA	Recommended Dietary Allowance
RDS	respiratory distress syndrome
REE	resting energy expenditure
RLL	right lower lobe
RLQ	right lower quadrant
ROM	range of motion
ROS	review of systems
RQ	respiratory quotient
RR	respiratory rate
RUL	right upper lobe
RUQ	right upper quadrant
Rx	take, prescribe, or treat
s̄	without
S/P	status post
SBGM	self blood glucose monitoring
SBO	small bowel obstruction
SBS	short bowel syndrome
SGOT	serum glutamic oxaloacetic transaminase
SGPT	serum glutamic pyruvic transaminase
SOB	shortness of breath
SQ	subcutaneous
stat	immediately
susp	suspension
T	temperature
T & A	tonsillectomy and adenoidectomy
T, tbsp	tablespoon
t, tsp	teaspoon
T_3	triiodothyronine
T_4	thyroxine
TB	tuberculosis
TEE	total energy expenditure
TF	tube feeding
TG	triglyceride
TIA	transient ischemic attack
TIBC	total iron binding capacity
TKO	to keep open
TLC	total lymphocyte count
TNM	tumor, node, metastasis
TPN	total parenteral nutrition
TSF	triceps skinfold

TSH	thyroid stimulating hormone	VLCD	very-low-calorie diet
TURP	transurethral resection of the prostate	VOD	venous occlusive disease
		VS	vital signs
UA	urinalysis	w.a.	while awake
UBW	usual body weight	WBC	white blood cell
UL	Tolerable Upper Intake Level	WNL	within normal limits
URI	upper respiratory infection	wt	weight
UTI	urinary tract infection	WW	whole wheat
UUN	urine urea nitrogen	yo	year old

Notations that Should Not Be Used[a]

Inappropriate Abbreviation(s) or Figures	Best Practices
U	Write "unit"
IU	Write "International Unit"
Q.D., QD, q.d., qd	Write "daily"
Q.O.D., QOD, q.o.d., qod	Write "every other day"
1.0	Write "1" (do not use a trailing zero with a whole number)
.5	Write "0.5" (use a leading zero when recording a value less than one)
MS, MSO_4	Write "morphine sulfate"
$MgSO_4$	Write "magnesium sulfate"

[a] Joint Commission (http://www.jointcommission.org/assets/1/18/dnu_list.pdf) Accessed 1/26/13.

NORMAL VALUES FOR PHYSICAL EXAMINATION

Vital Signs

Temperature

Rectal: C = 37.6°/F = 99.6°
Oral: C = 37°/F = 98.6° (± 1°)
Axilla: C = 37.4°/F = 97.6°

Blood Pressure: average 120/80 mmHg

Heart Rate (beats per minute)

Age	At Rest Awake	At Rest Asleep	Exercise or Fever
Newborn	100–180	80–160	≤220
1 week–3 months	100–220	80–200	<220
3 months–2 years	80–150	70–120	≤200
2–10 years	70–110	60–90	≤200
11 years–adult	55–90	50–90	≤200

Respiratory Rate (breaths per minute)

Age	Respirations
Newborn	35
1–11 months	30
1–2 years	25
3–4 years	23
5–6 years	21
7–8 years	20
9–10 years	19
11–12 years	19
13–14 years	18
15–16 years	17
17–18 years	16–18
Adult	12–20

Cardiac Exam: carotid pulses equal in rate, rhythm, and strength; normal heart sounds; no murmurs present

HEENT Exam (head, eyes, ears, nose, throat)
Mouth: pink, moist, symmetrical; mucosa pink, soft, moist, smooth
Gums: pink, smooth, moist; may have patchy pigmentation
Teeth: smooth, white, shiny
Tongue: medium red or pink, smooth with free mobility, top surface slightly rough
Eyes: pupils equal, round, reactive to light and accommodation
Ears: tympanic membrane taut, translucent, pearly gray; auricle smooth without lesions; meatus not swollen or occluded; cerumen dry (tan/light yellow) or moist (dark yellow/brown)
Nose: external nose symmetrical, nontender without discharge; mucosa pink; septum at the midline
Pharynx: mucosa pink and smooth
Neck: thyroid gland, lymph nodes not easily palpable or enlarged

Lungs: chest contour symmetrical; spine straight without lateral deviation; no bulging or active movement within the intercostal spaces during breathing; respirations clear to auscultation and percussion

Peripheral Vascular: normal pulse graded at 3+, which indicates that pulse is easy to palpate and not easily obliterated; pulses equal bilaterally and symmetrically

Neurological: normal orientation to people, place, time, with appropriate response and concentration

Skin: warm and dry to touch; should lift easily and return back to original position, indicating normal turgor and elasticity

Abdomen: umbilicus flat or concave, positioned midway between xyphoid process and symphysis pubis; bowel motility notes normal air and fluid movement every 5–15 seconds; graded as normal, audible, absent, hyperactive, or hypoactive

Appendix C

ROUTINE LABORATORY TESTS WITH NUTRITIONAL IMPLICATIONS[1]

This table presents a partial listing of some uses of commonly performed lab tests that have implications for nutritional problems.

Laboratory Test	Acceptable Range	Description
Hematology		
Red blood cell (RBC) count	4.2–5.4 × 10⁶/mm³ (women) 4.5–6.2 × 10⁶/mm³ (men)	Number of RBC; aids evaluation of anemias.
Hemoglobin (Hb)	12–15 g/dL (women) 14–17 g/dL (men)	Hemoglobin content of RBC; aids evaluation of anemias.
Hematocrit (Hct)	37–47% (women) 40–54% (men)	Percentage RBC in total blood volume; aids evaluation of anemias.
Mean corpuscular volume (MCV)	80–96 μm³	RBC size; helps to distinguish between microcytic and macrocytic anemias.
Mean corpuscular hemoglobin concentration (MCHC)	26–32 pg	Hb concentration within RBCs; helps to distinguish iron-deficiency anemia.
White blood cell (WBC) count	4.8–11.8 × 10³/mm³	Number of WBC; general assessment of immune function and/or presence of infection.
Blood Chemistry		
Serum Proteins		
Total protein	6–8 g/dL	Protein levels are not specific to disease or highly sensitive; they can reflect poor protein intake, illness or infections, changes in hydration or metabolism, pregnancy, or medications.
Albumin	3.5–5.0 g/dL	May reflect PEM; slow to respond to improvement or worsening of disease. Synthesis rate decreases during inflammation.
Transferrin	250–380 mg/dL (women) 215–365 mg/dL (men)	May reflect illness, PEM, or iron deficiency; slightly more sensitive to changes than albumin. Synthesis rate decreases during inflammation.
Prealbumin (transthyretin)	16–35 mg/dL	May reflect PEM; more responsive to health status changes than albumin or transferrin. Synthesis rate decreases during inflammation.
C-reactive protein	<1.00 mg/dL	Acute-phase protein—indicator of inflammation or disease.

[1] Adapted from Nelms MN, Sucher KP, Lacey K, Roth SL. *Nutrition Therapy & Pathophysiology*, 1st ed. (ISBN 053462154), Appendix B3, which was adapted from Rolfes SR, Pinna K, Whitney E. *Understanding Normal & Clinical Nutrition*, 7th ed. (ISBN 0534622089), Table 17-8, p. 594.

Fibrinogen	160–450 mg/dL	Acute-phase protein—indicator of inflammation or disease.
Lactate	0.3–2.3 mEq/L	Reflective of lactic acidosis—elevated during periods of critical illness.
Serum Enzymes		
Creatine kinase (CK, CPK)	0–145 IU/L 30–135 U/L (women) 55–170 U/L (men)	Different forms of CK are found in the muscle, brain, and heart. High blood levels may indicate heart attack, brain tissue damage, or skeletal muscle injury.
Lactate dehydrogenase (LDH)	208–378 IU/L	LDH is found in many tissues. Specific types may be elevated after heart attack, lung damage, or liver disease.
Alkaline phosphatase	30–120 U/L	Found in many tissues; often measured to evaluate liver function.
Aspartate aminotransferase (AST, formerly SGOT)	0–35 U/L	Usually monitored to assess liver damage; elevated in most liver diseases. Levels are somewhat increased after tissue damage.
Alanine aminotransferase (ALT, formerly SGPT)	4–36 U/L	Usually monitored to assess liver damage; elevated in most liver diseases. Levels are somewhat increased after tissue damage.
Serum Electrolytes		
Sodium	136–145 mEq/L	Helps to evaluate hydration status or neuromuscular, kidney, and adrenal functions.
Potassium	3.5–5.5 mEq/L	Helps to evaluate acid-base balance and kidney function; can detect potassium imbalances.
Chloride	95–105 mEq/L	Helps to evaluate hydration status and detect acid-base and electrolyte imbalances.
Other		
Glucose	70–110 mg/dL	Detects risk of glucose intolerance, diabetes mellitus, and hypoglycemia; helps to monitor diabetes treatment.
Glycosylated hemoglobin (HbA$_{1c}$)	3.9–5.2%	Used to monitor long-term blood glucose control (average over previous 120 days).
Blood urea nitrogen (BUN)	8–18 mg/dL	Primarily used to monitor renal function; value is altered by liver failure, dehydration, or shock.
Uric acid	2.8–8.8 mg/dL (women) 4.0–9.0 mg/dL (men)	Used for detecting gout or changes in renal function; levels affected by age and diet; varies among different ethnic groups.
Creatinine (serum or plasma)	0.6–1.2 mg/dL	Used to monitor renal function.

Note: μm = micrometer; dL = deciliter; pg = picogram; U/L = units per liter; mEq = milliequivalents

Appendix D

EXCHANGE LISTS FOR DIABETES MEAL PLAN FORM AND FOOD LISTS NUTRIENT INFORMATION [1]

Meal Plan for: _____

RD: _____

Date: _____

Phone: _____

Carbohydrate _____ (grams) _____ (% of calories)

Carbohydrate choices _____ (servings)

Protein _____ (% of calories)　　Fat _____ (% of calories)　　Calories _____

	Starches	Fruits	Milk	Nonstarchy Vegetables	Meat and Meat Substitutes	Fats	Menu Ideas
Breakfast Time: ___							
Snack Time: ___							
Lunch Time: ___							
Snack Time: ___							
Dinner Time: ___							
Snack Time: ___							

[1] From *Choose Your Foods: Exchange Lists for Diabetes*, published by the American Diabetes Association and the American Dietetic Association, 2007. Copyright © 2008 by the American Diabetes Association and the American Dietetic Association. Used by permission.

The Food Lists

The following chart shows the amount of nutrients in one serving from each list.

Food List	Carbohydrate (grams)	Protein (grams)	Fat (grams)	Calories
Carbohydrates				
Starch: breads, cereals and grains, starchy vegetables, crackers, snacks, and beans, peas, and lentils	15	0–3	0–1	80
Fruits	15	—	—	60
Milk				
Fat-free, low-fat, 1%	12	8	0–3	100
Reduced-fat, 2%	12	8	5	120
Whole	12	8	8	160
Sweets, desserts, and other carbohydrates	15	varies	varies	varies
Nonstarchy vegetables	5	2	—	25
Meat and Meat Substitutes				
Lean	—	7	0–3	45
Medium-fat	—	7	4–7	75
High-fat	—	7	8+	100
Plant-based proteins	varies	7	varies	varies
Fats	—	—	5	45
Alcohol	varies	—	—	100

INDEX

A

Abdominal pain, 104, 116, 142, 340
ABGs. *See* Arterial blood gases
Academy of Nutrition and Dietetics (AND), 1
Acute kidney injury (AKI)
  with burn injuries, 361
  case study of, 229–37
  overview of, 205
Acute lymphoblastic leukemia (ALL), 379–90
Acute pancreatitis, 127, 141–52
Acute respiratory distress, 313–23
Admitting history. *See also* Nutrition history
  for acute kidney injury, 230
  for acute pancreatitis, 142
  for adult type 1 diabetes mellitus, 170
  for adult type 2 diabetes mellitus, 194
  for Alzheimer's disease, 290
  for burn injuries, 352
  for celiac disease, 94
  for childhood obesity, 4
  for chronic kidney disease, 208, 220
  for chronic obstructive pulmonary disease, 302, 314
  for cirrhosis of the liver, 130
  for Crohn's disease, 116
  for esophageal cancer, 392
  for failed transplant, 220
  for gastric bypass surgery, 12, 366
  for gastroesophageal reflux disease, 72
  for heart failure, 60
  for hepatitis C, 130
  for hypertension, 36
  for irritable bowel syndrome, 104
  for malnutrition, 24
  for megoblastic anemia, 254
  for myocardial infarction, 48
  for nonhealing wounds, 290
  for open abdomen, 340
  for pediatric traumatic brain injury, 328
  for pediatric type 1 diabetes mellitus, 156
  for pediatric type 2 diabetes mellitus, 182
  for peptic ulcer disease, 82
  for pregnancy-related anemia, 242
  for respiratory failure, 314
  for sepsis, 366
  for stroke, 268
"Adult Starvation and Disease-Related Malnutrition"
  (Jensen et al.), 29–30
African Americans, 205
AGA antibodies, 98

AIDS/HIV, 401–12
AKI. *See* Acute kidney injury
Alcohol ingestion, 127, 135, 142, 145
ALL. *See* Acute lymphoblastic leukemia
Alzheimer's disease
  case study of, 289–98
  overview of, 265
American Thoracic Society, 299
Ancillary organs of digestion, 127
AND. *See* Academy of Nutrition and Dietetics
Anemia
  celiac disease and, 94–100
  classifications of, 249
  megoblastic type of, 253–64
  overview of, 239
  in pregnancy, 241–52
Angioplasty, 48–57
Antiretroviral therapy, 401
Arterial blood gases (ABGs)
  for acute pancreatitis, 148
  for adult type 1 diabetes mellitus, 175–76
  for burn injuries, 359
  for chronic obstructive pulmonary disease, 306–8,
    318, 322
  for respiratory failure, 318, 322
ASPEN. *See* Association of Parenteral and Enteral
  Nutrition
Aspiration, 275
Assessment
  for acute kidney injury, 230–36
  for acute lymphoblastic leukemia, 380–86, 388
  for acute pancreatitis, 142–48, 150
  for adult type 1 diabetes mellitus, 170–76, 178
  for adult type 2 diabetes mellitus, 194–200, 202
  for Alzheimer's disease, 290–96
  for burn injuries, 352–59, 361
  for cardiac cachexia, 65–66
  for celiac disease, 94–97
  for childhood overweight, 4–9
  for chronic kidney disease, 208–15, 220–26
  for chronic obstructive pulmonary disease, 302–6,
    308–9, 314–19, 321
  for cirrhosis of the liver, 130–34, 136–37
  for Crohn's disease, 116–21, 123
  for esophageal cancer, 392–99
  for failed transplant, 220–26
  for gastric bypass surgery, 12–17
  for gastroesophageal reflux disease, 72–76, 78
  for heart failure, 65–66

Assessment (*continued*)
  for HIV/AIDS, 404–11
  for hypertension, 37, 42–44
  for irritable bowel syndrome, 104–9, 111
  for malnutrition, 24–30
  for megoblastic anemia, 254–60, 262
  for myocardial infarction, 48–55
  for nonhealing wounds, 290–96
  for open abdomen, 340–49
  for Parkinson's disease, 280–84, 286
  for pediatric traumatic brain injury, 328–36
  for pediatric type 1 diabetes mellitus, 156–62, 164
  for pediatric type 2 diabetes mellitus, 182–85, 187
  for peptic ulcer disease, 82–85, 87–88
  for pregnancy-related anemia, 242–48, 250
  for respiratory failure, 321
  for sepsis, 366–74
  for stroke, 268–72, 274–75
Association for Parenteral and Enteral Nutrition (ASPEN), 1
Autoimmune disorders. *See specific disorders*

B
Bariatric surgery. *See* Gastric bypass surgery
Barichella, M., 285
Billroth II, 83
Blogs, 180
Blood glucose management
  for adult type 1 diabetes mellitus, 177–79
  for pediatric type 1 diabetes mellitus, 163, 165–66
  for pediatric type 2 diabetes mellitus, 186–87
  studies of, 153
Body mass index (BMI), 8, 188–89
Bronchitis. *See* Chronic obstructive pulmonary disease
B$_{12}$ deficiency, 253–64
Burn injuries, 351–63
Butterworth, C., 1

C
Cancer, 377–400. *See also specific types*
Canker sores, 94
Cardiac cachexia, 59–68
Cardiovascular disease
  heart failure case study in, 59–68
  hypertension case study in, 35–46
  myocardial infarction case study in, 47–57
  overview of, 33
  pediatric type 2 diabetes mellitus and, 186
Catabolism, 325
Celiac disease, 91, 93–101
Centers for Disease Control and Prevention, 265
Cerada, E., 285
Chemistry
  for acute kidney injury, 232–33

for acute lymphoblastic leukemia, 384–85
for acute pancreatitis, 146–47
for adult type 1 diabetes mellitus, 174–75
for adult type 2 diabetes mellitus, 198–99
for Alzeimer's disease, 293
for burn injuries, 358–59
for cachexia, 62–63
for celiac disease, 95–96
for childhood obesity, 6
for chronic kidney disease, 211–12, 222–23
for chronic obstructive pulmonary disease, 305–6, 317
for cirrhosis of the liver, 132–33
for Crohn's disease, 119–20
for esophageal cancer, 394–95
for failed transplant, 222–23
for gastric bypass surgery, 15–16
for gastroesophageal reflux disease, 74–75
for heart failure, 62–63
for HIV/AIDS, 407
for hypertension, 38–39
for irritable bowel syndrome, 107–8
for malnutrition, 27
for megoblastic anemia, 257–58
for myocardial infarction, 51–52
for nonhealing wounds, 293
for open abdomen, 344–45
for Parkinson's disease, 283
for pediatric traumatic brain injury, 332
for pediatric type 1 diabetes mellitus, 160–61
for pediatric type 2 diabetes mellitus, 183–84
for peptic ulcer disease, 84–85
for pregnancy-related anemia, 246–47
for respiratory failure, 317
for sepsis, 370–71
for stroke, 271
Chest tubes, 320
Chief complaint
  for acute lymphoblastic leukemia, 380
  for adult type 1 diabetes mellitus, 170
  for adult type 2 diabetes mellitus, 194
  for Alzheimer's disease, 290
  for celiac disease, 94
  for childhood obesity, 4
  for chronic kidney disease, 208, 220
  for chronic obstructive pulmonary disease, 302
  for cirrhosis of the liver, 130
  for Crohn's disease, 116
  for esophageal cancer, 392
  for failed transplant, 220
  for gastric bypass surgery, 12
  for gastroesophageal reflux disease, 72
  for heart failure, 60

for HIV/AIDS, 404
for irritable bowel syndrome, 104
for malnutrition, 24
for megoblastic anemia, 254
for nonhealing wounds, 290
for Parkinson's disease, 280
for pediatric type 1 diabetes mellitus, 156
for pediatric type 2 diabetes mellitus, 182
for pregnancy-related anemia, 242
for sepsis, 366
for stroke, 268
Childhood obesity, 1–9, 181–91
Choking, 281
Chronic kidney disease (CKD)
    case studies of, 207–28
    hemodialysis case for, 207–17
    overview of, 205
    peritoneal dialysis case for, 219–28
Chronic obstructive pulmonary disease (COPD)
    case studies of, 301–23
    malnutrition and, 299
    overview of, 299
Cirrhosis, 127, 130–39
CKD. See Chronic kidney disease
Closed head injuries, 325
Coagulation studies
    for acute kidney injury, 233
    for acute lymphoblastic leukemia, 385
    for acute pancreatitis, 147
    for Alzheimer's disease, 294
    for burn injuries, 359
    for cachexia, 63
    for celiac disease, 96
    for chronic kidney disease, 223
    for chronic obstructive pulmonary disease, 306, 317
    for cirrhosis of the liver, 133
    for Crohn's disease, 120
    for esophageal cancer, 395
    for failed transplant, 223
    for gastroesophageal reflux disease, 75
    for heart failure, 63
    for HIV/AIDS, 407
    for hypertension, 39
    for malnutrition, 27
    for megoblastic anemia, 258
    for myocardial infarction, 52
    for nonhealing wounds, 294
    for open abdomen, 345
    for pediatric traumatic brain injury, 332
    for peptic ulcer disease, 85
    for pregnancy-related anemia, 247
    for respiratory failure, 317
Cohen, D., 339

Collier, B., 349
Compher, C., 29–30
Complementary/alternative therapies, 112
Constipation, 104
Continuous renal replacement therapy (CRRT), 231, 235
COPD. See Chronic obstructive pulmonary disease
Coronary angioplasty, 48–57
Coughing, while eating, 280
Counterregulatory hormones, 325
Crohn's disease
    case study of, 115–26
    overview of, 91
CRRT. See Continuous renal replacement therapy
Cytokines, 325

D
DASH. See Dietary Approaches to Stop Hypertension
Dehydration, 201
Diabetes Control and Complications Trial (DCCT), 153
Diabetes mellitus
    in adults, 169–80, 193–203
    in children, 155–67, 181–91
    overview of, 153
    renal disorders and, 153, 205
Diabetic ketoacidosis (DKA), 177, 201
Diagnosis
    for acute kidney injury, 230–31, 236
    for acute lymphoblastic leukemia, 387, 389
    for acute pancreatitis, 146, 150
    for adult type 1 diabetes mellitus, 177–78
    for adult type 2 diabetes mellitus, 201–2
    for Alzheimer's disease, 295–96
    for burn injuries, 361
    for cardiac cachexia, 66
    for celiac disease, 99
    for childhood obesity, 9
    for chronic kidney disease, 215, 225–26
    for chronic obstructive pulmonary disease, 309, 321
    for cirrhosis of the liver, 132, 137
    for Crohn's disease, 123
    for esophageal cancer, 399
    for gastric bypass surgery, 18–20
    for gastroesophageal reflux disease, 73, 78
    for heart failure, 66
    for HIV/AIDS, 411
    for hypertension, 37, 44
    for irritable bowel syndrome, 105, 111
    for malnutrition, 29–30
    for megoblastic anemia, 261, 263
    for myocardial infarction, 55
    for nonhealing wounds, 295–96
    for open abdomen, 340, 349
    for Parkinson's disease, 285–86

Diagnosis (*continued*)
  for pediatric traumatic brain injury, 328, 336
  for pediatric type 1 diabetes mellitus, 159, 163–64
  for pediatric type 2 diabetes mellitus, 183, 186–87
  for peptic ulcer disease, 83, 88
  for pregnancy-related anemia, 245, 249–50
  for respiratory failure, 321
  for sepsis, 374
  for stroke, 275
Dialysis
  cost of, 205
  hemodialysis for, 207–17
  peritoneal dialysis for, 219–28
Diarrhea, 104, 116
Diastolic blood pressure, 33
Dietary Approaches to Stop Hypertension (DASH), 33
DKA. *See* Diabetic ketoacidosis
Dumping syndrome, 86
Duodenal ulcers, 81–82
Dysphagia diet, 274, 286–87
Dysphagia Outcome and Severity Scale (DOSS), 287

E
EMA antibodies, 98
Emphysema. *See* Chronic obstructive pulmonary disease
Endocrine disorders
  overview of, 153
  in pediatric type 1 diabetes mellitus, 155–67
  in pediatric type 2 diabetes mellitus, 181–91
  in type 1 adult diabetes mellitus, 169–80
Endotracheal intubation, 320, 322, 357
Enteral feeding, 67, 83, 149, 321–22, 398
Esophageal cancer, 391–400
Essential hypertension, 33
Evaluation/monitoring
  for acute kidney injury, 236
  for acute lymphoblastic leukemia, 389
  for acute pancreatitis, 151
  for adult type 1 diabetes mellitus, 179
  for adult type 2 diabetes mellitus, 202
  for burn injuries, 361–62
  for celiac disease, 100
  for childhood obesity, 9
  for chronic kidney disease, 227
  for chronic obstructive pulmonary disease, 309–10, 322
  for cirrhosis, 137–38
  for Crohn's disease, 124–25
  for esophageal cancer, 399
  for gastric bypass surgery, 20
  for HIV/AIDS, 411
  for irritable bowel syndrome, 112
  for malnutrition, 30
  for megoblastic anemia, 263

  for myocardial infarction, 56
  for open abdomen, 349
  for Parkinson's disease, 287
  for pediatric traumatic brain injury, 336
  for pediatric type 1 diabetes mellitus, 165–66
  for pediatric type 2 diabetes mellitus, 190
  for pregnancy-related anemia, 251
  for respiratory failure, 322
  for sepsis, 374
  for stroke, 275–77
Examination
  for acute lymphoblastic leukemia, 380–81, 383
  for acute pancreatitis, 142–43, 145–46
  for adult type 1 diabetes mellitus, 170–71, 173
  for adult type 2 diabetes mellitus, 194–95, 197
  for Alzheimer's disease, 290–91
  for burn injuries, 352–53, 357
  for celiac disease, 94
  for childhood overweight, 4–5
  for chronic kidney disease, 208–9, 220–21
  for chronic obstructive pulmonary disease, 302–3, 314–15
  for Crohn's disease, 116–17
  for failed transplant, 220–21
  for gastric bypass surgery, 14
  for gastroesophageal reflux disease, 73
  for HIV/AIDS, 404–5
  for hypertension, 36–37
  for irritable bowel syndrome, 104–5
  for malnutrition, 24–25
  for megoblastic anemia, 254–56
  for myocardial infarction, 48–49
  for nonhealing wounds, 290–91
  for open abdomen, 340–41
  for Parkinson's disease, 280–81
  for pediatric traumatic brain injury, 328, 331
  for pediatric type 1 diabetes mellitus, 156–57, 159
  for pediatric type 2 diabetes mellitus, 182–83
  for peptic ulcer disease, 82
  for pregnancy-related anemia, 242–43, 245
  for respiratory failure, 314–15
  for sepsis, 366–67, 369
  for stroke, 268–69

F
FODMAP assessment, 106–7, 111
Folate deficiency, 253–64
Food diaries, 111
Functional disorders, 110

G
Gastric bypass surgery
  case study of, 11–20

megoblastic anemia and, 253–64
metabolic stress and, 365–75
Gastric residual volume (GRV), 322
Gastroesophageal reflux disease (GERD)
  case study of, 71–80
  overview of, 69
  peptic ulcer disease and, 81–89
Gastrointestinal (GI) disorders
  celiac disease in, 91, 93–101
  Crohn's disease in, 115–26
  gastroesophageal reflux disease in, 71–80
  irritable bowel syndrome in, 103–14
  overview of, 69
  peptic ulcer disease in, 81–89
Gastrojejunostomy, 83
GERD. *See* Gastroesophageal reflux disease
Gliadin, 91
Glomerular filtration rate (GFR), 214
Glomerulonephritis, 219–28
Gluconeogenesis, 325
Glucose management
  for adult type 1 diabetes mellitus, 177–79
  for pediatric type 1 diabetes mellitus, 163, 165–66
  for pediatric type 2 diabetes mellitus, 186–87
  studies of, 153
Gluten, 91, 98
Glycemic index, 179
Glycogen, 325
Graft-versus-host disease (GVHD), 379–89
Growth charts, 8, 188–89
GRV. *See* Gastric residual volume
Gunshot wound, 339–40

H
HAART, 404–12
Health care costs
  of diabetes mellitus, 153
  of renal disorders, 205
Heart failure (HF), 33
Helicobacter pylori, 69, 82, 86
Hematological disorders. *See specific disorders*
Hematology
  for acute kidney injury, 233–34
  for acute lymphoblastic leukemia, 385–86
  for acute pancreatitis, 147–48
  for adult type 1 diabetes mellitus, 175
  for adult type 2 diabetes mellitus, 199–200
  for Alzheimer's disease, 294
  for burn injuries, 359
  for cachexia, 63–64
  for celiac disease, 96–97
  for childhood obesity, 6
  for chronic kidney disease, 223–24

for chronic obstructive pulmonary disease, 306, 318
for cirrhosis of the liver, 133–34
for Crohn's disease, 120–21
for esophageal cancer, 395–96
for failed transplant, 223–24
for gastric bypass surgery, 16
for gastroesophageal reflux disease, 75–76
for heart failure, 63–64
for HIV/AIDS, 408
for hypertension, 39–40
for irritable bowel syndrome, 108–9
for malnutrition, 27–28
for megoblastic anemia, 258–59
for myocardial infarction, 52
for nonhealing wounds, 294
for open abdomen, 345
for Parkinson's disease, 283–84
for pediatric traumatic brain injury, 332–33
for pediatric type 1 diabetes mellitus, 161–62
for pediatric type 2 diabetes mellitus, 184–85
for peptic ulcer disease, 85
for pregnancy-related anemia, 247
for respiratory failure, 318
for sepsis, 371
for stroke, 271–72
Hematopoietic cell transplantation, 377–90
Hemiparesis, 268
Hemodialysis, 205, 208–17
Hemorrhagic edema, 329–34
Hepatitis C virus, 127, 130
Hepatobiliary disease
  cirrhosis of the liver in, 127, 129–39
  overview of, 127
  pancreatitis in, 127, 141–52
HF. *See* Heart failure
History, medical. *See also* Nutrition history
  for acute kidney injury, 230
  for acute pancreatitis, 142
  for adult type 1 diabetes mellitus, 170
  for adult type 2 diabetes mellitus, 194
  for Alzheimer's disease, 290
  for burn injuries, 352
  for celiac disease, 94
  for childhood obesity, 4
  for chronic kidney disease, 208, 220
  for chronic obstructive pulmonary disease, 302, 314
  for cirrhosis of the liver, 130
  for Crohn's disease, 116
  for esophageal cancer, 392
  for failed transplant, 220
  for gastric bypass surgery, 12, 366
  for gastroesophageal reflux disease, 72
  for heart failure, 60

History, medical (*continued*)
   for hepatitis C, 130
   for hypertension, 36
   for irritable bowel syndrome, 104
   for malnutrition, 24
   for megoblastic anemia, 254
   for myocardial infarction, 48
   for nonhealing wounds, 290
   for open abdomen, 340
   for pediatric traumatic brain injury, 328
   for pediatric type 1 diabetes mellitus, 156
   for pediatric type 2 diabetes mellitus, 182
   for peptic ulcer disease, 82
   for pregnancy-related anemia, 242
   for respiratory failure, 314
   for sepsis, 366
   for stroke, 268
HIV/AIDS, 401–12
Hyperglycemic hyperosmolar syndrome (HHS), 197, 201
Hypertension
   case study in, 35–46
   overview of, 33
Hypochromic microcytic anemia, 241–52
Hypoglycemia, 135

I
IBS. *See* Irritable bowel syndrome
Indirect calorimetry, 124
Inflammatory bowel disease. *See* Crohn's disease
Insulin
   for adult type 1 diabetes mellitus, 171, 173, 177
   for adult type 2 diabetes mellitus, 193–203
   lack of production of, 153
   metabolic stress and, 325
   for pediatric type 1 diabetes mellitus, 157
Insulin:CHO ratio, 164
Intake/output records
   for acute kidney injury, 232
   for acute pancreatitis, 145
   for adult type 1 diabetes mellitus, 173
   for adult type 2 diabetes mellitus, 198
   for burn injuries, 357–58
   for cachexia, 64
   for chronic kidney disease, 211, 224
   for chronic obstructive pulmonary disorder, 319
   for Crohn's disease, 119
   for esophageal cancer, 396
   for failed transplant, 224
   for gastric bypass surgery, 15
   for heart failure, 64
   for malnutrition, 26
   for megoblastic anemia, 257

   for open abdomen, 343
   for Parkinson's disease, 282
   for pediatric traumatic brain injury, 331
   for pediatric type 1 diabetes mellitus, 160
   for peptic ulcer disease, 84
   for pregnancy-related anemia, 248
   for respiratory failure, 319
   for sepsis, 369
Integumentary system, 289–98
International Consensus Guideline Committee, 29
Internet resources
   for acute kidney injury, 237
   for acute lymphoblastic leukemia, 390
   for acute pancreatitis, 152
   for adult type 1 diabetes mellitus, 180
   for Alzheimer's disease, 298
   for burn injuries, 363
   for celiac disease, 101
   for childhood obesity, 10
   for chronic kidney disease, 217, 228
   for chronic obstructive pulmonary disease, 311, 323
   for cirrhosis, 139
   for Crohn's disease, 126
   for esophageal cancer, 400
   for gastric bypass surgery, 21
   for heart failure, 68
   for hepatitis, 139
   for HIV/AIDS, 412
   for hypertension, 46
   for irritable bowel syndrome, 114
   for malnutrition, 31
   for megoblastic anemia, 264
   for myocardial infarction, 57
   for open abdomen, 350
   for Parkinson's disease, 288
   for pediatric traumatic brain injury, 337
   for pediatric type 1 diabetes mellitus, 167
   for pediatric type 2 diabetes mellitus, 191
   for pregnancy-related anemia, 252
   for respiratory failure, 323
   for sepsis, 375
   for stroke, 277
   for upper gastrointestinal disorders, 80, 89
   for wounds, 298
Intervention
   for acute kidney injury, 236
   for acute lymphoblastic leukemia, 382, 384, 389
   for acute pancreatitis, 144, 146, 150
   for adult type 1 diabetes mellitus, 171, 173, 178–79
   for adult type 2 diabetes mellitus, 195, 197, 202
   for Alzheimer's disease, 292, 297
   for burn injuries, 354–57, 361–62
   for cardiac cachexia, 62, 67

for celiac disease, 95, 99
for childhood obesity, 9
for chronic kidney disease, 210, 215–16, 222,
    226–27
for chronic obstructive pulmonary disease, 304, 309,
    316, 321
for cirrhosis of the liver, 132, 137
for Crohn's disease, 118, 124
for esophageal cancer, 394, 399
for failed transplant, 222
for gastric bypass surgery, 14, 20
for gastroesophageal reflux disease, 73, 78–79
for heart failure, 62, 67
for HIV/AIDS, 406, 411
for hypertension, 37, 42, 44–45
for irritable bowel syndrome, 105, 111–12
for malnutrition, 25–26, 30
for megoblastic anemia, 255, 257, 263
for myocardial infarction, 54–56
for nonhealing wounds, 292, 297
for open abdomen, 340, 349
for Parkinson's disease, 282, 286–87
for pediatric traumatic brain injury, 329, 331, 336
for pediatric type 1 diabetes mellitus, 157, 159, 164–65
for pediatric type 2 diabetes mellitus, 183, 187, 190
for peptic ulcer disease, 83, 88
for pregnancy-related anemia, 244–45, 250–51
for respiratory failure, 316, 321
for sepsis, 368–69, 374
for stroke, 270, 275
Intubation, 320, 322, 357
Iron deficiency, 249
Irritable bowel syndrome (IBS)
    case study of, 103–14
    overview of, 91
Ischemic stroke
    case study of, 267–78
    overview of, 265
Isosource HN, 321, 398
Itchy rash, 94

J
Jejunostomy, 83
Jejunostomy tubes, 341
Jensen, G.L., 29–30
Joint pain, 94

K
Kidney disorders
    acute type of, 205, 229–37, 361
    chronic type of, 205, 207–28
    diabetes mellitus and, 153, 205
Kidney transplant, 219–28

L
Laboratory results
    for acute kidney injury, 232–34
    for acute lymphoblastic leukemia, 384–86
    for acute pancreatitis, 146–48
    for adult type 1 diabetes mellitus, 174–76
    for adult type 2 diabetes mellitus, 198–200
    for Alzheimer's disease, 293–94
    for burn injuries, 358–59
    for cachexia, 62–64
    for celiac disease, 95–97
    for childhood obesity, 6
    for chronic kidney disease, 211–12, 222–24
    for chronic obstructive pulmonary disease, 305–6,
        317–18
    for cirrhosis of the liver, 132–34
    for Crohn's disease, 119–21
    for esophageal cancer, 394–96
    for failed transplant, 222–24
    for gastric bypass surgery, 15–17
    for gastroesophageal reflux disease, 74–76
    for heart failure, 62–64
    for HIV/AIDS, 407–8
    for hypertension, 38–40
    for irritable bowel syndrome, 107–9
    for malnutrition, 27–28
    for megoblastic anemia, 257–60
    for myocardial infarction, 51–53
    for nonhealing wounds, 293–94
    for open abdomen, 344–46
    for Parkinson's disease, 283–84
    for pediatric traumatic brain injury, 332–33
    for pediatric type 1 diabetes mellitus, 160–62
    for pediatric type 2 diabetes mellitus, 183–85
    for peptic ulcer disease, 84–85
    for pregnancy-related anemia, 246–47
    for respiratory failure, 317–18
    for sepsis, 370–72
    for stroke, 271–72
Lacerations, 289–98
Latent autoimmune diabetes of adult
    (LADA), 173
LES incompetence. *See* Lower esophageal sphincter
    incompetence
Leukemia, 377–90
Levodopa, 285
Lipolysis, 325
Liver
    cirrhosis of, 127, 129–39
    metabolism and, 135
    overview of, 127
Living will, 67
Lower esophageal sphincter (LES) incompetence, 69

Lower gastrointestinal disorders
  celiac disease in, 91, 93–101
  Crohn's disease in, 115–26
  irritable bowel syndrome in, 103–14
  overview of, 91
Lung disorders. *See* Pulmonary disorders

M
Malnutrition
  cancer and, 377
  case study of, 23–31
  cirrhosis and, 136
  in hepatobiliary disease, 127
  overview of, 1
  pulmonary disorders and, 299
Meal planning
  for chronic kidney disease, 215
  for diabetes mellitus, 153
Mechanical soft diet, 336
Medical history. *See also* Nutrition history
  for acute kidney injury, 230
  for acute pancreatitis, 142
  for adult type 1 diabetes mellitus, 170
  for adult type 2 diabetes mellitus, 194
  for Alzheimer's disease, 290
  for burn injuries, 352
  for celiac disease, 94
  for childhood obesity, 4
  for chronic kidney disease, 208, 220
  for chronic obstructive pulmonary disease, 302, 314
  for cirrhosis of the liver, 130
  for Crohn's disease, 116
  for esophageal cancer, 392
  for failed transplant, 220
  for gastric bypass surgery, 12, 366
  for gastroesophageal reflux disease, 72
  for heart failure, 60
  for hepatitis C, 130
  for hypertension, 36
  for irritable bowel syndrome, 104
  for malnutrition, 24
  for megoblastic anemia, 254
  for myocardial infarction, 48
  for nonhealing wounds, 290
  for open abdomen, 340
  for pediatric traumatic brain injury, 328
  for pediatric type 1 diabetes mellitus, 156
  for pediatric type 2 diabetes mellitus, 182
  for peptic ulcer disease, 82
  for pregnancy-related anemia, 242
  for respiratory failure, 314
  for sepsis, 366
  for stroke, 268

Megoblastic anemia, 253–64
Membranoproliferative glomerulonephritis, 219–28
Metabolic panel
  for acute kidney injury, 232–33
  for acute lymphoblastic leukemia, 384–85
  for acute pancreatitis, 146–47
  for adult type 1 diabetes mellitus, 174–75
  for adult type 2 diabetes mellitus, 198–99
  for Alzeimer's disease, 293
  for burn injuries, 358–59
  for cachexia, 62–63
  for celiac disease, 95–96
  for childhood obesity, 6
  for chronic kidney disease, 211–12, 222–23
  for chronic obstructive pulmonary disease, 305–6, 317
  for cirrhosis of the liver, 132–33
  for Crohn's disease, 119–20
  for esophageal cancer, 394–95
  for failed transplant, 222–23
  for gastric bypass surgery, 15–16
  for gastroesophageal reflux disease, 74–75
  for heart failure, 62–63
  for HIV/AIDS, 407
  for hypertension, 38–39
  for irritable bowel syndrome, 107–8
  for malnutrition, 27
  for megoblastic anemia, 257–58
  for myocardial infarction, 51–52
  for nonhealing wounds, 293
  for open abdomen, 344–45
  for Parkinson's disease, 283
  for pediatric traumatic brain injury, 332
  for pediatric type 1 diabetes mellitus, 160–61
  for pediatric type 2 diabetes mellitus, 183–84
  for peptic ulcer disease, 84–85
  for pregnancy-related anemia, 246–47
  for respiratory failure, 317
  for sepsis, 370–71
  for stroke, 271
Metabolic stress
  of burn injuries, 351–63
  of morbid obesity, 365–75
  of open abdomen, 339–50
  overview of, 325
  of pediatric brain injuries, 327–37
  phases of, 347
  of sepsis, 365–75
MI. *See* Myocardial infarction
Microcytic anemia, 241–52
Mirtallo, J., 29–30
Monitoring/evaluation
  for acute kidney injury, 236
  for acute lymphoblastic leukemia, 389

for acute pancreatitis, 151
for adult type 1 diabetes mellitus, 179
for adult type 2 diabetes mellitus, 202
for burn injuries, 361–62
for celiac disease, 100
for childhood obesity, 9
for chronic kidney disease, 227
for chronic obstructive pulmonary disease, 309–10, 322
for cirrhosis, 137–38
for Crohn's disease, 124–25
for esophageal cancer, 399
for gastric bypass surgery, 20
for HIV/AIDS, 411
for irritable bowel syndrome, 112
for malnutrition, 30
for megoblastic anemia, 263
for myocardial infarction, 56
for open abdomen, 349
for Parkinson's disease, 287
for pediatric traumatic brain injury, 336
for pediatric type 1 diabetes mellitus, 165–66
for pediatric type 2 diabetes mellitus, 190
for pregnancy-related anemia, 251
for respiratory failure, 322
for sepsis, 374
for stroke, 275–77
Morbid obesity. *See* Obesity
Motor vehicle accidents, 327–37
Mouth sores, 94
Myeloablative transplant, 387
Myocardial infarction (MI)
    case study of, 47–57
    definition of, 33

N
Nasogastric tubes, 149
National Dysphagia Diet, 274
National Institutes of Health, 205
Native Americans, 205
Neoplasm, 377
Neoplastic disease, 377–400
Neurological disorders
    of Alzheimer's disease, 265, 289–98
    overview of, 265
    of Parkinson's disease, 265, 279–88
    of stroke, 265, 267–78
Neuropathy, 94
Nitrogen balance, 125, 335–36, 349
Non-myeloablative transplant, 387
Nursing assessment
    for acute kidney injury, 231
    for acute lymphoblastic leukemia, 381–82
    for acute pancreatitis, 143–44

for adult type 1 diabetes mellitus, 171–72
for adult type 2 diabetes mellitus, 195–96
for Alzheimer's disease, 291–92
for burn injuries, 353–54
for chronic kidney disease, 209–10, 221–22
for chronic obstructive pulmonary disease, 303–4, 315–16
for cirrhosis of the liver, 131–32
for Crohn's disease, 117–18
for esophageal cancer, 393
for failed transplant, 221–22
for gastric bypass surgery, 13
for heart failure, 61
for HIV/AIDS, 405–6
for malnutrition, 25
for megoblastic anemia, 255
for myocardial infarction, 49–50
for nonhealing wounds, 291–92
for open abdomen, 342–43
for Parkinson's disease, 281
for pediatric traumatic brain injury, 330
for pediatric type 1 diabetes mellitus, 157–58
for pregnancy related anemia, 243–44
for respiratory failure, 315–16
for sepsis, 367–68
for stroke, 269
Nutrition assessment. *See* Assessment
Nutrition history. *See also* Patient history
    for acute pancreatitis, 144–45
    for adult type 2 diabetes mellitus, 196–97
    for Alzheimer's disease, 292
    for cachexia, 62
    for childhood overweight, 5
    for chronic kidney disease, 210–11, 222
    for chronic obstructive pulmonary disease, 304–5, 316
    for cirrhosis of the liver, 132
    for Crohn's disease, 118–19
    for esophageal cancer, 394
    for failed transplant, 222
    for gastroesophageal reflux disease, 73–74
    for heart failure, 62
    for HIV/AIDS, 406
    for hypertension, 37–38
    for irritable bowel syndrome, 105–7
    for malnutrition, 26
    for megoblastic anemia, 256
    for myocardial infarction, 50–51
    for nonhealing wounds, 292
    for open abdomen, 342
    for Parkinson's disease, 282
    for pediatric traumatic brain injury, 329
    for pediatric type 1 diabetes mellitus, 158–59
    for pediatric type 2 diabetes mellitus, 183

Nutrition history (*continued*)
   for pregnancy-related anemia, 244–45
   for respiratory failure, 316
   for stroke, 270–71
Nutrition intervention. *See* Intervention
Nutrition monitoring and evaluation
   for acute kidney injury, 236
   for acute lymphoblastic leukemia, 389
   for acute pancreatitis, 151
   for adult type 1 diabetes mellitus, 179
   for adult type 2 diabetes mellitus, 202
   for Alzheimer's disease, 297
   for burn injuries, 361–62
   for celiac disease, 100
   for childhood obesity, 9
   for chronic kidney disease, 227
   for chronic obstructive pulmonary disease, 309–10, 322
   for cirrhosis, 137–38
   for Crohn's disease, 124–25
   for esophageal cancer, 399
   for gastric bypass surgery, 20
   for HIV/AIDS, 411
   for irritable bowel syndrome, 112
   for malnutrition, 30
   for megoblastic anemia, 263
   for myocardial infarction, 56
   for nonhealing wounds, 297
   for open abdomen, 349
   for Parkinson's disease, 287
   for pediatric traumatic brain injury, 336
   for pediatric type 1 diabetes mellitus, 165–66
   for pediatric type 2 diabetes mellitus, 190
   for pregnancy-related anemia, 251
   for respiratory failure, 322
   for sepsis, 374
   for stroke, 275–77
Nutrition therapy
   for acute kidney injury, 235
   for acute lymphoblastic leukemia, 387–88
   for acute pancreatitis, 149
   for adult type 1 diabetes mellitus, 178
   for adult type 2 diabetes mellitus, 202
   for Alzheimer's disease, 295
   for burn injuries, 360
   for cardiac cachexia, 65
   for celiac disease, 98–99
   for childhood obesity, 7
   for chronic kidney disease, 213, 225
   for chronic obstructive pulmonary disease, 308, 320–21
   for cirrhosis, 136
   for Crohn's disease, 123
   for diabetes mellitus, 153
   for esophageal cancer, 397

   for gastric bypass surgery, 19
   for gastroesophageal reflux disease, 78
   for heart failure, 65
   for HIV/AIDS, 409
   for hypertension, 42
   for irritable bowel syndrome, 110–11
   for malnutrition, 29
   for megoblastic anemia, 262
   for myocardial infarction, 54
   for nonhealing wounds, 295
   for open abdomen, 347
   for Parkinson's disease, 285–86
   for pediatric traumatic brain injury, 334
   for pediatric type 1 diabetes mellitus, 164
   for pediatric type 2 diabetes mellitus, 187
   for peptic ulcer disease, 69, 86–87
   for pregnancy-related anemia, 249–50
   for respiratory failure, 320–21
   for sepsis, 373
   for stroke, 273–74

O
Obesity
   in children, 1–9, 181–91
   gastric bypass surgery for, 11–20
   health consequences of, 1
   metabolic stress and, 325, 365–75
   prevalence of, 1
Obstructive sleep apnea (OSA), 4–5
Oncology. *See* Cancer
Online resources
   for acute kidney injury, 237
   for acute lymphoblastic leukemia, 390
   for acute pancreatitis, 152
   for adult type 1 diabetes mellitus, 180
   for Alzheimer's disease, 298
   for burn injuries, 363
   for celiac disease, 101
   for childhood obesity, 10
   for chronic kidney disease, 217, 228
   for chronic obstructive pulmonary disease, 311, 323
   for cirrhosis, 139
   for Crohn's disease, 126
   for esophageal cancer, 400
   for gastric bypass surgery, 21
   for heart failure, 68
   for hepatitis, 139
   for HIV/AIDS, 412
   for hypertension, 46
   for irritable bowel syndrome, 114
   for malnutrition, 31
   for megoblastic anemia, 264
   for myocardial infarction, 57

for open abdomen, 350
for Parkinson's disease, 288
for pediatric traumatic brain injury, 337
for pediatric type 1 diabetes mellitus, 167
for pediatric type 2 diabetes mellitus, 191
for pregnancy-related anemia, 252
for respiratory failure, 323
for sepsis, 375
for stroke, 277
for upper gastrointestinal disorders, 80, 89
for wounds, 298
Open abdomen, 339–50
Orben, K., 379
Osteoporosis, 262
Output/intake records
for acute kidney injury, 232
for acute pancreatitis, 145
for adult type 1 diabetes mellitus, 173
for adult type 2 diabetes mellitus, 198
for burn injuries, 357–58
for cachexia, 64
for chronic kidney disease, 211, 224
for chronic obstructive pulmonary disorder, 319
for Crohn's disease, 119
for esophageal cancer, 396
for failed transplant, 224
for gastric bypass surgery, 15
for heart failure, 64
for malnutrition, 26
for megoblastic anemia, 257
for open abdomen, 343
for Parkinson's disease, 282
for pediatric traumatic brain injury, 331
for pediatric type 1 diabetes mellitus, 160
for peptic ulcer disease, 84
for pregnancy-related anemia, 248
for respiratory failure, 319
for sepsis, 369

P
Palliative care, 67
Pancreas, 127
Pancreatitis
case study of, 141–52
overview of, 127
Pancytopenia, 253–64
Parenteral nutrition, 67, 124
Parkinson's disease
case study of, 279–88
overview of, 265
Parkland formula, 360
Pathophysiology
for acute kidney injury, 235

for acute lymphoblastic leukemia, 387
for acute pancreatitis, 149
for adult type 1 diabetes mellitus, 177–78
for adult type 2 diabetes mellitus, 201
for Alzheimer's disease, 295
for burn injuries, 360
for cardiac cachexia, 65
for celiac disease, 98
for childhood obesity, 7
for chronic kidney disease, 213, 225
for chronic obstructive pulmonary disease, 307–8, 320
for cirrhosis, 135–36
for Crohn's disease, 122–23
for esophageal cancer, 397
for gastric bypass surgery, 18–19
for gastroesophageal reflux disease, 77
for heart failure, 65
for HIV/AIDS, 409
for hypertension, 41–42
for irritable bowel syndrome, 110
for malnutrition, 29
for megoblastic anemia, 261–62
for myocardial infarction, 54
for nonhealing wounds, 295
for open abdomen, 347
for Parkinson's disease, 285
for pediatric traumatic brain injury, 334
for pediatric type 1 diabetes mellitus, 163–64
for pediatric type 2 diabetes mellitus, 186
for peptic ulcer disease, 86
for pregnancy-related anemia, 249
for respiratory failure, 320
for sepsis, 373
for stroke, 273
Patient history. *See also* Nutrition history
for acute kidney injury, 230
for acute pancreatitis, 142
for adult type 1 diabetes mellitus, 170
for adult type 2 diabetes mellitus, 194
for Alzheimer's disease, 290
for burn injuries, 352
for celiac disease, 94
for childhood obesity, 4
for chronic kidney disease, 208, 220
for chronic obstructive pulmonary disease, 302, 314
for cirrhosis of the liver, 130
for Crohn's disease, 116
for esophageal cancer, 392
for failed transplant, 220
for gastric bypass surgery, 12, 366
for gastroesophageal reflux disease, 72
for heart failure, 60
for hepatitis C, 130

Patient history (*continued*)
  for hypertension, 36
  for irritable bowel syndrome, 104
  for malnutrition, 24
  for megoblastic anemia, 254
  for myocardial infarction, 48
  for nonhealing wounds, 290
  for open abdomen, 340
  for pediatric traumatic brain injury, 328
  for pediatric type 1 diabetes mellitus, 156
  for pediatric type 2 diabetes mellitus, 182
  for peptic ulcer disease, 82
  for pregnancy-related anemia, 242
  for respiratory failure, 314
  for sepsis, 366
  for stroke, 268
Pediasure, 335
Pediatric cases
  of obesity, 1–9
  of traumatic brain injuries, 327–37
  of type 1 diabetes mellitus, 155–67
  of type 2 diabetes mellitus, 181–91
Peptic ulcer disease (PUD)
  case study of, 81–89
  overview of, 69
Perforated duodenal ulcers, 81–82
Peritoneal dialysis, 205, 219–28
Pezzoli, G., 285
Physical exam
  for acute lymphoblastic leukemia, 380–81, 383
  for acute pancreatitis, 142–43, 145–46
  for adult type 1 diabetes mellitus, 170–71, 173
  for adult type 2 diabetes mellitus, 194–95, 197
  for Alzheimer's disease, 290–91
  for burn injuries, 352–53, 357
  for celiac disease, 94
  for childhood overweight, 4–5
  for chronic kidney disease, 208–9, 220–21
  for chronic obstructive pulmonary disease, 302–3,
      314–15
  for Crohn's disease, 116–17
  for failed transplant, 220–21
  for gastric bypass surgery, 14
  for gastroesophageal reflux disease, 73
  for HIV/AIDS, 404–5
  for hypertension, 36–37
  for irritable bowel syndrome, 104–5
  for malnutrition, 24–25
  for megoblastic anemia, 254–56
  for myocardial infarction, 48–49
  for nonhealing wounds, 290–91
  for open abdomen, 340–41
  for Parkinson's disease, 280–81

  for pediatric traumatic brain injury, 328, 331
  for pediatric type 1 diabetes mellitus, 156–57, 159
  for pediatric type 2 diabetes mellitus, 182–83
  for peptic ulcer disease, 82
  for pregnancy-related anemia, 242–43, 245
  for respiratory failure, 314–15
  for sepsis, 366–67, 369
  for stroke, 268–69
Pima Indians, 205, 207, 215
Pivot 1.5, 348
Pneumonia, 401–12
Powell, N. J., 349
Pregnancy, 241–52
Pressure ulcers, 289–98
Probiotics, 105
Protein catabolism, 325
Protein-energy malnutrition, 136
PUD. *See* Peptic ulcer disease
Pulmonary disorders
  case studies of, 301–23
  overview of, 299
Pulmonary function tests, 307

R
Radiation therapy, 377–400
Rash, 94
Reduced-intensity conditioning, 387
Renal disorders
  acute type of, 205, 229–37, 361
  chronic type of, 205, 207–28
  diabetes mellitus and, 153, 205
Renal replacement therapy
  continuous type of, 231, 235
  cost of, 205
  hemodialysis for, 207–17
  peritoneal dialysis for, 219–28
Resources
  for acute kidney injury, 237
  for acute lymphoblastic leukemia, 390
  for acute pancreatitis, 152
  for adult type 1 diabetes mellitus, 180
  for Alzheimer's disease, 298
  for burn injuries, 363
  for celiac disease, 101
  for childhood obesity, 10
  for chronic kidney disease, 217, 228
  for chronic obstructive pulmonary disease, 311, 323
  for cirrhosis, 139
  for Crohn's disease, 126
  for esophageal cancer, 400
  for gastric bypass surgery, 21
  for heart failure, 68
  for hepatitis, 139

for HIV/AIDS, 412
for hypertension, 46
for irritable bowel syndrome, 114
for malnutrition, 31
for megoblastic anemia, 264
for myocardial infarction, 57
for open abdomen, 350
for Parkinson's disease, 288
for pediatric traumatic brain injury, 337
for pediatric type 1 diabetes mellitus, 167
for pediatric type 2 diabetes mellitus, 191
for pregnancy-related anemia, 252
for respiratory failure, 323
for sepsis, 375
for stroke, 277
for upper gastrointestinal disorders, 80, 89
for wounds, 298
Respiratory disorders. *See* Pulmonary disorders
Respiratory failure, 313–23
Roux-en-Y gastric bypass surgery, 11–20, 365–75

S
School, problems in, 4
Sepsis, 365–75
"The Skeleton in the Hospital Closet" (Butterworth), 1
Sleep disturbances, 4–5
Slurred speech, 268
Social networks, 180
Sodium, 65
Speech-language therapists, 285
Stent placement, 54
Stress response, 325
Stroke
    case study of, 267–78
    overview of, 265
Supplements, 286
Swallowing evaluation, 329–30
Swallowing, phases of, 273
Systolic blood pressure, 33

T
Terminal illnesses, 399
Texture, of foods, 275
Therapeutic Lifestyle Changes (TLC), 33
Thomas, S., 351
Total body irradiation, 377–90
Total parenteral nutrition (TPN), 341–42, 389
Trauma
    of burn injuries, 351–63
    of open abdomen, 339–50
    overview of, 325
    of pediatric brain injuries, 327–37
Traumatic brain injuries, 327–37

Treatment. *See* Intervention
Trickle feeds, 348
tTG antibodies, 98
Tube feeding, 67, 83
Type 1 diabetes mellitus
    in adults, 169–80
    overview of, 153
    in pediatrics, 155–67
Type 2 diabetes mellitus
    in adults, 193–203
    in children, 181–91
    chronic kidney disease and, 208
    overview of, 153

U
Ulcerative colitis, 91
Ulcers. *See* Peptic ulcer disease
Unified Parkinson's Disease Rating Scale (UPDRS), 285
Upper gastrointestinal disorders
    gastroesophageal reflux disease in, 71–80
    overview of, 69
    peptic ulcer disease in, 81–89
Urinalysis
    for acute lymphoblastic leukemia, 386
    for acute pancreatitis, 148
    for adult type 1 diabetes mellitus, 175
    for adult type 2 diabetes mellitus, 200
    for chronic kidney disease, 212
    for cirrhosis of the liver, 134
    for gastric bypass surgery, 16
    for hypertension, 40
    for irritable bowel syndrome, 109
    for malnutrition, 28
    for myocardial infarction, 53
    for open abdomen, 345–46
    for pediatric type 1 diabetes mellitus, 161–62
    for pediatric type 2 diabetes mellitus, 185
    for sepsis, 371–72
Urinary nitrogen, 125

V
Ventilation, 320, 356
Viral hepatitis, 127, 130
Viral load, 401
Viscosity, of fluids, 275
Vitamin B_{12} deficiency, 253–64

W
Wasting syndrome, 401
Websites
    for acute kidney injury, 237
    for acute lymphoblastic leukemia, 390
    for acute pancreatitis, 152

Websites (*continued*)
  for adult type 1 diabetes mellitus, 180
  for Alzheimer's disease, 298
  for burn injuries, 363
  for celiac disease, 101
  for childhood obesity, 10
  for chronic kidney disease, 217, 228
  for chronic obstructive pulmonary disease, 311, 323
  for cirrhosis, 139
  for Crohn's disease, 126
  for esophageal cancer, 400
  for gastric bypass surgery, 21
  for heart failure, 68
  for hepatitis, 139
  for HIV/AIDS, 412
  for hypertension, 46
  for irritable bowel syndrome, 114
  for malnutrition, 31
  for megoblastic anemia, 264
  for myocardial infarction, 57
  for open abdomen, 350
  for Parkinson's disease, 288
  for pediatric traumatic brain injury, 337
  for pediatric type 1 diabetes mellitus, 167
  for pediatric type 2 diabetes mellitus, 191
  for pregnancy-related anemia, 252
  for respiratory failure, 323
  for sepsis, 375
  for stroke, 277
  for upper gastrointestinal disorders, 80, 89
  for wounds, 298
Weight loss, 397–98
Wounds
  from gunshots, 339–50
  nonhealing types of, 289–98

X
Xie equation, 361

Y
Yogurt, 105

Z
Zawacki equation, 361